AF393801

P. Rathert   S. Roth   M. S. Soloway

URINARY CYTOLOGY

Manual and Atlas

Second Edition, Fully Revised and Expanded

In Collaboration with
A. Böcking, R. Friedrichs, F. Hofstädter, J.-D. Hoppe,
E. Huland, H. Huland, C. Hunold, R. Nafe, S. Peter,
P. Röttger, H. Rübben, and B. J. Schmitz-Dräger

Translated by Terry C. Telger

With 196 Figures in 279 Separate Illustrations,
Mostly in Color

Springer-Verlag
Berlin Heidelberg New York
London Paris Tokyo
Hong Kong Barcelona
Budapest

Prof. Dr. Peter Rathert
Klinik für Urologie und Kinderurologie
Krankenhaus Düren
Akademisches Lehrkrankenhaus
Roonstrasse 30
W-5160 Düren, FRG

Dr. Stephan Roth
Klinik und Poliklinik für Urologie
Westfälische Wilhelms-Universität Münster
Albert-Schweitzer-Strasse 33
W-4400 Münster, FRG

Prof. Mark S. Soloway, M. D.
University of Miami
Department of Urology M-814
P. O. Box 016960
Miami, FL 33101, USA

Translator:
Terry C. Telger
6112 Waco Way
Fort Worth, TX 76133, USA

Title of the German Edition
P. Rathert · St. Roth, Urinzytologie – Praxis und Atlas
© Springer-Verlag Berlin Heidelberg 1991

Library of Congress Cataloging/in/Publication Data. Rathert, Peter. [Urinzytologie. English] Urinary cytology : manual and atlas / P. Rathert, S. Roth, M. S. Soloway ; in collaboration with A. Böcking ... [et al.] ; translated by Terry C. Telger. – 2nd ed., fully rev. and expanded. p.   cm. Rev. ed. of: Urinary cytology / H. J. de Voogt, P. Rathert, M. E. Beyer–Boon. 1977. Includes bibliographical references and index.

ISBN-13: 978-3-642-76186-7   e-ISBN-13: 978-3-642-76184-3
DOI: 10.1007/978-3-642-76184-3

1. Urine – Examination – Atlases.   2. Diagnosis, Cytologic – Atlases.   3. Urine – Examination – Handbooks, manuals, etc.   4. Diagnosis, Cytologic – Handbooks, manuals, etc. I.Roth, S.(Stephan), 1957– . II. Soloway, Mark S.   III. Voogt, H. J. de, 1925 – Urinary cytology. IV. Title. [DNLM: 1. Bladder Neoplasms – diagnosis – atlases. 2. Cytodiagnosis – methods – atlases. 3. Urine – cytology – atlases. WJ 17 R234u]   RB53. R.3713 1993
616.99 '262'07582–dc20   DNLM/DLC

Reproduction of figures: Scantrans Pte Ltd., Singapur

10/3130-5 4 3 2 1 0 – Printed on acid-free paper

Contributors

Professor Dr. A. Böcking
Institut für Pathologie der Medizinischen Fakultät der Rheinisch-Westfälischen
Technischen Hochschule, Pauwelsstr. 30, W-5100 Aachen, FRG

Dr. R. Friedrichs
Urologische Klinik, Universitätsklinikum Rudolf Virchow, Standort
Charlottenburg, Spandauer Damm 130, W-1000 Berlin 19, FRG

Professor Dr. F. Hofstädter
Institut für Pathologie der Universität Regensburg, Universitätsstr. 31,
W-8400 Regensburg, FRG

Dr. J.-D. Hoppe
Institut für Pathologie, Krankenhaus Düren, Akademisches Lehrkrankenhaus,
Roonstr. 30, W-5160 Düren, FRG

Dr. E. Huland
Urologische Klinik und Poliklinik, Universitätsklinik, Martinistr. 52,
W-2000 Hamburg 20, FRG

Professor Dr. H. Huland
Direktor der Urologischen Klinik und Poliklinik, Universitätsklinik,
Martinistr. 52, W-2000 Hamburg 20, FRG

Dr. C. Hunold
Urologische Klinik und Poliklinik der Medizinischen Einrichtungen der
Universität, Gesamthochschule Essen, Hufelandstr. 55, W-4300 Essen 1, FRG

Dr. R. Nafe
Institut für Pathologie, Medizinische Hochschule Hannover, Konstanty-
Gutschow-Str. 8, W-3000 Hannover 61, FRG

Professor Dr. S. Peter
Urologische Klinik Darmstadt, Grafenstr. 9, W-6100 Darmstadt, FRG

Professor Dr. P. Röttger
Institut für Pathologie, Krankenhaus Düren, Akademisches Lehrkrankenhaus,
Roonstr. 30, W-5160 Düren, FRG

Professor Dr. H. Rübben
Urologische Klinik und Poliklinik der Medizinischen Einrichtungen der
Universität, Gesamthochschule Essen, Hufelandstr. 55, W-4300 Essen 1, FRG

Privatdozent Dr. B. J. Schmitz-Dräger
Urologische Klinik der Universität, Moorenstr. 5, W-4000 Düsseldorf 1, FRG

Preface to the Second Edition

*"Urine cytology, no doubt, deserves a much more
central position in urologic diagnosis."*
L. ANDERSON, 1978

After an initial period of relative stagnation following the work of V. D. Lambl
(1854), oncologic urinary cytology has evolved at an explosive pace since the
work of G. V. Papanicolaou and V. F. Marshall (1945). This led the
cytopathologist L. G. Koss to state in 1979 that urinary cytology should be made
an essential diagnostic tool of the urologist.

The first edition of this book, written in collaboration with Drs. H. J. de Voogt
and M. E. Boon, already combined the disciplines of pathology and urology in
recognition of the fact that both the clinical and pathologic features of urothelial
tumors must be understood in order to select patients for urinary cytology and
comprehend its limitations and especially its interpretive criteria.

Since the first edition was published, basic research has given us new insights
into the ultrastructure of the urothelium, and many new preparatory techniques
have been developed for the collection, concentration, fixation, staining, and
analysis of cellular material. Not all of them could be included in this book (i.e.
thin layer prep.). In particular, the scientific accuracy of urinary cytology has
been enhanced by DNA single-cell cytometry, flow cytophotometry, and
immunocytology. Terminology, moreover, has been refined and standardized.

Besides oncologic urothelial cellular analysis, urinary erythrocyte morphology also has assumed clinical importance.

We could not do justice to these developments simply by reworking the 1979
text. We are grateful to Springer-Verlag for preparation of the new edition and
especially to Dr. U. HEILMANN and Mrs. I. C. LEGNER for their competent advice
and to E. KIRCHNER, J. SCHAUBEL, and W. BISCHOFF for their meticulous
production work. Through their help we were able to add numerous authorities
from the fields of pathology and urology to our list of contributors. We are
grateful to these colleagues for their spontaneous willingness to assist in the
creation of this monograph and atlas. Their authoritative input was essential for
providing up-to-date, comprehensive coverage of our subject matter. Special
emphasis has been placed on conveying reproducible, practice-oriented
methods while also presenting basic scientific principles in a comprehensible
form. We hope that the book will stimulate interest in urinary tract cytology
among cytophatologists and urologists, deepen their awareness of its importance
and limitations, and assist them in the preparation and analysis of cytologic
material.

In the practice of urinary cytology, close cooperation between the pathologist
and urologist leads to progress in the early diagnosis and follow-up of patients
with urothelial malignancies. Besides cystoscopy, sonography, and urography,
exfoliative oncologic urinary cytology is an essential component in the
management of these patients.

PETER RATHERT, Düren
STEPHAN ROTH, Münster
MARK S. SOLOWAY, Miami

Foreword to the First Edition

The cytologic diagnosis of cancer has its roots in clinical microscopy as it was shaped during the first half of the nineteenth century. In reviewing some of the early writing on this subject, one is amazed at the accuracy of the descriptions and soundness of the observations. Cytology of the urine is no exception: In 1864 Sanders described fragments of cancerous tissue in the urine of a patient with bladder cancer (*Edinburgh Medical Journal* 111, 273). This observation was confirmed by Dickinson in 1869 (*Transactions of the Pathological Society London* 20, 233). It is a source of special pride to me that in 1892 a New York pathologist, FRANK FERGUSON, advocated the examination of the urinary sediment as the best means of diagnosing bladder cancer, short of cystoscopy. PAPANICOLAOU freely acknowledged these contributions while establishing sound scientific bases for continuation and spread of this work. PAPANICOLAOU's work in the area of the urinary tract did not fall on deaf ears. He documented to several urologists who were within his sphere of personal influence, mainly Dr. VICTOR MARSHALL, Professor of Urology at Cornell University Medical School, that urinary tract cytology was a reliable tool in the diagnosis of urothelial carcinoma. Some of us who have attempted to spread the master's word had their share of success in institutions with which we were associated. Perhaps the most important contribution of urinary tract cytology has been in the identification of nonpapillary carcinoma in situ, a key lesion in the assessment or prognosis of urothelial neoplasia. Yet, the authors of this fine book on urinary cytology are quite right when they imply that the vast majority of urologists are either unaware or skeptical of this method of diagnosis. There are many reasons for this, perhaps the most important of which are its limitations. Well-differentiated papillary lesions of the bladder, such as papilloma and grade 1 papillary carcinoma, are unlikely to yield diagnostic cells. Thus, the expectation of the urologists that *any* bladder tumor will be reliably diagnosed by cytology is false. Similar mistakes are committed by pathologists and cytopathologists who fail to recognize the limitations of the method and, in attempting to diagnose too much, make major errors of judgment that arouse the mistrust of their clinical colleagues. Urinary tract cytology is difficult and is full of pitfalls and distressing sources of diagnostic mistakes. It cannot be learned casually but requires many years of experience and close cooperation between the pathologist and the urologist. This atlas should contribute to the popularization of this important method of diagnosis, which admirably complements but does not replace clinical judgment and the biopsy. The goal of these efforts is a relatively simple one: to offer the patient with cancer of the lower urinary tract the best possible chance for an early diagnosis resulting in a cure or at least containment of the disease and as comfortable a life as possible if a cure is not attainable. Urinary cytology can contribute significantly to this goal by identifying the patients at very high risk for invasive cancer whose urinary sediment contains obvious cancer cells. For such patients, radical treatment of the diseased urothelium prior to the development of metastases may be the best and sometimes only chance of salvation.

Drs. BEYER-BOON, DE VOOGT, and RATHERT should be congratulated on this fine atlas. It should contribute substantially to the clarification and education of both urologists and pathologists who are interested in cancer of the lower urinary tract.

LEOPOLD G. KOSS
Professor and Chairman
Department of Pathology
Albert Einstein College of Medicine
at Montefiore Hospital and Medical Center
Bronx, New York 10467

Contents

1 History of Urinary Cytology

P. RATHERT

CONTENTS

1.1 Introduction

Even today the "matula", a flask for inspection of the urine, is depicted in the official emblems of the American Urological Association, the German Society for Urology, and the Professional Association of German Urologists. It symbolizes the central importance of urinalysis as a medical and especially a urologic method of investigation. The color and odor of urine were of keen diagnostic interest to the physicians of the Middle Ages (Figs. 1.1, 1.2). But due to lack of alternative diagnostic methods, uroscopy soon evolved into a mystical uromancy in which the color of a urine specimen was subject to fantastic interpretations (Fig. 1.3). Urinalysis was not taken seriously as a diagnostic tool until the nineteenth century, when chemical and microscopic techniques were applied scientifically to the examination of the urine.

1.2 History of Oncologic Urinary Cytology

The general acceptance and evolution of urinary cytology, like that of many other subjective diagnostic methods, has taken place over a long period of time (Rathert 1986). Nearly a century passed before urinary cytology became established as a diagnostic technique.

In 1843 Dr. Julius Vogel of Göttingen first reported on a method that became popular a century later as *exfoliative cytology* (Grunze and Spriggs 1983). This method, which predated the histologic examination of tissue, involved the microscopic examination of cells desquamated from tissue surfaces. In one patient with a palpable mass in the mandibular angle region and a draining retroauricular sinus tract, the cytologic examination of exfoliated cells in the discharge raised suspicion of a malignant process, which was confirmed by the subsequent course of the disease.

V. D. Lambl (Fig. 1.4) was the first, in 1856, to apply this technique to the *detection of cancer cells in urine*. We must credit Grunze and Spriggs (1983) with identifying Lambl as the inaugurator of oncologic urinary cytology. It was previously thought that Sanders (1864) was the first to detect bladder cancer cells microscopically in urine; Ferguson (1892) also has been incorrectly named as the founder of urine cytology.

Vilém Dusan Lambl, born in Letina (near Pilsen) in 1824, became known for a species of infectious protozoan that he discovered and was subsequently named for him, *Giardia lamblia* (formerly known as *Lamblia intestinalis*). His historic paper on the cytologic diagnosis of bladder cancer (Fig. 1.5a,b) stands alone in his general body of work and was not pursued by other authors. This was probably due to the lack of therapeutic options at that time in patients diagnosed with bladder cancer (Rathert 1987).

Lambl's publication is interesting not only because it establishes his precedence in the field of oncologic urinary cytology but also because it includes a reference to "bedside diagnosis" (Fig. 1.5 a), a concept that would become important many years later. Lambl also notes various methods of specimen collection, most notably bladder washing, and the possibility of preserving urine by acidification and cooling. Lambl's work was also prophetic in its acknowledgement of the importance of microscopy: "The microscope is founding a science that is not necessary for the clinician in all its details; but if ever the assistance of a microscope were urgently needed, it is in the area

Fig. 1.1. **Page from a medieval manuscript (1580) on the physical examination of urine in a flask (matula).** (J. A. Benjamin Collection. Medical Library, University of California, Los Angeles)

Fig. 1.2. **"Uroscopy"** as portrayed in a German painting from the eighteenth century (Museo Nat. Art Sanitaria, Rome)

of urinalysis, where a confident diagnosis must precede the selection of an appropriate therapy" (Lambl 1856).

The decisive *breakthrough* in oncologic urinary cytology came later with the 1945 publication by *Papanicolaou* and *Marshall*, which led to a progressive integration of urinary cytology into routine urologic diagnosis.

1.3 Landmark Technical Developments

Clinical cytology and urinary cytology were introduced to medicine in the mid-1800s. This had been preceded by more than 200 years of scientific inquiry and innovation, which laid the necessary instrumental foundation in the form of the microscope.

Although Francesco Fontana claimed in 1646 to have built the *first microscope* in 1618 (Singer 1914), today this distinction is generally credited to the lens grinder Hans Hanssen and his son Zacharias of Middelburg (Turner 1980; Grunze and Spriggs 1983). Their instrument, which followed the basic design of Galileo's telescope, had a biconcave eyepiece a biconvex objective lens that provided a magnification of 60 diameters. According to Dittrich (1971), the term *microscopium* was coined by Demesianos, a member of the Accademia dei Lincei in Rome.

Besides optical magnification, another important prerequisite for effective visual cellular analysis is sufficient contrast of the object being examined. This led very early to the use of *stains* to bring out structural details. Anthony van Leeuwenhoek

Fig. 1.3. **"The Urine Prophet"**, title page from a medieval manuscript dated 1637 (Thomas Brian)

Fig. 1.4 **Vilém Dusan Lambl (1824–1895).** First description of tumor cells in urine

(1632–1723), one of the great authorities of early microscopy who first described spermatozoa with his pupil Johan Ham (1677), used saffron in the staining of microscopic preparations (Grunze and Spriggs 1983).

The discovery of the polychromatic staining properties of *methylene blue* by Romanowsky (1891) had far-reaching effects. In the early twentieth century, Quensel (1918) first used methylene blue stain for the immediate examination of urine sediment and successfully diagnosed seven of 12 papillomas and eight of 13 bladder cancers by urinary cytology (Fig. 1.7a,b). Several years earlier *Giemsa* (1910) had described the technique for staining air-dried specimens that was named for him and is still widely used today. The stain introduced by Feulgen in 1924 is still commonly used for automated image analysis, as it provides a homogeneous staining of nucleic acid that is ideal for instrumented measurements. The major impetus for the establishment of urinary cytology as a routine diagnostic study occurred in 1942, when *Papanico-*

laou (Fig. 1.8) introduced his alcohol stain that is still accepted today as the standard stain for urinary cytology.

Despite advances in staining techniques, light microscopy had reached a state of relative methodological stagnation by the early 1930s (Breinl 1979). This ended dramatically in 1934 with the invention of *phase contrast microscopy* by the Dutch physicist Fritz Zernike, for which he received the Nobel Prize in 1953. In his acceptance speech for the award, Zernike described his unexpected discovery:

"In about 1930 our laboratory received a large concave grating for installation in a Runge-Paschen system. The lined appearance of the surface was soon found. Because the grating was placed 6 m from the eye, I set up a small telescope to view it. Then the unexpected happened. The lines were clearly visible to the naked eye but vanished as soon as the telescope was focused precisely on the grating surface!" (Zernike 1953).

The phase contrast effect (see Chap. 8) had been discovered. Although phase microscopy is not

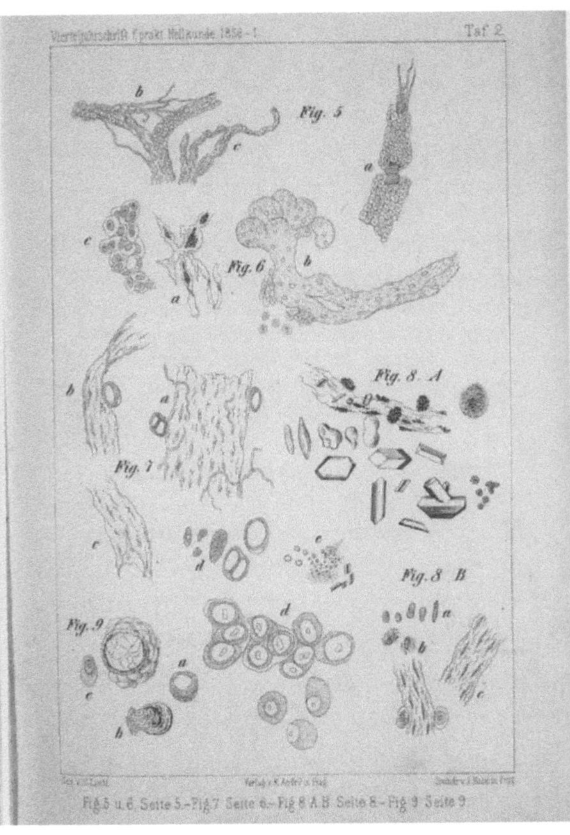

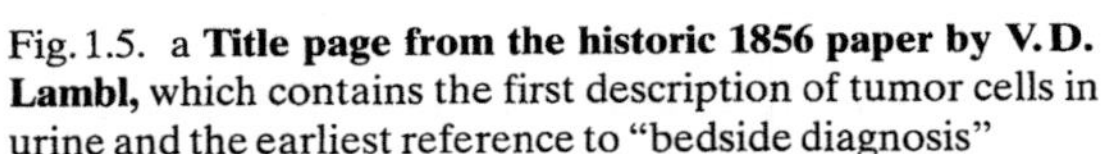

Fig. 1.5. a **Title page from the historic 1856 paper by V. D. Lambl,** which contains the first description of tumor cells in urine and the earliest reference to "bedside diagnosis"

b Lithographic plate from original drawings made by Lambl in 1856, which demonstrate not only his artistic talent but also the meticulous care with which he documented his observations

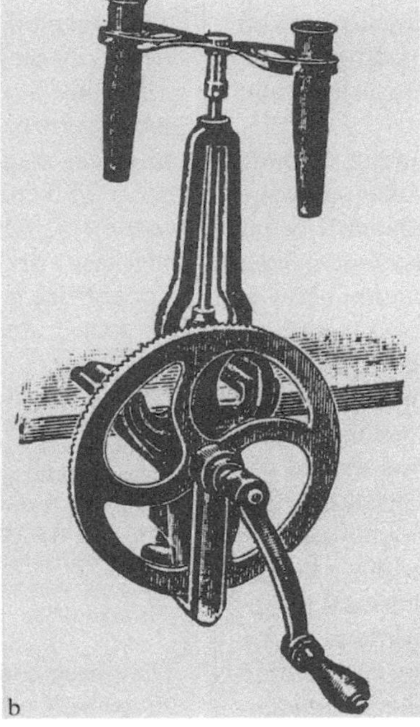

Fig. 1.6 a, b. **Early devices for cell concentration.** a Sedimentation flask (from Daiber 1906). b Manually operated centrifuge (from Laache 1914)

Fig. 1.7 a, b. **First micrographs of bladder tumor cells stained with methylene blue, published by Quensel in 1918**

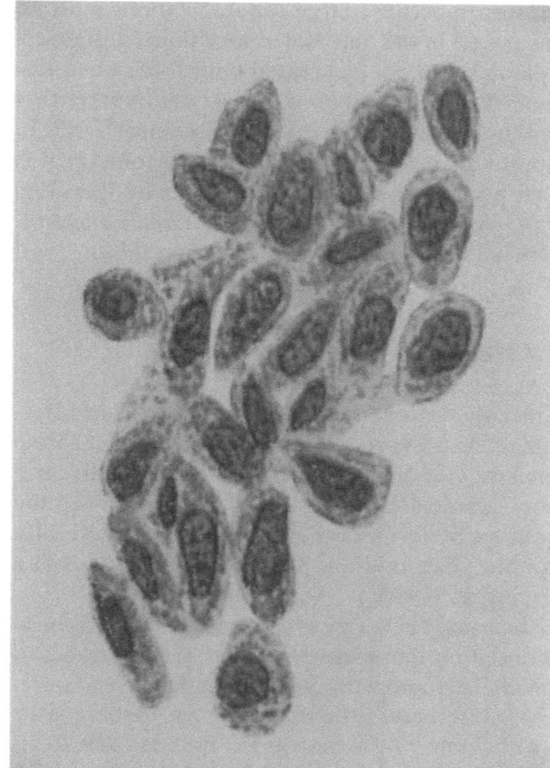

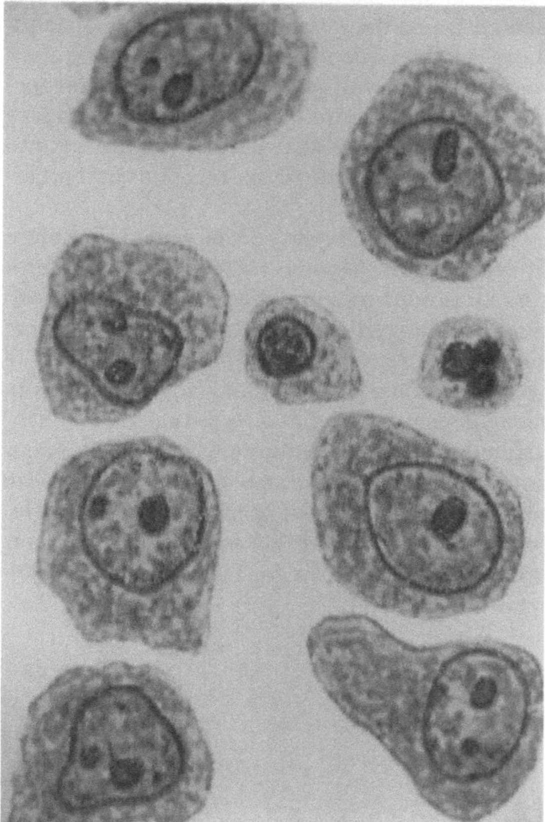

Fig. 1.8. **Postage stamp honoring Dr. G. Papanicolaou,** inaugurator of the Papanicolaou alcohol stain, still recognized today as the standard stain for urine cytology

widely used today for oncologic urinary cytology, it continues to be important in the evaluation of erythrocyte morphology (see Chap. 13).

Because of the small numbers of urothelial cells that are desquamated into the urine and the correspondingly low cell yields due to dilution, it was necessary to find a practical, effective way to *concentrate the cells* in a sample. Toward the end of the nineteenth century, sedimentation flasks were replaced by centrifuges, which were first manually operated and then driven by an electric motor (Fig. 1.6a,b). Solomon et al. (1958) greatly advanced the cytologic preparation of urine by introducing the *membrane-filter technique,* a negative-pressure filtration method in which cells are trapped on the surface of a filter as urine is passed through it. The filter is then mounted directly on a glass slide for further processing and examination. Subsequently a number of practical alternatives have been developed (see Chap. 8).

1.4 Innovations in Urinary Cytology

1.4.1 Automated Image Analysis

In conventional cytology as in most medical investigations, the examiner works descriptively with estimated values that are subjectively influenced and are based on personal knowledge and experience. A more objective and reproducible analysis can be achieved by *image analysis systems* that automatically evaluate specific, quantifiable properties of the collected cells.

Unlike histopathology, in which a semiautomated planimetric analysis is performed by outlining areas manually with a stylet, urinary cytology has focused on the development of *fully automated systems* (Nafe and Frohneberg 1989) that automatically quantify the DNA content of cells by the densitometric principle. These systems exploit the fact that the malignant transformation of cells is associated with a characteristic change in the distribution pattern of nuclear DNA.

Current research is focusing on two different types of systems. While some of the systems (flow cytometry, microscopic photometric systems) examine a whole cell population and yield a statistically determined central tendency as their end result, other systems work interactively, i.e., the qualitative experience of the cytologist is combined with an automated measuring process. In these systems the examiner, guided by his own criteria, sorts out suspicious cells which are then subjected to a more objective, quantifiable analysis (TV-image cytometry, quantitative fluorescence image analysis).

Although instrumented cell diagnosis per se was introduced by the Freiburg pathologist W. Sandritter (Sandritter et al. 1960), the foundations of *cytophotometry* were laid in the first half of the twentieth century. Sandritter cites, among others, the biophysical studies of Köhler (1904), who performed microphotographic measurements with ultraviolet light, and Feulgen's results with his biochemical nucleic acid stain (Feulgen and Rossenbeck 1924). Quantitative cytometry was introduced in 1936 when *Caspersson* synthetized these separate discoveries by performing measurements of prestained nucleic acids.

The development of flow cytometry, which has been employed in urinary cytology for more than 10 years, chiefly as a research tool, dates back to *Lagercrantz*, who first used a photodetector in 1948 to count cells suspended in a fluid medium. However, because large cell populations must be measured in the automated technique, a practical system could not be created until 1965, when Kamentsky developed an ultra-fast photometer that within minutes could quantitate nuclear DNA in approximately 30 000 cells. Flow cytometry was first performed in the United States by *Van Dilla* et al. in 1969, and in Germany by *Göhde and Dittrich* in 1971 (Tribukait and Gustafson 1980).

1.4.2 Immunocytology

Since the discovery of radioactivity by Henri Becquerel in 1896 and the studies of the Hungarian chemist George C. de Hevesky on radioactive tracers, scientists and physicians have explored the concept of developing specially labeled molecules for the diagnosis and treatment of diseases (La France et al. 1985).

In urinary cytology as well, much effort has been focused on the discovery of antigenic markers of urothelial tumors that would specifically and selectively discriminate tumor cells from nonneoplastic urothelium. *Pressman* was the first, in 1949, to attempt isotopic labeling with antibodies and later demonstrated their localization in mammalian tumors. In subsequent decades heterologous antisera were used in an attempt to identify tumor-specific antigens, but results were disappointing due to the inadequate specificity of these polyvalent antisera and the associated cross-reactions that occurred.

The most revolutionary development took place in 1975 when *Köhler and Milstein* described a process for producing antibodies of predefined specificity in unlimited quantities. *Monoclonal antibody technology* represented a major advance in oncologic urinary cytology and led to the development of many antibodies directed against urothelial tumors. But while various tumor-associated antigens have been found in the urine of patients with bladder cancer, a tumor-specific antibody has remained elusive, and further historic developments are needed in the quest for Paul Ehrlich's proclaimed "magic bullet."

References

Breinl H (1979) Die Phasenkontrastmikroskopie als morphologische Untersuchungsmethode in Biologie und Medizin. In: Witte S, Ruch F (eds) Moderne Untersuchungsmethoden in der Zytologie, 2nd edn. Witzstrock, Baden-Baden, pp 3–19

Caspersson O (1936) Quantitative cytochemical studies on normal malignant, premalignant and atypical cell populations from the human uterine cervix. Scand Arch Physiol 73: 8

Daiber A (1906) Mikroskopie der Harnsedimente, 2nd edn. Bergmann, Stuttgart

Dittrich M (1971) Hauptetappen der mikroskopischen Forschung. Jenaer Rundsch 4: 211

Ferguson F (1892) The diagnosis of tumors of the bladder by microscopical examinations. Proc NY Pathol Soc, p 71

Feulgen F, Rossenbeck H (1924) Der mikroskopisch-chemische Nachweis einer Nukleinsäure vom Typus der Thymusnukleinsäure und darauf beruhende elektive Färbung von Zellkernen in mikroskopischen Präparaten. Hoppe-Seylers Z Physiol Chem 135: 203

Giemsa G (1910) Über eine neue Schnellfärbung mit meiner Azur-Eosin-Lösung Muench Med Wochenschr 47: 2476

Göhde W, Dittrich W (1971) Impulsfluorometrie – ein neuartiges Durchflußverfahren zur ultraschnellen Mengenbestimmung von Zellinhaltsstoffen. Acta Histochem [Suppl] 10: 42

Grunze H, Spriggs AI (1983) History of clinical cytology: a selection of documents, 2nd edn. G-I-T Verlag, Darmstadt

Kamentsky LA (1965) Spectrophotometer: new instruments for ultrarapid cell analysis. Science 150: 630

Köhler A (1904) Mikrophotographische Untersuchungen mit ultraviolettem Licht. Z Wiss Mikr 21: 129, 273

Köhler G, Milstein C (1975) Continuous cultures of fused cells secreting antibody of predefined specificity. Nature 265: 495

Laache S (1914) Klinsk Urin-Analyse. Steen'ske Bogtrykkeri OG Forlag

LaFrance ND, Donner MW, Larson S, Scheffel U (1985) Diagnostische Anwendung monoklonaler Antikörper. Dtsch Med Wochenschr 110: 651

Lagercrantz C (1948) Photo electric counting of individual microscopic plant and animal cells. Nature 161: 25

Lambl VD (1856) Über Harnblasenkrebs. Ein Beitrag zur mikroskopischen Diagnostik am Krankenbette. Prager Vierteljahresschr Heilk 49: 1–32

Nafe R, Frohneberg D (1989) Automatische histologisch-zytologische Bildanalyseverfahren an den Organen des Urogenitaltraktes. Urologe [A] 28: 163

Papanicolaou GN (1942) A new procedure for staining vaginal smears. Science 95: 438

Papanicolaou GN, Marshall V (1945) Urine sediment smears as diagnostic procedure in cancer of the urinary tract. Science 101: 500–520

Pressman D (1949) The zone of activity of antibodies as determined by the use of radioactive tracers. Ann NY Acad Sci 11: 203

Quensel U (1918) Untersuchungen über die Morphologie des organisierten Harnsediments bei Krankheiten der Nieren und der Harnwege und über die Entstehung von Harnzylindern. Reprint from Nord Med Ark 50: 319. Nordstedt & Söner, Stockholm

Rathert P (1986) VD Lambl (1824–1895) Begründer der onkologischen Urinzytologie. Niere – Blase – Prostata 1: 15

Rathert P (1987) Onkologische Urinzytologie 1854: VD Lambl (1824–1895) Mitteilungen Dtsch Ges Urol 3: 41

Romanowsky D (1891) Zur Frage der Parasitologie und Therapie der Malaria. Med Wochenschr (St. Petersburg) 16: 297

Sanders WR (1864) Cancer of the bladder. Edinburgh Med J 10: 273

Sandritter W, Cramer H, Mondorf W (1960) Zur Krebsdiagnostik an vaginalen Zellausstrichen mittels cytophotometrischer Messungen. Arch gynecol 192: 293

Singer C (1914) Notes on the early history of microscopy. Proc Roy Soc Med 7/2: 247

Solomon C, Amelar RD, Hyman RM, Chaiban R, Europa DL (1958) Exfoliated cytology of the urinary tract: a new approach with reference to the isolation of cancer cells and the preparation of slides for study. J Urol 80: 374

Tribukait B, Gustafson H (1980) Impulscytophotometrische DNS-Untersuchungen bei Blasenkarzinomen. Onkologie 6: 278

Turner GL'E (1980) Essays on the history of the microscope. Senecio, Oxford

Van Dilla MA, Trujillo TT, Mullaney PF, Coulter JR (1969) Cell micro-fluorometry: a method for rapid fluorescence measurements. Science 163: 1213

Zernike F (1934) Beugungstheorie des Schneidenverfahrens und seiner verbesserten Form, der Phasenkontrastmethode. Physika [I] 18: 689

Zernike F (1953) Wie ich den Phasenkontrast entdeckte. Nobelvortrag 1953. Phys Blatter [II] 159

2 Indications for Urinary Cytology

P. RATHERT and S. ROTH

CONTENTS

2.1 Introduction

The cancer statistics clearly demonstrate a *rising
incidence* of urothelial tumors, especially in the industrialized regions (for the Federal Republic of
Germany, see, for example, Federal Public Health
Office 1987). While bladder cancer among males
ranked sixth in 1977 with 6100 new cases (Federal
Minister 1983), it rose to third place in 1986 with
10000 reported primary occurrences (see Table
5.1, p. 36).

The effective treatment of these cancers is not
possible without early diagnosis of both the primary tumor and of recurrent disease (Table 2.1).
Urinary cytology is of paramount importance in
this setting as the only *noninvasive* method of examination available.

Table 2.1. **Indications for urinary cytology**

Dysuria and alguria of unknown causes
Micro- and macrohematuria
Preoperative staging of urothelial tumors (CA in situ?)
Postoperative treatment of urothelial tumors
 - After transurethral tumor resection
 - Monitoring of instillation treatment
 - Lavage cytology of the urethral stump after cystectomy
Screening in risk groups
 - Analgesic (phenacetin) abuse
 - Occupational exposure to carcinogenic substances
Rare indications
 - Suspicious for vesicoenteral fistula
 - Extraurologic tumor invasion

2.2 Reasons for the Broad Range of Indications for Urinary Cytology

2.2.1 Broad Diagnostic Coverage of Urothelial Surfaces

Because more than 95% of all urothelial cancers
arise on the mucosal surface, they are detectable
by the presence of tumor cells that are desquamated (exfoliated) into the urine. This broad-coverage
aspect of urinary cytology is especially important
in the diagnosis of dysplasias and flat tumors,
which can escape detection by endoscopy, uroradiology, and needle biopsy (Fig. 2.1).

2.2.2 High Diagnostic Accuracy of Conventional Urinary Cytology

Conventional urinary cytology can detect approximately 70% of all urothelial cancers (Esposti et al.
1978; Murphy et al. 1986; Rübben et al. 1989).

This *numerically low accuracy rate* is due to the
relative insensitivity of urinary cytology to *highly
differentiated urothelial tumors* of low biological
aggressiveness, which are detected in less than
50% of cases. The cells shed from these tumors often show a lack of pathologic anatomy in that they
display minimal morphologic criteria of malignan-

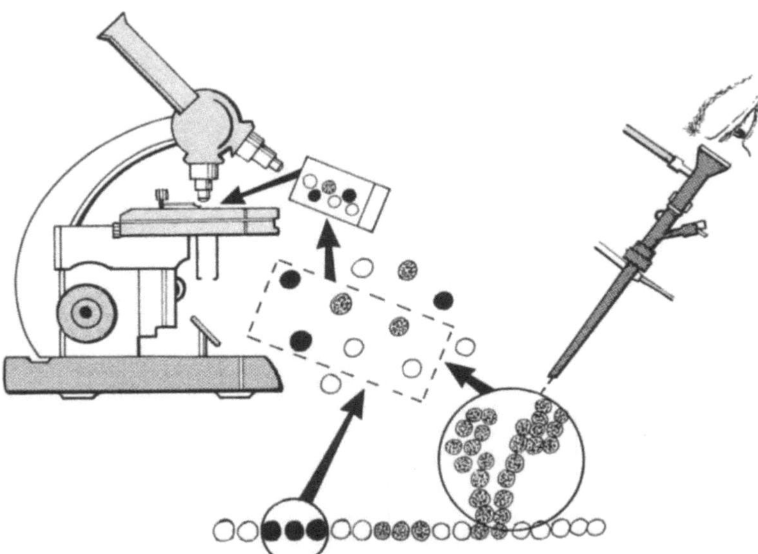

Fig. 2.1. While only exophytic lesions can be detected by cystoscopy, urinary cytology can also detect cells sloughed from flat areas of dysplasia or carcinoma. (Modified from Hofstädter)

cy. Also, they are as difficult to distinguish from reactive cell changes due to infection or lithiasis as they are from mild urothelial dysplasias, the latter having no pathologic significance for the patient (Jakse et al. 1986). Because less than 2% of highly differentiated urothelial tumors exhibit an invasive type of growth (Rübben et al. 1989), the low accuracy rate of the conventional cytologic diagnosis of these tumors has *little clinical significance*.

On the other hand, conventional urinary cytology has demonstrated a very high sensitivity, or tumor detection rate, for the frequently invasive moderately differentiated (grade 2) and poorly differentiated (grade 3) urothelial tumors. It is between 65% and 80% sensitive in the detection of grade 2 cancers, and 85%–90% sensitive in detecting poorly differentiated urothelial neoplasms (Jakse et al. 1986; Koss et al. 1985; Murphy et al. 1986, Rübben et al. 1989). Its specificity, or rate of true-negative findings, is between 78% and 95% (Rübben et al. 1989).

The clinical relevance of the *high sensitivity* of conventional urine cytology to *poorly differentiated urothelial tumors* is well illustrated by carcinoma in situ (see Fig. 2.1). This growth variant of urothelial cancer can occur in association with highly differentiated tumors and can have serious prognostic implications (Althausen et al. 1976). Since carcinoma in situ is virtually asymptomatic and is rarely detected by cystoscopy, the 90% detection rate by urinary cytology is critically important. For example, a positive cytologic diagnosis prior to a tumor resection would confirm the need to take multiple biopsies of the bladder mucosa separate from the visible area of exophytic tumor growth. A routine biopsy performed as part of the tumor resection cannot replace adjunctive urinary cytology, because the biopsy covers only a small fraction of the urothelium and lacks the broad coverage of the cytologic examination. Harving et al. (1988) confirmed that urinary cytology is more sensitive than multiple biopsies in detecting an accompanying carcinoma in situ.

2.2.3 Follow-Up of Surgical Patients

The follow-up of surgically treated urothelial tumors varies in accordance with the diversity of surgical options. Possible studies include urinary cytology performed after a transurethral bladder tumor resection to check for residual or recurrent tumor in the bladder and upper urinary tract, urethral irrigation cytology performed after radical cystectomy, and the cytologic follow-up of a supravesical urinary diversion or ileal conduit.

2.2.3.1 Follow-Up of Transurethral Tumor Resections

A transurethral tumor resection leads to significant reactive and degenerative urothelial changes that can complicate cytologic evaluation. Although residual tumor cells can be detected as early as 3 days after the incomplete resection of a bladder tumor (Müller et al. 1985), an interval of at least 7 days should pass between surgery and cytologic follow-up to optimize the readability of the specimens and improve diagnostic accuracy.

2.2.3.2 Monitoring of the Postcystectomy Urethral Remnant

The reported incidence of recurrent carcinoma of the urethral remnant after radical cystectomy ranges from 4% (Cordonnier and Spjut 1962) to 18% (Gowing 1960). Schellhammer and Whitemore (1976) report a 7% incidence of urethral stump carcinoma (24/348), while Hickey et al. (1986) report a 10% incidence (7/72).

Due to the slim chance of curing a recurrent tumor of the urethral remnant by the time clinical symptoms are noted (Schellhammer and Whitemore 1976), there have been repeated calls to combine a prophylactic urethrectomy with the original cystectomy. But because this procedure has become established only for special indications such as carcinoma in situ, the problem of *effective aftercare* becomes a challenge. The inadequate sensitivity of urethroscopy as an exclusive follow-up study was demonstrated by Schellhammer and Whitemore (1976), who detected only one-half of 24 recurrent tumors by endoscopic examination.

This underscores the potential importance of *urethral irrigation cytology*, which offers the additional advantage of low invasiveness. Although cytologic specimens are sometimes difficult to interpret because of the reactive changes induced by the irrigation, cytology has nevertheless proven to be a valuable technique (Hermansen et al. 1988).

For purely technical reasons, the urethra should be irrigated through a thin-gauge disposable catheter rather than through an olive applied externally to the urethral meatus to ensure that sufficient cellular material is collected from the relevant portion of the proximal urethra (see Chap. 8).

2.2.3.3 Monitoring Response to Intravesical Chemoimmunotherapy

The *reactive cytomorphologic changes* induced by intravesical chemotherapy and immune therapy can pose a significant obstacle to cytologic follow-up (Roth and Rathert 1989). Nevertheless, this therapy represents an important indication for oncologic urinary cytology, owing in part to a lack of other cost-effective noninvasive options.

In a follow-up study of 65 patients treated by the intravesical instillation of BCG, Bretton et al. (1989) showed that conventional cytology correctly detected the absence of tumor (specificity) in 29 of 36 patients (81%) 3 months after therapy. Parallel automated flow cytometry achieved a specificity of only 56% (20/36). Tumor recurrence was detected by cytology with a sensitivity of "only" 55% (19/29) compared with 69% (20/29) by flow cytometry, but the purely numerical inferiority of conventional urine cytology is qualified by other considerations. On the one hand, the *correct identification of tumor-free patients (specificity)* is *just as important as the detection of recurrence* in terms of avoiding unnecessary biopsy; on the other, there is the problem of practicality. Because oncologic follow-up is largely provided in clinical and office settings that do not have sophisticated image analysis equipment, there is a need for the optimum utilization of conventional urinary cytologic techniques.

2.2.4 Urinary Cytology and Hematuria

One of the *most common diagnostic problems* confronting the urologist is hematuria. This applies less to gross hematuria in older patients, which is an ominous sign necessitating a careful exclusion of malignancy, than to persistent microhematuria in younger individuals.

Several factors support the use of urinary cytology in the diagnostic evaluation of hematuria:

– Only a microscopic analysis can *distinguish* "*true*" hematuria from "*apparent*" hematuria caused, for example, by dyes, medications, or hemolysis.
– Hematuria, whether gross or microscopic, is the *cardinal symptom of urothelial cancer*. The sloughing of tumor cells into the urine and the high accuracy rate of their cytologic detection should make urinary cytology a standard procedure in the investigation of hematuria.
– A *glomerular origin* of the bleeding can be established with more than 90% sensitivity and specificity based on characteristic changes in urinary erythrocyte morphology. This can be done within the framework of oncologic urinary cytology without using special stains or microscopic equipment (see Chap. 12).

2.2.5 Urinary Cytology and Drug Abuse

The initial reports from Sweden in 1969 on the link between chronic analgesic abuse and urothelial tumors of the renal pelvis were received with skepticism (Rathert et al. 1975). However, numerous pharmacologic, epidemiologic, and case studies have shown that excessive *phenacetin consumption* (more than 5 kg) not only leads to renal papillary necrosis but also may be responsible for the *induction of urothelial cancer* after an average latent period of 22 years (Rathert et al. 1975; Porpaczy and Schramek 1981).

Although phenacetin has since been banned as an ingredient in analgesic products, there will continue to be a need for the urinary cytologic monitoring of patients with a prior history of abuse. Moreover, it remains to be seen whether the substitution of acetaminophen for phenacetin is a real solution, for nearly all the metabolic pathways of phenacetin are also accessible to acetaminophen (Rathert 1987).

2.2.6 Urinary Cytology and Exposure to Carcinogens

Since 1895 when Ludwig Rehn pointed to the link between papillary bladder tumors and exposure to aniline dyes, there has been interest in testing for potential exogenous and endogenous carcinogens in the urine.

Known exogenous urothelial carcinogens include the *aromatic amines* α-naphthylamine and paraaminodiphenyl, which occur as intermediate products in the manufacture of dyes, textiles, leather, and rubber (Zingg 1982). In 503 men who had long-term occupational exposure to paraaminodiphenyl, 35 (7%) developed bladder cancer (Koss et al. 1969). The duration of exposure was approximately 2 years, and the latent period ranged from 18 to 45 years.

This fact has led to the recognition of bladder cancer as an *occupational disease* among workers in certain industries, and trade associations in Germany have made it *mandatory* for workers regularly exposed to aromatic nitro or amino compounds to undergo *routine medical examinations* that include a cytologic examination of the urinary sediment. This is repeated at intervals of 6–12 months, depending on previous findings.

Although a relationship between *urothelial cancer and cigarette smoking* has not yet been definitely proven (Zingg 1982), it is strongly supported by epidemiologic evidence. On the one hand, a direct exogenous carcinogenic action is ascribed to the small amounts of α- and β-naphthylamine present in cigarette smoke. Smoking also appears to inhibit the breakdown of various tryptophan metabolites into niacin, leading to a buildup of endogenous carcinogens (O'Flynn et al. 1975). Overall, it is estimated that the risk of urothelial carcinogenesis is two to five times higher in cigarette smokers than in nonsmokers (Zingg 1982). In a randomized epidemiologic study of more than 6000 postmortem bladder biopsies performed in 282 persons, a striking correlation was noted between the degree of dysplasia and cigarette consumption (Auerbach and Garfinkel 1989). Nuclear atypias and epithelial structural changes were found in only 4.3% of all nonsmokers, compared with 72.9% of persons smoking 20–39 cigarettes per day and 88.4% of persons smoking more than 40 cigarettes per day.

In addition, animal studies have proven the carcinogenic effect of certain *cytostatic drugs*, especially cyclophosphamide (Endoxan), although there are only isolated case reports on the occurrence of urothelial cancer in humans secondary to cytostatic use.

Especially in persons exposed chronically to aromatic amines, urinary cytology provides a rapid, elegant, effective, and economical method of cancer screening.

2.2.7 Other Indications for Urinary Cytology

2.2.7.1 Vesicoenteric Fistulas

The diagnosis of vesicoenteric fistulas is frequently difficult. Despite sophisticated uroradiologic techniques (Roth and Rathert 1988), urinary cytology can be a useful adjunct, since even cystoscopy can locate the fistula in no more than 40% of cases. Cytology typically demonstrates many coliform bacteria and fibrous residues from undigested vegetable matter in addition to general inflammatory urinary constituents.

2.2.7.2 Extraurologic Tumors Invading the Urinary Tract

Although urinary cytology is of little value for the differentiation of nonurothelial malignant cells, it can detect cells that are at least pathologically suspicious in cases where an extraurologic tumor has penetrated the urothelium. In selected cases this can provide information useful for the planning of therapy.

References

Althausen AF, Prout GR, Daly JJ (1976) Non-invasive papillary carcinoma of the bladder associated with carcinoma in situ. J Urol 116: 575

Auerbach O, Garfinkel L (1989) Histologic changes in the urinary bladder in relation to cigarette smoking and use of artifical sweeteners. Cancer 64: 983

Berufsgenossenschaftliche Grundsätze für arbeitsmedizinische Vorsorgeuntersuchungen, 2nd edn. (1981). Genter, Stuttgart

Betton PR, Herr HW, Kimmel M, Fair WR, Whitemore WF Jr, Melamed MR (1989) Flow cytometry as a predictor of response and progression in patients with superficial bladder cancer treated with bacillus calmette-guerin. J Urol 141: 1332

Bundesgesundheitsamt (1987) Krebsstatistik der Bundesrepublik Deutschland 1986. Berlin

Bundesminister für Forschung und Technologie (1983) Krebsfrüherkennung. Bonn, pp 20–21

Cordonnier JJ, Spjut HJ (1962) Urethral occurrence of bladder carcinoma following cystectomy. J Urol 87: 398

Esposti PL, Edsmyr B, Tribukait B (1978) The role of exfoliative cytology in the management of bladder carcinoma. Urol Res 6: 197

Gowing NFC (1960) Urethral carcinoma associated with cancer of the bladder. Br J Urol 32: 428

Harving N, Wolf H, Melsen F (1988) Positive urinary cytology after tumor resection: an indicator for concomitant carcinoma in situ. J Urol 140: 495

Hermansen DK, Badalament RA, Whitemore WF Jr, Fair WF, Melamed MR (1988) Detection of carcinoma in the post-cystectomy urethral remnant by flow cytometric analysis. J Urol 139: 304

Hickey DP, Soloway MS, Murphy WM (1986) Selective urethrectomy following cystprostatectomy for bladder cancer. J Urol 136: 828

Jakse G, Hufnagel B, Hofstädter F, Rübben H (1986) Sequentielle Blasenschleimhautbiopsie beim Urothelkarzinom der Harnblase. Verh Dtsch Ges Urol 37: 200

Koss LG, Melamed MR, Kelly RE (1969) Further cytologic and histologic studies of bladder lesicns in workers exposed to paraaminodiphenyl: progress report. J Natl Cancer Inst 43: 233

Koss LG, Deitch D, Ramanathan AB, Sherman AB (1985) Diagnostic value of cytology of voided urine. Acta Cytol (Baltimore) 29: 810

Müller F, Kraft R, Zingg E (1985) Exfoliative cytology after transurethral resection of superficial bladder tumors. Br J Urol 57: 530

Murphy WM, Emerson LD, Chandler RW, Moinuddin SM, Soloway MS (1986) Flow cytometry versus urinary cytology in the evaluation of patients with bladder cancer. J Urol 136: 815

O'Flynn JD, Smith JD, Hanson JS (1975) Transurethral resection of the assessment and treatment of vesical neoplasma. Eur Urol 1: 38

Porpaczy P, Schramek P (1981) Analgesic nephropathy and phenacetin-induced transitional cell carcinoma – Analysis of 300 patients with long-term consumption of phenacetin-containing drugs. Eur Urol 7: 349

Rathert P (1987) Paracetamol. Dtsch Med Wochenschr 112: 40

Rathert P, Melchior HJ, Lutzeyer W (1975) Phenacetin: a carcinogen for the urinary tract? J Urol 113: 653

Roth St, Rathert P (1988) Vesiko-enterale Fisteln: Lösungswege eines diagnostischen Dilemmas. Urologe [A] 27: 142

Roth St, Rathert P (1989) Cytological surveillance of carcinoma in situ during and after intravesikal chemotherapy. In: Therapeutic progress in urological cancers. Liss, New York, p 523

Rübben H, Rathert P, Roth S, Hofstädter F, Giani G, Terhorst B, Friedrichs R (1989) Exfoliative Urinzytologie. Harnwegstumorregister, 4th edn. Fort- und Weiterbildungskommission der Deutschen Urologen, Arbeitskreis Onkologie, Sektion Urinzytologie

Schellhammer PF, Whitemore WF Jr (1976) Transitional cell carcinoma of the urethra in men having cystectomy for bladder cancer. J Urol 115: 56

Zingg EJ (1982) Maligne Tumoren der Harnblase. In: Hohenfellner R, Zingg EJ (eds) Urologie in Klinik und Praxis, vol 1. Thieme, Stuttgart, p 520

3 Non-neoplastic Transitional Epithelium of the Urinary Tract

J.-D. HOPPE and P. RÖTTGER

CONTENTS

3.1 Introduction

Most of the excretory portion of the urinary tract – the renal pelvis and ureter, the bladder except for the trigone, and the proximal part of the urethra – is lined by urothelium. This regional variant of transitional epithelium is distinguished by a superficial layer of "umbrella cells," large caplike cells that typically cover the intermediate and basal layers of urothelial cells (see Fig. 3.1). The umbrella cells themselves are covered on the luminal side by a mucopolysaccharide layer containing sialic acid and, on scanning electron microscopy, exhibit a network of prominent surface ridges that disappear following certain types of urothelial injury (see Chap. 4).

The thickness of the urothelium gradually increases from the calices of the renal pelvis (two to three layers of cells) to the ureters (four to five layers of cells) to the bladder and urethra (six to seven layers of cells). Beyond this, there are no appreciable differences in the cell spectra of these urinary tract segments, and urinary cytology of exfoliated cells shows no differentiating features as to their origin. The contention that umbrella cells can arise only from normal urothelium, and that corresponding conclusions can be drawn from their presence in cytologic material, is no longer justified since these cells have also been demonstrated on (benign) urothelial papillomas.

The cells of the intermediate and basal layers of the urothelium do not differ substantially from the transitional epithelial cells in other body regions. The basement membrane of the urothelium is only faintly visible by light microscopy and relates closely to the capillaries and to processes from nerve fibers in the variably thick subepithelial stroma. This relatively thin basement membrane and its connection with the capillary bloodstream account for the rapid appearance of granulocytes, monocytes, and extravasated red cells in the urothelium, and their subsequent passage into the urinary stream, in response to various insults ("hemorrhagic cystitis").

Unlike the epithelium of the small bowel, for example, the intact urothelium contains no intraepithelial representatives from the local population of lymphocytes or monocytes.

At the trigone of the urinary bladder in women and in older men, the urothelial lining is replaced by a noncornifying squamous epithelium having about the same structure and thickness as the vaginal epithelium ("trigonal metaplasia"). This tissue also lines the distal portion of the urethra in both sexes. Thus, isolated superficial cells from this epithelium are part of the normal cell spectrum observed in spontaneously voided urine.

The changes that can affect the non-neoplastic urothelium are highly diverse and shall be treated here only by reference to selected examples that illustrate their effects on the cell spectrum in centrifuged urine sediment.

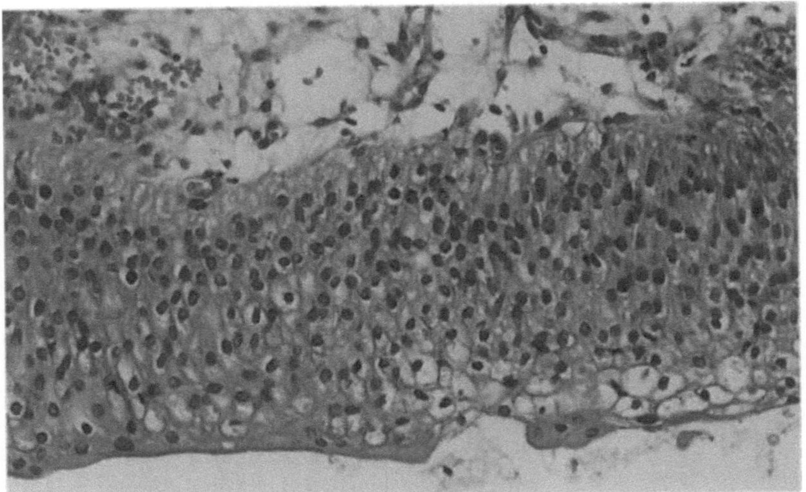

Fig. 3.1. **Urinary bladder,**
H & E, × 316. Simple numerical
hyperplasia of the urothelium in
a 52-year-old male. The urothelial structure is preserved, but
there is widening of the intermediate cell layer with detachment
of a superficial umbrella cell

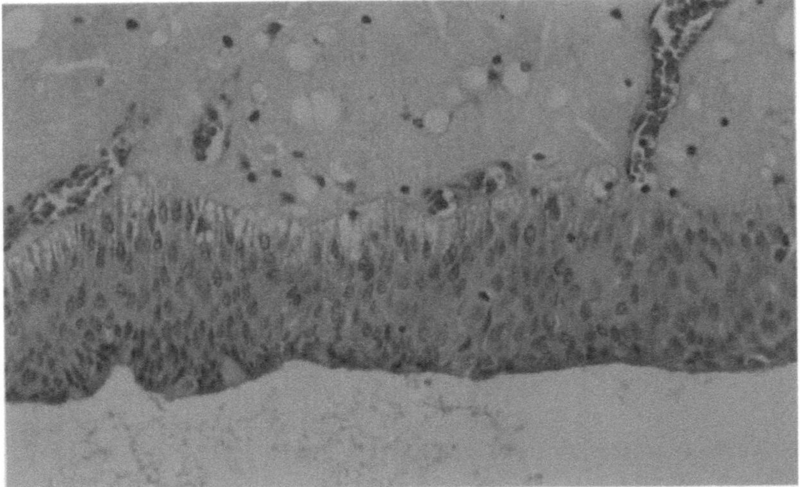

Fig. 3.2. **Urinary bladder,**
H & E, × 316. Pure mechanical
alteration of the bladder by an
indwelling catheter in a 79-year-
old male. The stroma is markedly hyperemic and edematous,
and the urothelium shows increased cellularity with epidermoid transformation of the umbrella cells, which are barely
identifiable as such

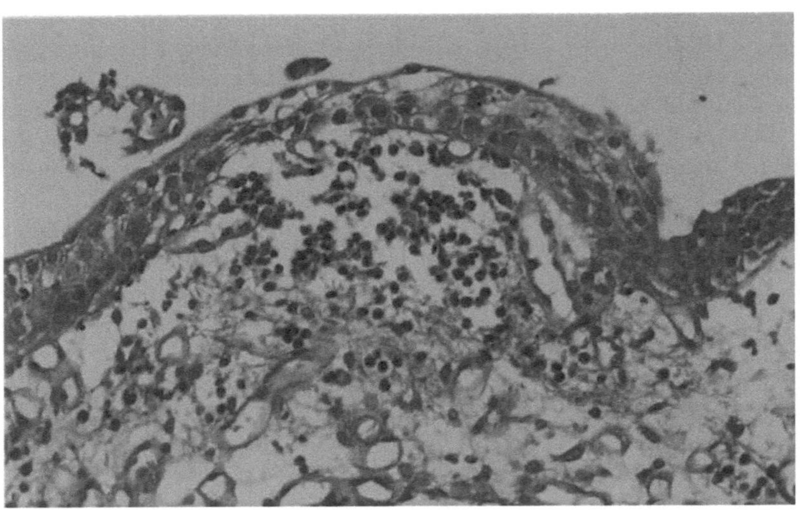

Fig. 3.3. **Urinary bladder,**
H & E, × 316. Acute flareup of
Hunner's cystitis in a 20-year-old
female. The urothelium is pro-
tuberant and edematous and is
permeated by granulocytes and
markedly dilated capillaries.
The cell spectrum is normal

3.2 Simple Urothelial Hyperplasia

Simple (numerical) urothelial hyperplasia is caused by an insult, usually localized, that increases the number of cells in the urothelium without altering its basic stratification or structure. This reactive change in the urothelium does not affect the cell spectrum observed in the urine (Fig. 3.1).

3.3 Adaptation of the Urothelium to Mechanical Stresses

The most notable adaptation involves the urothelial changes induced by urinary lithiasis, although the basic process underlying these adaptive changes is illustrated by a very simple and well-known mechanism: the effect of an indwelling catheter on the bladder urothelium (Fig. 3.2). There is conspicuous hyperemia and high-grade edema of the subepithelial stroma, but most significant is an early increase of cellularity and a transformation of the umbrella cells, whose nuclei become smaller and more compact and whose cytoplasm becomes denser and eosinophilic, giving rise to a kind of epidermoid metaplasia that becomes apparent within a few days. These altered cells can be recognized in the urine sediment.

3.4 Acute Inflammation

Banal findings are often difficult to demonstrate morphologically, for they generally do not require histologic evaluation. This is true of most acute urinary tract infections. Hunner's cystitis (Fig. 3.3), because of its episodic course, requires histologic investigation to establish a diagnosis or detect a recurrent. The associated superficial changes basically match the features of other acute inflammations: marked capillary dilatation, the exudation of granulocytes, and edematous expansion of the urothelium, so that the urine sediment may contain cells from all layers of the transitional epithelium.

3.5 Chronic Inflammation

TUR cystitis, which has become a clinically important entity, is associated not just with ulcerations and foreign body reactions, which occasionally may be apparent in the sediment, but also with nonspecific chronic inflammatory states in the intact urothelium (Fig. 3.4). The subepithelial stroma shows heavy lymphohistiocytic infiltration and increased vascularity. The urothelium is markedly thickened, its cellularity is increased, and its layered arrangement is preserved. Besides inflammatory cells, which also spread intraepithelially, the sediment contains increased numbers of cells from the middle and upper layers and fewer from the basal cell layer of the urothelium. Atypias do not occur except in association with a recurrence of the underlying disease that necessitated the original procedure. There are no cytomorphologic features that distinguish this component of TUR cystitis from other chronic nonsuppurative urinary tract infections.

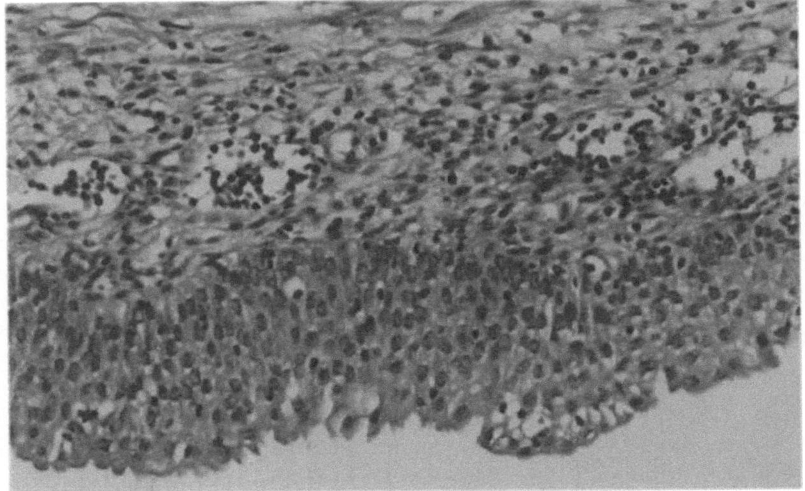

Fig. 3.4. Urinary bladder,
H & E, × 316. Chronic cystitis as
a nonulcerous component of
TUR cystitis in a 63-year-old female. Basally there is perivascu-
lar lymphohistiocytic infiltration
and capillary proliferation, and
there is marked general epithelial thickening with heavy superficial vacuolation and an otherwise normal cell spectrum

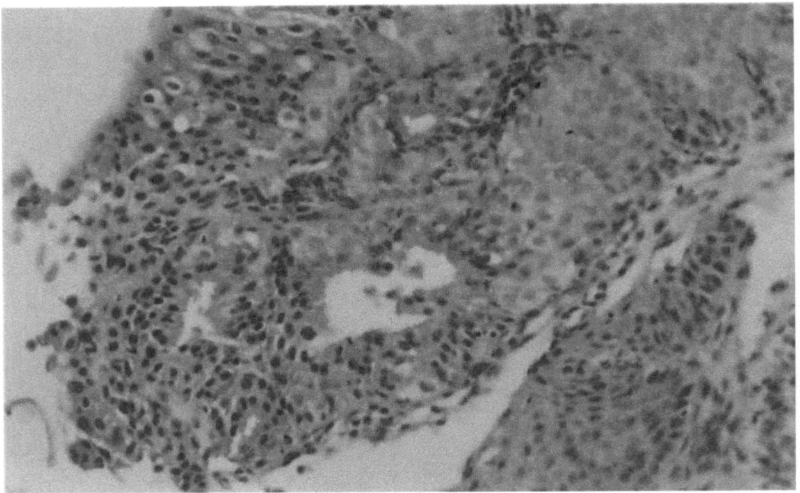

Fig. 3.5. Urinary bladder,
H & E, × 316. Cystitis glandularis (cystica) in a 66-year-old female. The urothelium is permeated by columnar epithelial
cells with superficial tubular
structures; otherwise the cellular pattern is normal

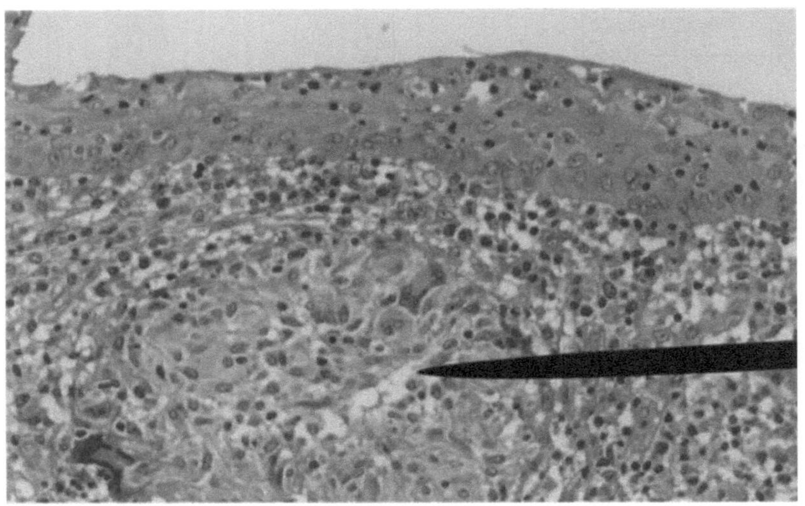

Fig. 3.6. Urinary bladder,
H & E, × 316. Tuberculous cystitis in a 62-year-old female. The
arrow points to an epithelioid
cell granuloma within the stroma. The urothelium is flattened,
shows increased cellularity with
superficial sloughing, and is permeated by inflammatory cells

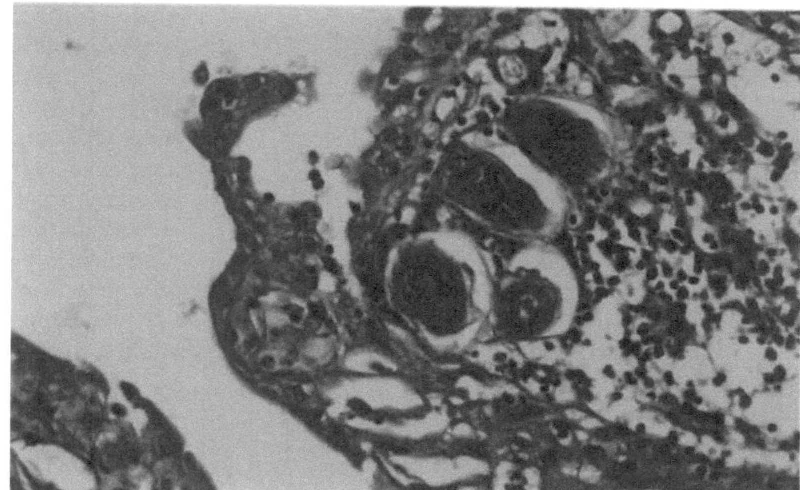

Fig. 3.7. **Urinary bladder,**
H & E, × 316. Bilharzial cystitis in
a 29-year-old male. There are
parasite eggs and heavy inflammatory infiltrate in the superficial
stroma. There is sloughing of
large urothelial fragments (ero-
sion) and a generally variegated
cellular pattern

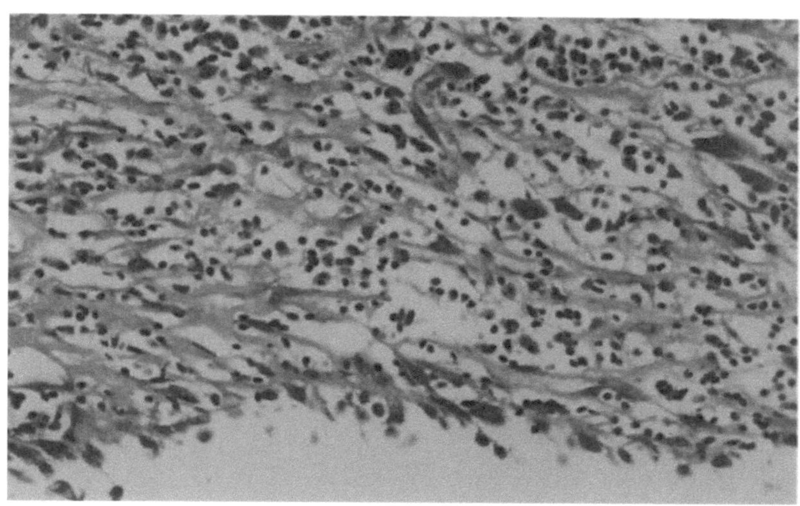

Fig. 3.8. **Urinary bladder,**
H & E, × 316. Cystitis in a 64-year-
old female secondary to radiation
injury 3 years before. There is
epithelial atrophy with nuclear
atypias and corresponding nuclear changes in the sclerosed stroma, which shows diffuse histiocy-
tic infiltration. The cellular
pattern is very atypical

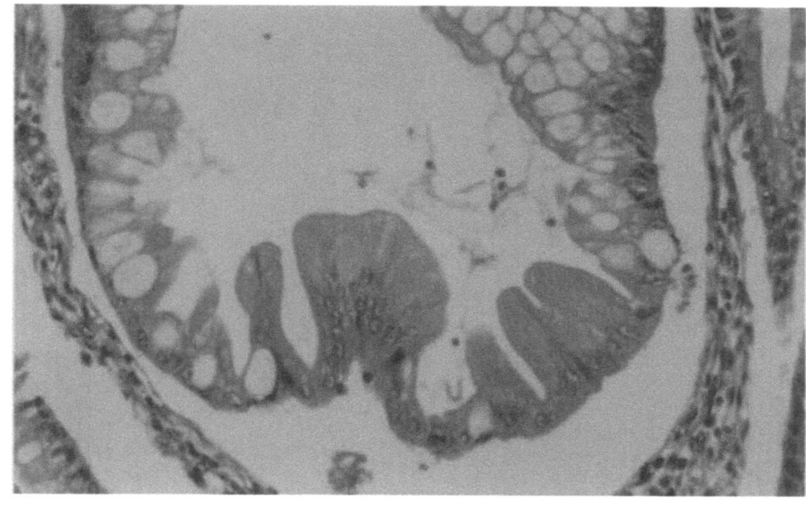

Fig. 3.9. **Ileal conduit,**
H & E, × 316. Crypt region of the
substitute bladder in a 45-year-old
female 3 months after urinary diversion. The crypt lumen is dilated, and there is proliferation of
enterocytes with abundant cytoplasm. There is an increase of mucus secretion at abnormal sites
without the typical goblet cell pattern

3.6 Cystitis Glandularis (Cystica)

Follicular and glandular (cystic) chronic inflammations of the urothelium are relatively common. The epithelial changes associated with follicular inflammation correspond largely to those in nonspecific chronic inflammations, whereas the glandular form (Fig. 3.5) may lead to the appearance of columnar cells in the sediment if the glandular metaplasia is not just manifested by cyst formation in the subepithelial stroma but also involves the epithelium itself.

3.7 Tuberculous Cystitis

Tuberculosis of the urinary tract, which today occurs chiefly in immune-compromised individuals, can have manifestations similar to those of TUR-cystitis. Besides ulcerations due to caseous necrosis, the superficial stroma may contain typical epithelioid cell granulomas (Fig. 3.6) while the epithelium itself is markedly edematous and permeated by inflammatory cells. An abnormal, though not "specific," cell spectrum is found in the urine sediment.

3.8 Bilharzial Cystitis

Schistosomiasis is a (sub)tropical disease whose practical relevance, even among urinary tract infections, has increased with changes in modern living conditions. The coexistence of chronic and acute inflammatory changes leads to a heavy exfoliation of cells (Fig. 3.7). The frequent erosive defects can allow the subepithelial parasite eggs to enter the urine, where they are accessible to cytologic detection (see Fig. 9.93a–c). The variegated cytologic features of this chronic recurrent erosive cystitis call for close surveillance due to the significant potential for degeneration to squamous cell carcinoma of the bladder.

3.9 Radiation Cystitis

Even early radiation-induced damage to the urothelium is manifested by a flattening of the epithelium along with epithelial defects and a relatively mild inflammatory infiltrate. In the late stage (Fig. 3.8) the cell picture is generally atrophic due to the impaired regenerative capacity of the urothelium, with nuclear atypias and corresponding nuclear changes appearing in the mesenchymal cells of the stroma. The histologic interpretation of these findings is facilitated by the prominent atrophic features, whereas radiogenic changes in the urine sediment are very difficult to identify cytologically unless the prior history of radiation exposure is known (see Fig. 9.84 ff.).

3.10 Appendix: Epithelial Changes in the Ileal Conduit

In addition to the urothelial changes mentioned above, there is another situation of potential practical relevance: changes relating to the metamorphosis of the ileal mucosa in a substitute bladder. These structural adaptations include a shortening of the villi, a deepening of the crypts, and changes in the intestinal epithelium. Eight distinct cell types are still present in the mucosa, but the predominant cells, the enterocytes, tend to shift their nuclear–cytoplasmic ratio in favor of the (nonhydropic) cytoplasm, especially in the expanded crypts, with some cells acquiring a structure resembling that of umbrella cells (Fig. 3.9). In all the crypts there is intensified mucus secretion at atypical sites from proliferated goblet cells. The foregoing cell structures are easily differentiated from malignant cells in the urine sediment following, say, the spread of recurrent cancer to the ileal conduit.

3.11 Conclusion

This morphologic examination of nonneoplastic changes in the excretory portion of the urinary tract, and especially the urinary bladder, differs from traditional descriptions in that the main emphasis has been placed on changes in the urothelium. Conventional viewpoints usually relegate these findings to the background as "associated reactions." But the urothelial reactions associated with almost all lower urinary tract disorders generate cytologic abnormalities that can permit a differential diagnosis of the underlying disease at an earlier and hence more favorable point in time.

Thus, we have not only examined the normal appearance of the urothelium but have also considered the somewhat broader spectrum of non-neoplastic urothelia and their impact on findings in the urine sediment.

References

Petersen RO (1992) Urologic pathology. Lippincott, Philadelphia, 2nd Edition

Schubert GE (1984) Niere und ableitende Harnwege. In: Remmele W (ed) Pathologie, vol 3. Springer, Berlin Heidelberg New York Tokyo, p 1

4 Ultrastructure of the Urothelium [*]

S. PETER

CONTENTS

4.1 Physiology of the Bladder Epithelium

The composition of the urine in mammals is determined by the interaction of filtration pressures and concentration gradients and by a hormonally controlled enzyme transport system in the kidneys. The urine collected in the bladder is usually hypertonic and differs substantially from blood plasma in its organic and inorganic constituents. Normally the osmotic pressure of urine is two to four times higher than the osmotic pressure of blood plasma: Urine contains 100 times more urea and creatinine, 30 times more phosphate, and 60 times more sulfate ions (Valtin 1978).

The blood capillaries of the urinary bladder are located in the lamina propria below the epithelium that lines the bladder. Only the thin urothelium separates the space containing hypertonic urine from the vessels with their isotonic blood plasma, and thus sustains the high chemical gradient between the urine and plasma. In this sense the bladder epithelium forms an impermeable barrier for water and electrolytes (Englund 1956), comparable to a plastic bag.

In amphibians, by contrast, the urine can be chemically altered while still inside the bladder. In animals that hatch in fresh water, this provides an extra mechanism for the conservation of salt. The capacious urinary bladder in these animals can reabsorb sodium ions from the urine against a high electrochemical gradient. Because this absorption is influenced partly by aldosterone, the toad bladder provides an excellent biological model for the electrophysiologic investigation of transepithelial transport processes (Leaf 1966). The bladder epithelia of mammals and amphibians thus perform different physiologic tasks, and one must be careful in drawing morphologic comparisons. On the other hand, Nelson et al. (1975) report that the brown bear (*Ursus americanus*) is able during hibernation to concentrate bladder urine by the reabsorption of water; this protective mechanism greatly curtails fluid loss during winter sleep while eliminating urine odor that might attract predators. These findings are particularly interesting in the light of in vitro studies by Lewis and Diamond (1975) and Lewis et al. (1976a,b), which demonstrated a transepithelial transport of ions in the bladder epithelium of rabbits. Schütz (1980) made similar findings in the urothelium of the human renal pelvis.

These qualifications aside, the bladder epithelium in mammals and humans may be regarded as an essentially impermeable layer.

Another distinctive property of the bladder epithelium is its ability to adapt morphologically to contraction and expansion of the bladder wall. This transformational ability of the urothelial cells with bladder distention lends a special meaning to the term "transitional epithelium" proposed by Jakob Henle. This property is necessary because the epithelium closely overlies the muscular coat, separated from it only by a submucosal layer.

4.2 The Urothelium: A Multilayered Epithelium

Since the work of Petry and Amon (1966), the transitional epithelium was viewed histologically in German-language textbooks as a unilayered pseudostratified epithelium. The Anglo-American and pathohistologic literature, however, continued to regard the transitional epithelium as a multilayered structure. We were able to show in ultrathin electron microscopic sections that the superficial cells of the urothelium have no connections with the basement membrane (Peter 1985). Thus,

---

[*] With funding from the Deutsche Forschungsgemeinschaft.

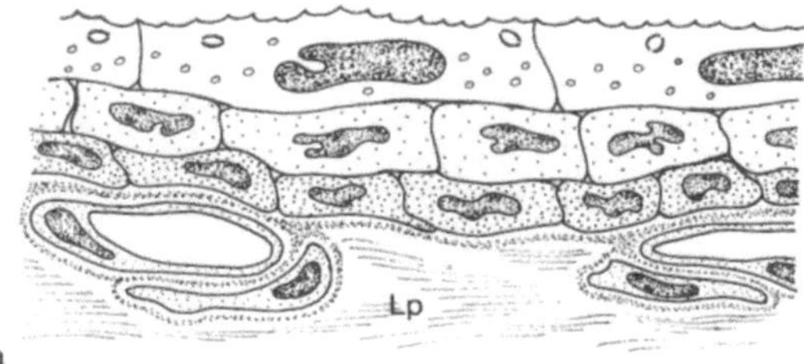

a

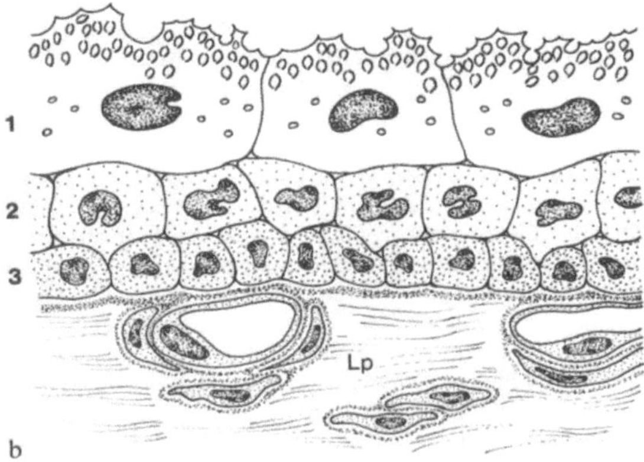

b

Fig. 4.1. **Schematic diagram of the distended** (a) **and nondistended** (b) **urinary bladder epithelium.** The urothelial cells increase in size from the basal layer toward the lumen. Luminal (*1*), intermediate (*2*), and basal (*3*) cell layers. *Lp*, Lamina propria

the mammalian urothelium should continue to be regarded as a multilayered epithelium.

Three cell layers – a superficial (luminal) layer, an intermediate layer, and a basal layer – can be distinguished in the moderately distended urinary bladder.[1] The size of the cells increases from the basal layer toward the bladder lumen (Figs. 4.1, 4.2).

The cytoplasmic structures of the basal cells display no unusual features. The cells contain many free ribosomes, abundant filaments, some mitochondria, dense bodies, and scant endoplasmic reticulum. The Golgi apparatus is poorly developed. The basal cells of the urothelium lie upon a lamina propria that consists largely of fibrocytes and collagenous connective tissue permeated by blood vessels and lymphatics (Fig. 4.2). Deep to the lamina propria is the muscular coat, a layer of smooth muscle cells reinforced by strands of connective tissue.

In the cells of the intermediate layer as well, the cell nuclei, endoplasmic reticulum, mitochondria and ribosomes present no unusual features. The intermediate cells differ from the basal cells, however, in their very well developed Golgi apparatus. Scattered fusiform vacuoles occur in proximity to the Golgi apparatus. Lysosomes are rarely seen.

The most distinctive feature of the cells in the superficial layer is the abundance of fusiform vacuoles in their cytoplasm. These vacuoles appear less numerous in the distended bladder than in the contracted or nondistended bladder, where they tend to be crowded along the luminal side of the cells (Fig. 4.3). The membranes of the fusiform vacuoles are peculiar in that they are composed of two layers, an inner layer approximately 80 Å thick and an outer shell approximately 40 Å thick (Fig. 4.4). Many a fusiform vacuole is in direct contact with the luminal cell membrane, touching it with one of its tapered extremities (Fig. 4.5).

Hicks (1986) showed by electron microscopy that Golgi cisterns, fusiform vacuoles, and the apical cell membranes have the same membranous structure, suggesting that the fusiform vacuoles form from the membranes of the Golgi apparatus. Hicks also showed that the fusiform vacuoles serve as a "substitute membrane" for the apical cell membrane. Labeling studies by Porter et al. (1965) and Hicks (1966) showed that the fusiform vacuoles also form after invagination of the apical cell membrane. This process is thought to occur during or at least immediately following evacuation of the bladder. Invaginations or infoldings of the luminal cell membrane occur during bladder contraction to adapt the luminal surface to the reduced bladder volume. The reverse process likely occurs during bladder expansion: Membranes of the fusiform vacuoles and the invaginated areas are incorporated to enlarge the surface area of the luminal membrane (Minsky and Chlapowski 1978).

4.3 Ultrastructure of Urothelial Carcinoma

The cells of undifferentiated urothelial tumors show a loss of the specialized apical cell membrane. The new apical cell membranes are no longer distinguishable by electron microscopy from normal membranes of other cells or from the lateral and basolateral membranes of healthy urothelial cells. There is a characteristic *loss of the double membrane* and the appearance of stubby *cellular processes*, called *microvilli*, radiating toward the lumen. As dedifferentiation progresses, the microvilli become more numerous in each section through the membrane and are no longer confined to its luminal side. As the cancer cells become in-

---

[1] For the terminology of urothelial cells, see also Chap. 7, p. 53.

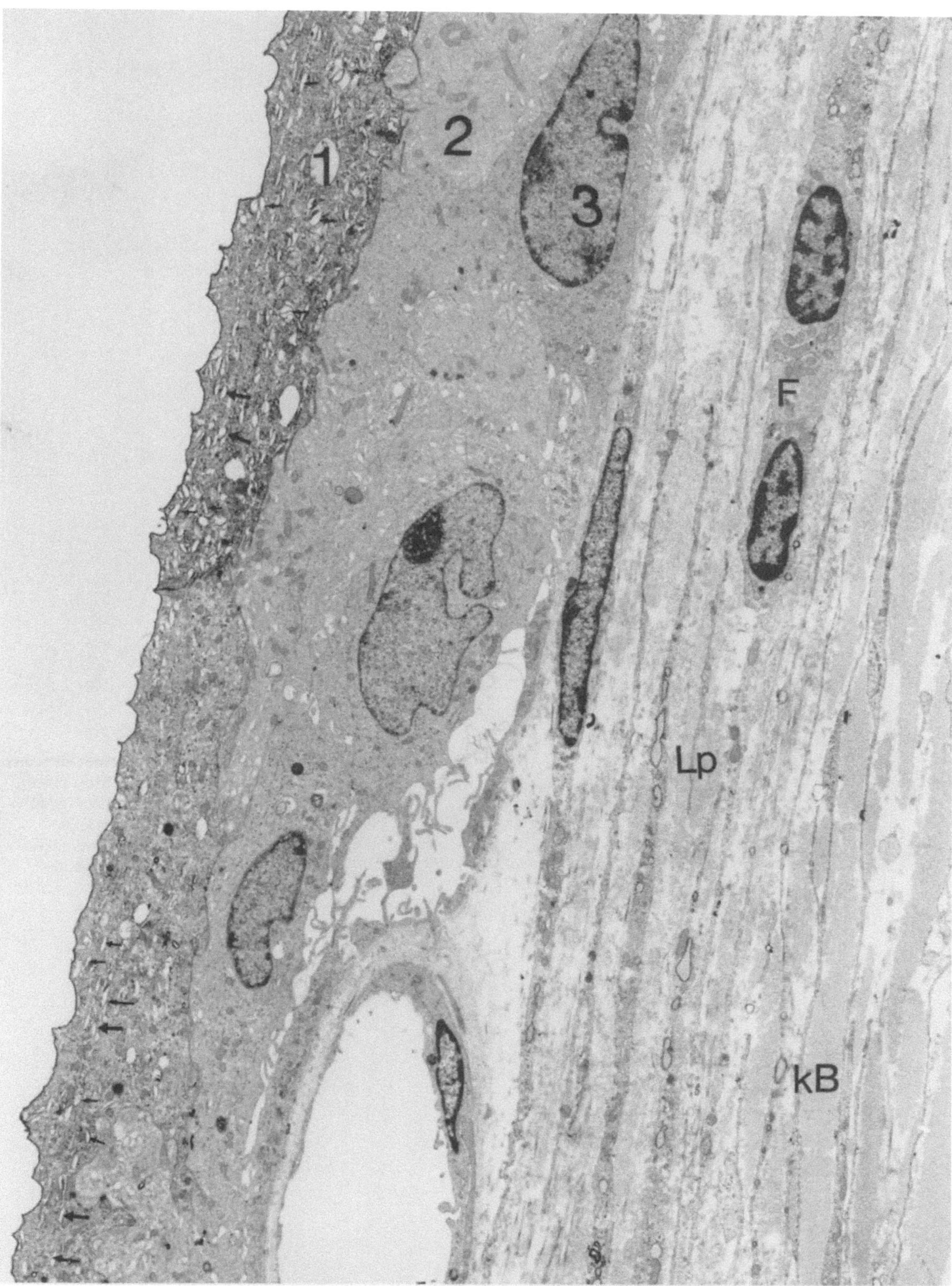

Fig. 4.2. **Low-power electron micrograph of the bladder epithelium showing the luminal** (*1*) **and the intermediate** (*2*) **and basal** (*3*) **cell layers.** The epithelium rests on the lamina propria (*Lp*) containing fibrocytes (*F*) and collagenous connective tissue (*kB*). The abundant fusiform vacuoles (← ←) are visible even in the low-power view

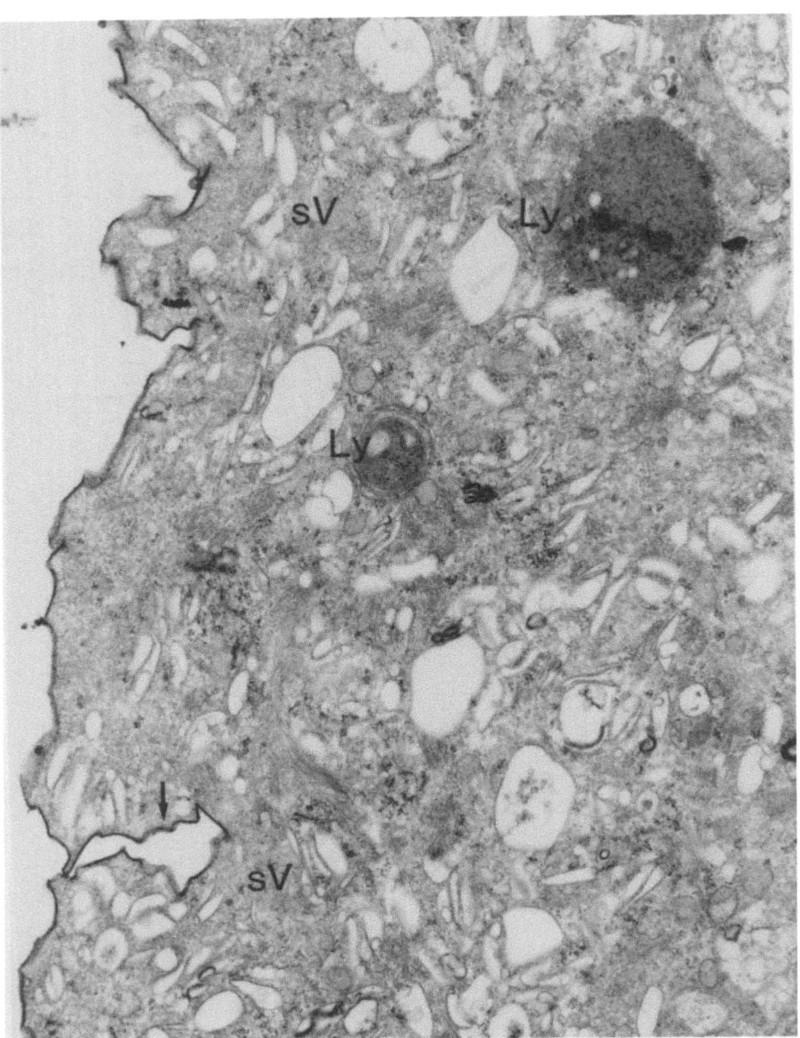

Fig. 4.3. **Cytoplasmic structures of a luminal cell.** Most prominent are the fusiform vacuoles (*sV*). At left (←) is an invagination of the luminal cell membrane. *Ly*, Lysosome

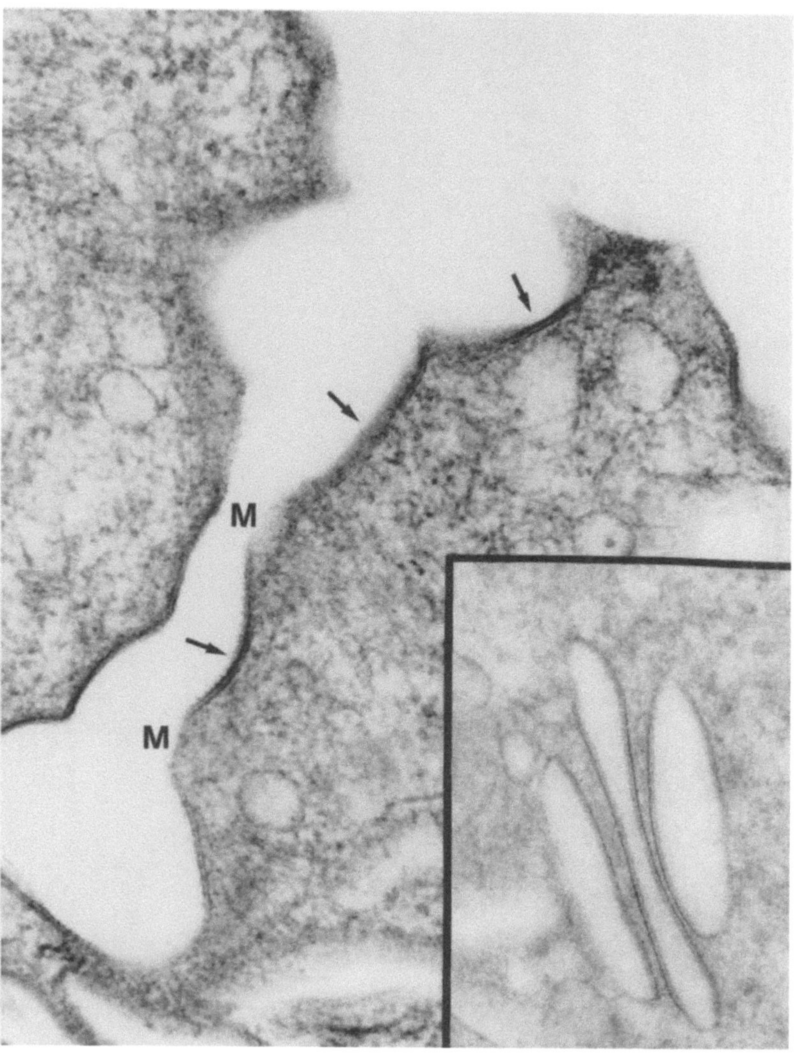

Fig. 4.4. a **High-power view of the luminal cell membrane** clearly demonstrates the two-layered structure of the concave membrane segments (← ←). The membrane bridges (*M*) do not display this bilaminarity.
b High-power view of fusiform vacuoles, whose membranes exhibit a bilaminar structure like the luminal cell membranes

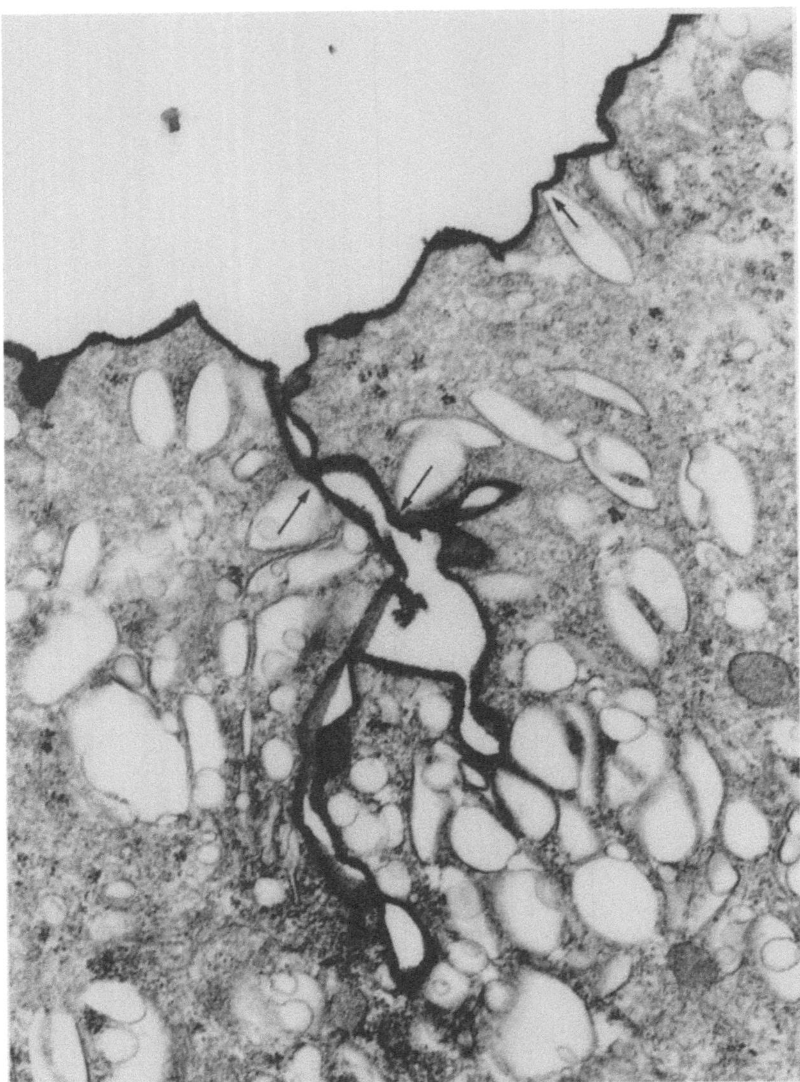

Fig. 4.5. **The fusiform vacuoles in the urothelium of the non-distended bladder are frequently in direct contact with the luminal cell membrane ($\leftarrow \leftarrow$).** The apical cell membrane is stained with horseradish peroxidase (*black*)

creasingly detached from the surrounding tissue, greater numbers of microvilli appear all around their periphery (Fig. 4.6).

4.4 Intercellular Connections in Urothelium

Disregarding special forms or subgroups, we can distinguish three types of intercellular connections or bridges in the urothelium:

1. Tight junctions
2. Gap junctions
3. Desmosomes

Tight junctions occur between adjacent lateral cell membranes in the apical cell layers (Fig. 4.7). Each tight junction encircles each luminal cell close to its luminal surface, effectively sealing off the luminal compartment from the basolateral compartment. The adjacent cells are in such intimate contact at a tight junction that even high-power electron microscopy cannot define a space between them. The height of the tight junction, identifiable by its electron-dense structure (i.e., dark in the contrast-enhanced section), measures approximately 1.5 μm in the urothelium. The tight junctions appear in freeze-fracture specimens as strands of membrane particles that encircle the cell like a belt and fuse with the "particle belts" of adjoining cells (Fig. 4.8). The actual intercellular attachment sites are con-

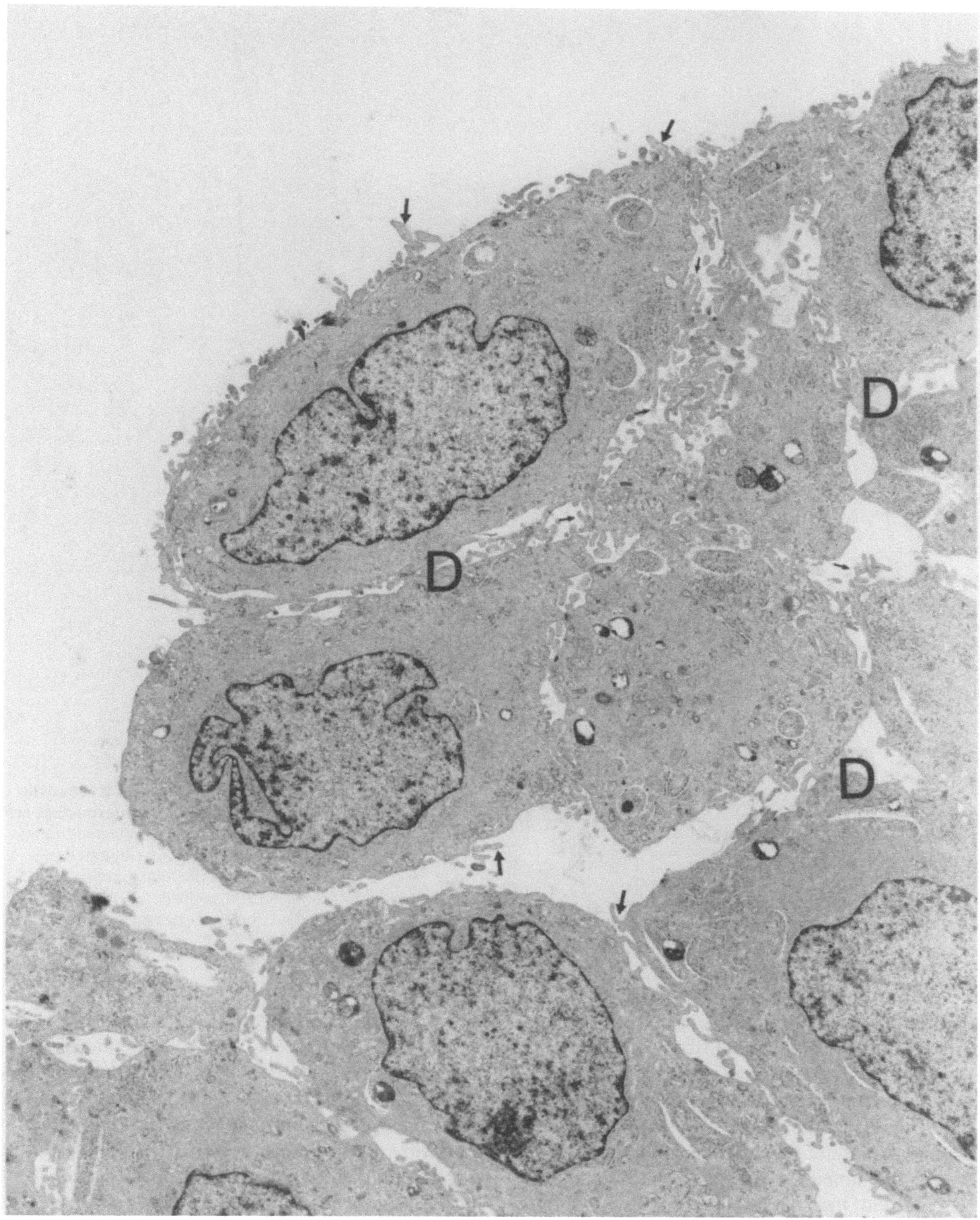

Fig. 4.6. **These urothelial cancer cells are interconnected only at single points by desmosomes** (*D*). Microcilli (← ←) appear on all sides of the cells

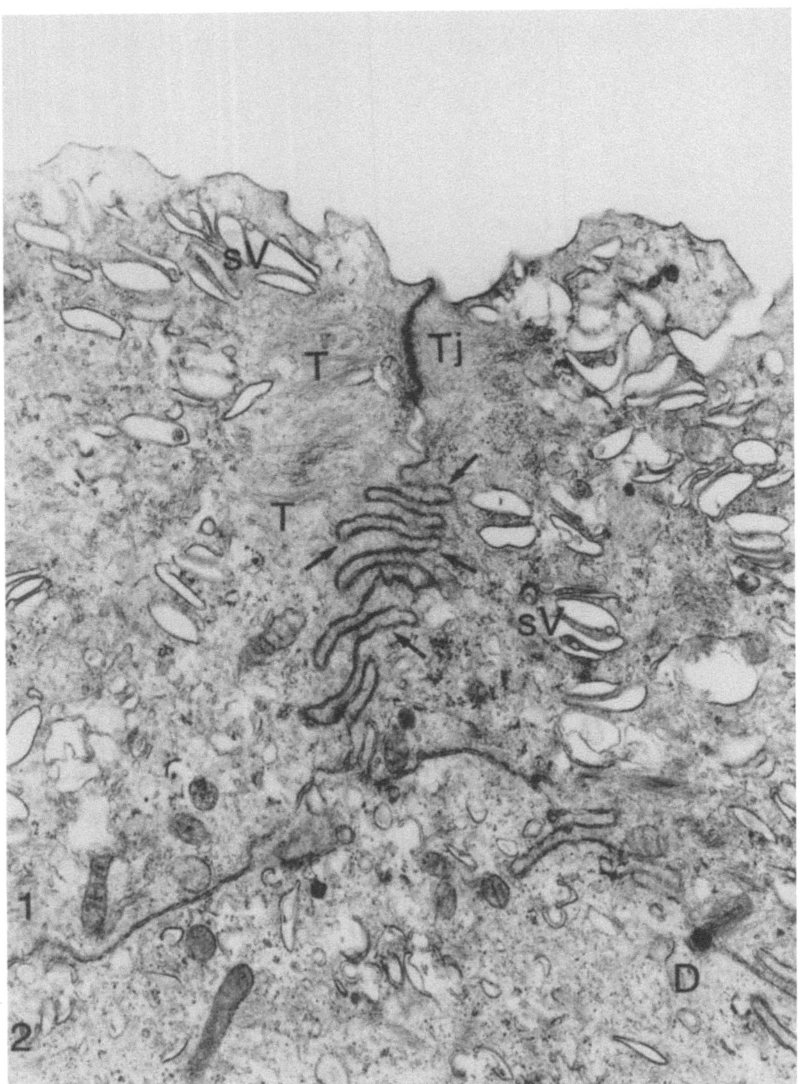

Fig. 4.7. **Contiguous cells of the luminal (*1*) and intermediate cell layers** (*2*). The cell membranes are deeply interdigitated (← ←) below the tight junction (*Tj*). *T*, tonofibrils; *sV*, fusiform vacuoles; *D*, desmosomes

fined to points along the linear strands that encircle the cells. The tight junctions in the urothelium of the moderately dilated bladder are never located directly on the epithelial surface but are always recessed from it at the base of small infoldings. In the distended bladder, however, the tight junctions approach the lumen very closely. The infoldings may be regarded, then, as reserve luminal membranes that help provide the increased epithelial surface necessary during bladder distention. Tight junctions are not found outside the apical zones.

Gap junctions, which are not visible in electron microscopic thin sections of the urothelium, show a variety of shapes and sizes on freeze-fracture replicas. They exhibit specific features in terms of loca-tion and internal structure. They appear to be confined to urothelial cells of the intermediate and basal layers (Peter 1978). The membrane particles themselves may be loosely distributed or densely arranged. The membrane particles seen in freeze-fracture preparations are the collapsed cavities of canals that interconnect the cytoplasm of adjacent cells. This junction, then, can provide direct electrical coupling of adjacent cells through an unimpeded flow of ions, as well as mechanical coupling by amino acids (Fig. 4.9). The only limiting factor is the diameter of the canals, which consistently measures approximately 20 Å.

In theory, the canal sizes are sufficient to allow for cell-to-cell transmission of *oncogenes* or oncogene products through the gap junctions, permitting their spread throughout the bladder epithelium and enabling a multifocal pattern of growth.

Desmosomes perform the physiologic task of establishing mechanical couplings between individual cells in a given tissue. Thus, desmosomes are particularly well developed in tissues that are subjected to large mechanical stresses. The desmosomes join with the tonofilaments to form a cytoskeleton which prevents excessive stretching and deformation of the cells. As in other cell systems, the desmosomes form attachment sites that ensure the mechanical coupling of one urothelial cell to an adjoining urothelial cell. Desmosomes appear in sections as electron-dense, finely fibrous thickenings of the cell membrane (Fig. 4.10). The opposing membrane surfaces are never directly adherent to each other, and the intercellular space is occupied by electron-dense material.

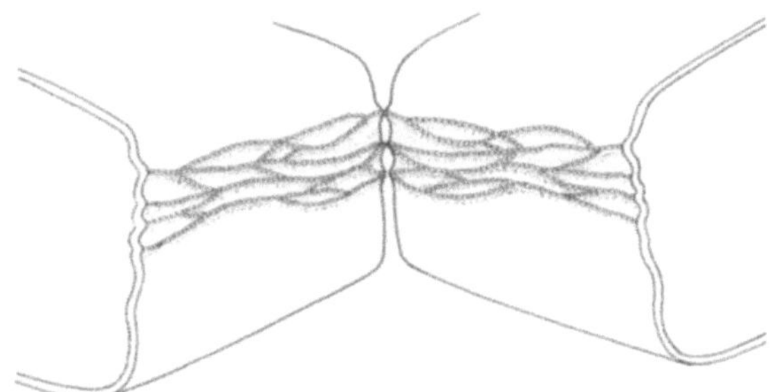

Fig. 4.8. **Schematic diagram of an opened tight junction, which encircles the cells like a belt.** The width of the tight junctions is variable and depends on the number of particle strands

A *reduction of intercellular attachments* ranging to a complete loss of intercellular cohesiveness is one of many characteristics of poorly differentiated bladder tumors. At least in some tumors, the reduction of tight junctions correlates directly with the degree of anaplasia, the increased transepithelial permeability giving carcinogens easier access to the basal cell layer (Simani et al. 1974).

In the cervical epithelium as well, McNutt et al. (1971) noted a correlation between the number of gap junctions (nexuses) and the grade of malignan-

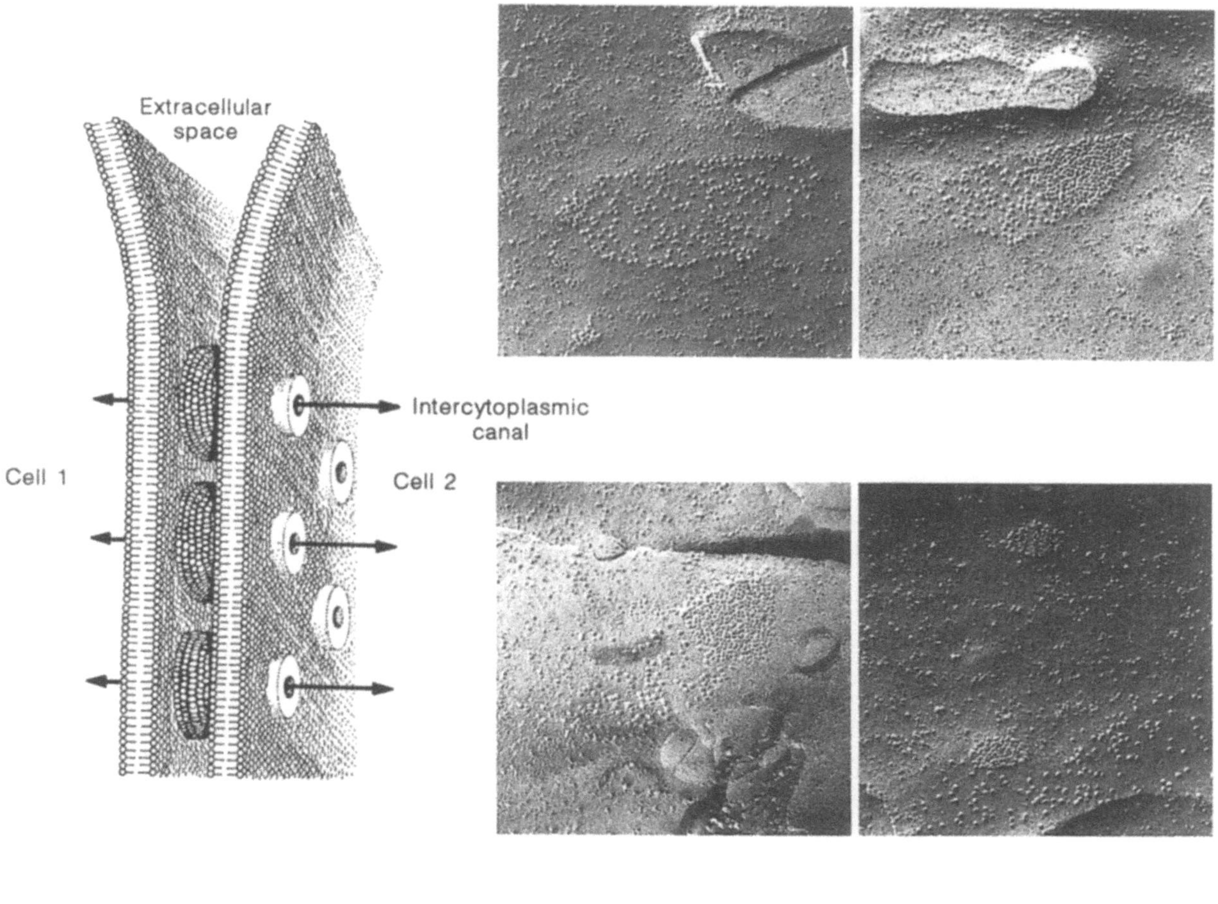

Fig. 4.9. **Schematic and electron microscopic depictions of gap junctions.** The electron micrographs illustrate the various patterns of particle distribution in gap junctions. The particles represent the collapsed canals that interconnect the cytoplasm of adjacent cells

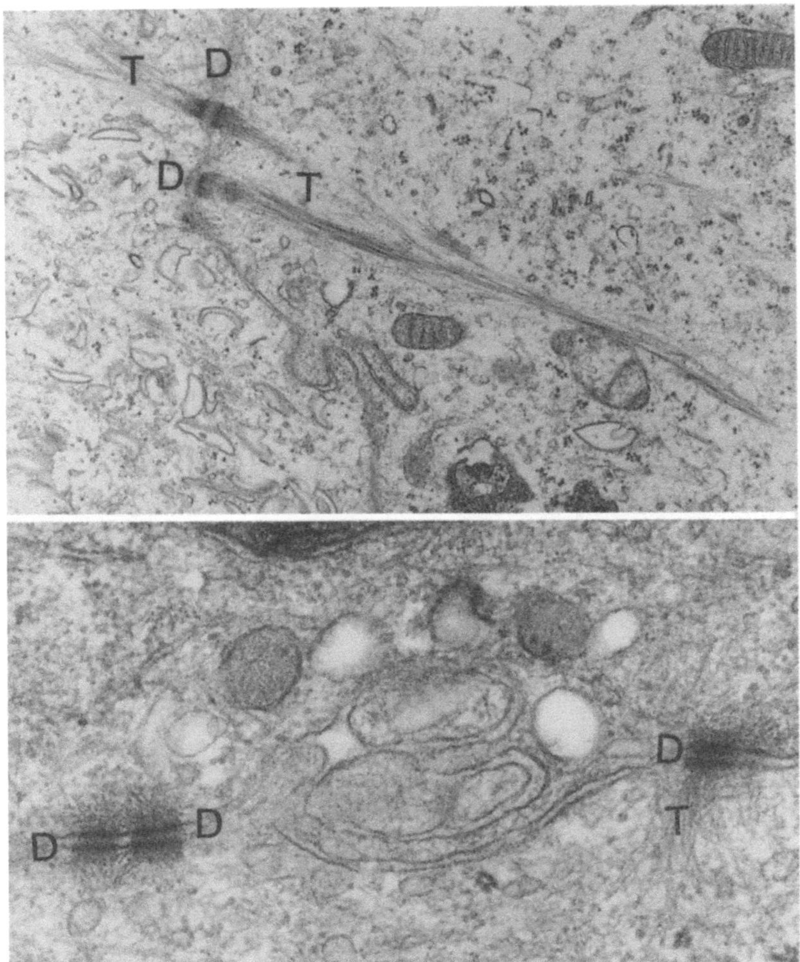

Fig. 4.10. **Desmosomes** (*D*),
from which tonofibrils (*T*)
radiate into the cells

cy. Gap junctions are abundant in cervical epithelium that is normal or mildly dysplastic, but they are sparse in malignant and invasive cervical cancers. Since the decrease in gap junctions was observed prior to invasive growth, the authors suggest that deficient cell coupling promotes disordered, invasive growth, perhaps following the cell-to-cell propagation of malignant growth via the gap junctions.

The loss of desmosomes in dedifferentiated urothelial cells is associated with diminished intercellular cohesiveness. The reduced cohesiveness of tumor cells leads to a copious sloughing of tumor cells from bladder carcinoma and thus assumes practical importance in cytologic diagnosis.

References

Englund SE (1956) Observations on the migration of some labelled substances between the urinary bladder and the blood in rabbit. Review of the literature. Acta Radiol Suppl (Stockholm) 135: 1

Hicks RM (1966) The function of the Golgi complex in transitional epithelium. J Cell Biol 30: 623

Leaf A (1966) On the functional structure of the transport system in the toad bladder. Proceedings of the third international congress on nephrology, Washington. Karger, Basel

Lewis SA, Diamond JM (1975) Active sodium transport by mammalian urinary bladder. Nature 253: 747

Lewis SA, Eaton DC, Diamond JM (1976a) Na^+transport by rabbit urinary bladder, a tight epithelium. J Membr Biol 28: 1

Lewis SA, Eaton DC, Diamond JM (1976b) The mechanism of Na^+transport by rabbit urinary bladder. J Membr Biol 28: 41

McNutt NS, Hershberg RA, Weinstein RS (1971) Further observations on the occurrence of nexuses in benign and malignant human cervical epithelium. J Cell Biol 51: 805

Minsky BD, Chlapowski FJ (1978) Morphometric analysis of the translocation of lumenal membrane between cytoplasm and cell surface of transitional epithelial cells during the expansion-contraction cycles of mammalian urinary bladder. J Cell Biol 77: 685

Nelson RA, Jones JD, Wahner HW, McGill DB, Code CF (1975) Nitrogen metabolism in bears: urea metabolism in summer starvation and in winter sleep and role of urinary bladder in water and nitrogen conservation. Mayo Clin Proc 50: 141

Peter S (1978) The junctional connections between the cells of the urinary bladder in the rat. Cell Tiss Res 187: 439

Peter S (1985) Das Übergangsepithel der Harnblase: Ein mehrschichtiges Epithel. In: Harzmann et al. (eds) Experimentelle Urologie. Springer, Berlin Heidelberg New York Tokyo

Petry G, Amon H (1966) Licht- und elektronenmikroskopische Studie über Struktur und Dynamik des Übergangsepithels. Z Zellforsch 69: 587

Porter KR, Kenyon KR, Badenhausen S (1965) Origin of discoidal vesicles in cells of transitional epithelium. Anat Rec 151: 401

Schütz W (1980) Aktiver transepithelialer Natriumtransport am isolierten Warmblüterurothel. Habilitationsschrift, University of Munich

Simani AS, Inoue S, Hogg JC (1974) Penetration of respiratory epithelium of guinea pigs following exposure to cigarette smoke. Lab Invest 31: 75

Valtin H (1973) Renal function mechanisms: Preserving fluid and solute balance in health. Boston, Little, Brown and Co.

5 Epidemiology, Etiology, and Classification of Urothelial Tumors

H. Rübben and C. Hunold

CONTENTS

5.1 Introduction

Knowledge about the etiology and epidemiology of urothelial tumors provides a rationale for the selective use of urinary cytology in the screening of high-risk individuals and population groups.

The etiologic relationship between exposure to harmful agents and malignant neoplasia was noted as early as 1775 by Percival Pott, who described scrotal tumors in chimney sweeps. Since then, countless studies have been published on the increased risk of malignant disease linked with exposure to environmental factors.

Like Ludwig Rehn, who first postulated aromatic amines as a cause of bladder carcinoma in 1895, clinicians usually recognized the causal relationship between agent and tumor when they noted unusually high incidences of disease occurrence in certain population groups.

Today, epidemiologic studies have become so complex that they are reserved for epidemiologists. Nevertheless, there are several reasons why etiology is still a matter of concern for clinicians:

1. For the discovery and reporting of tumor cases caused by occupational exposure to known carcinogens, because

- Not all carcinogens are known.
- All known offending substances have not been withdrawn from use despite laws prohibiting the use and manufacture of the most potent urothelial carcinogens.
- The latent period between carcinogen exposure and disease onset may be many years, and we see occupationally related urothelial cancers in individuals exposed decades previously.

2. For patient education. The physician is obliged to inform the patient and his family about potential exposure so that they can proceed with possible compensation claims following an epidemiologic risk assessment.

3. For the identification of new carcinogens. Many new potentially carcinogenic substances are placed into circulation each year. Reports on findings submitted to the epidemiologist by the clinician can aid in the early identification of these carcinogens.

4. For the identification of etiologic factors that influence the further course of the disease. Most bladder tumor patients remain under the care and supervision of a urologist for many years. During this time new tumors may develop and spread by infiltration or metastasis. There is reason to believe that urothelial carcinogenesis is based on a multifactorial process in which the initiating event may have no therapeutic relevance, but the persistence of carcinogenic exposure will adversely affect the course of the disease.

5.2 Epidemiology

The basis for epidemiologic studies is the observation that diseases do not occur randomly. The essential goal of epidemiology is to identify these nonrandom events. This can be done descriptively or analytically.

Descriptive studies document the morphology, stage, and differentiation of tumors based on population statistics, mortality statistics, and special tumor registries, and they supply information on incidence, prevalence, and mortality taking into account the factors of age, sex, race, geographic distribution, and temporal development.

Analytic studies can supply considerably more information than descriptive studies because they are directed toward a specific inquiry.

In a *cohort study*, the group under study is identified by its exposure to a certain carcinogen over a designated period of time. In a *case control study*, the study group is defined by the disease itself and is compared with a corresponding control group that matches the study group in all respects (age, sex, etc.) except that it is unaffected by the neoplastic disease.

5.2.1 Incidence

The occurrence of bladder tumors in a population can be stated as either an age-specific incidence or an age-normalized incidence for the population as a whole and is expressed as the number of new cases per 100 000 inhabitants per year. The age-specific incidence of bladder cancer is the same in the United States and Western Europe; the incidence data are listed in Table 5.1.

The relative incidence of bladder cancer rises sharply during the sixth decade of life. Incidence by gender in the Western industrialized nations shows an approximately 3:1 preponderance of males.

Studies by the National Cancer Institute show that, between 1939 and 1971, the incidence of bladder cancer rose from 14.1 to 21.3 in white males and from 3.8 to 9.8 in black males.

The incidence of bladder cancer varies significantly among different continents, countries, and even among different regions of the same country. For example, the mortality rate is 7.9 in South Africa, approximately 5.5 in Central Europe, but only 2.4 in Japan. The apparent causal significance of environmental factors in bladder cancer prevalence is further supported by a comparison of urban and rural populations, indicating a two times higher incidence among city dwellers (Morrison and Cole 1976).

Table 5.1. **Age-specific and age-normalized incidence rates for bladder cancer in the white population of the United States** (after Cutler and Young 1975)

Age	Males (%)	Females (%)
0–19	0,2	0,2
20–29	1,0	0,3
30–39	2,5	1,0
40–49	10,3	2,7
50–59	32,4	10,5
60–69	91,2	25,9
70–79	172,3	42,2
<80	201,5	62,5

5.3 Etiology

5.3.1 Aromatic Amines (Arylamine)

The first report by Rehn (1895) on the development of bladder tumors in four chemical dye workers has been followed by numerous reports from Europe and the U.S. Hueper et al. (1938) supplied the first experimental proof that aromatic amines cause bladder tumors.

Transitional cell tumors of the bladder developed in dogs that had been fed 2-naphthylamine. This finding prompted the chemical industry to conduct extensive epidemiologic studies (Case et al. 1954). The results in Great Britain showed that workers who were exposed to 1-naphthylamine, 2-naphthylamine, auramine, fuchsin, or benzidine had an increased risk of developing a bladder tumor, and that this risk was proportional to the intensity and duration of the exposure. The data did not demonstrate an increased risk associated with the manufacture or use of aniline (Case and Pearson 1954). Rehn's original theory had been that aniline was the agent responsible for the development of bladder tumors, and even today aniline dyes are still used as a synonym for occupationally induced bladder cancer.

On the basis of these findings, manufacture of the offending agents was banned in subsequent years. Japan was the last nation, in 1972, to suspend the manufacture of 2-naphthylamine (Ohkawa et al. 1982).

Melick identified 4-aminobiphenyl as a urothelial carcinogen that had caused bladder tumors in workers at two factories in the U.S. A 17-year follow-up demonstrated a bladder cancer rate of 16.1%–18.5% in exposed workers (Melick et al. 1971).

Chemical urothelial carcinogens generally have comparable chemical structures and frequently belong to the class of aromatic amines. They are absorbed via the gastrointestinal tract, lung, and skin.

Known human bladder carcinogens include:

- 2-Naphthylamine
- Benzidine
- 4-Aminobiphenyl
- Dichlorobenzidine
- Orthodianisidine
- Orthotolidine
- Phenacetin
- Chlornaphazine
- Cyclophosphamide

Many occupations are known that have involved worker exposure to aromatic amines. Despite stringent surveillance and controls, it is inevitable that some workers will still develop occupationally induced bladder tumors, because a latent period of 10–40 can elapse between exposure to the agent and the development of cancer.

Occupational groups in the following industries are at particularly high risk:

- Dye industry
- Rubber processing industry (cables, etc.)
- Gas production in the coal industry
- Exterminators
- Laboratory employees
- Aluminum industry
- Textile industry, including textile dyeing
- Printing industry
- Kimono painters, hairdressers

In the case of 2-naphthylamine, for example, a seven times higher incidence of bladder cancer has been demonstrated in exposed workers compared with nonexposed workers (Schulte et al. 1986). Of interest are azo dyes, which release benzidine when degraded by bacterial action. This is one reason for the high incidence of bladder tumors among Japanese kimono painters, who ingest the azo dyes when they lick their brushes.

The metabolic pathway by which aromatic amines are transformed into urothelial carcinogens is largely known and has been described by Lower (1982). After hydroxylation of the aromatic amines in the N position within the liver (Radomski and Brill 1970; Cramer et al. 1960), the N-hydroxy metabolites are glucoronated and excreted in the urine. The rate of inactivation of aromatic amines by this enzyme system varies among different species, but fast and slow inactivation types are also observed within the same individual. This could account for the different sensitivities of different persons to carcinogen exposure (Cartwright et al. 1982), although a clear correlation between a slow acetylation system and the development of bladder tumors has not yet been

demonstrated in man. The conjugated N-hydroxy metabolites must be deconjugated in the urine before they can interact with the urothelial cell. Urinary enzymes such as β-glucoronidase that originate from the kidney or from the urothelium itself can effect this decoupling and thereby release the active carcinogen. Attempts to block β-glucoronidase in experiments on bladder tumor induction (Boyland et al. 1960) and in the treatment of bladder tumor patients (Boyland et al. 1964) have been unsuccessful in terms of preventing recurrence. N-hydroxylamines lose water in acidic urine, leading to the formation of electrophilic aryl nitrite ions that can react with the cellular nucleic acid (Kadlubar et al. 1978). This produces the DNA alteration that is the basic prerequisite for carcinoma induction.

5.3.2 Cigarette Smoking

Many retrospective and prospective studies have demonstrated an increased risk of bladder cancer in cigarette smokers. The relative risk compared with nonsmokers is from 2:1 to 6:1 (Cole 1971; Kunze et al. 1986). It has been estimated that nicotine has some degree of etiologic significance in 30%–40% of bladder tumor patients (Cole 1973; Wynder and Goldsmith 1977). Armstrong and Doll (1974) attributed the rise in bladder cancer mortality among males since 1870 to an increase in cigarette consumption. Augustine et al. (1988) documented the influence of the duration and quantity of cigarette smoking, finding that the risk in a 40-year smoker is twice that in a 20-year smoker. Apparently the quality of the tobacco also affects risk, since the use of dark grades of tobacco is associated with an approximately three times higher risk of bladder cancer. This is attributed to the higher concentration of aromatic amines in dark tobacco, resulting in higher blood levels of 4-aminobiphenyl. So far an increased risk has not been established in pipe and cigar smokers.

The analysis of cigarette smoke has identified 2-naphthylamine as the principal offending agent (Hoffman et al. 1969; Hecht et al. 1976; Patrianakos and Hoffman 1979). Little has been learned about the effect of continued cigarette smoking in known bladder tumor patients, although it is reasonable to expect that the cessation of nicotine abuse in patients with superficial tumors would favorably influence the course of the disease.

5.3.3 Urinary Tract Infections

5.3.3.1 Nonspecific Chronic Urinary Tract Infection, Bladder Stone

An increased incidence of bladder cancer has been found in patients with chronic urinary tract infections, especially when accompanied by bladder stones or the prolonged use of an indwelling catheter (Wynder and Goldsmith 1977). Most of the tumors are squamous cell carcinomas. In paraplegic patients with permanent urinary catheter diversion, Olson and de Vere White (1979) detected diffuse squamous cell bladder cancers in five of 100 patients, and Kaufman et al. (1977) in six of 62 patients. Five of the six patients had used an indwelling catheter for more than 10 years.

The apparent carcinogenic agents in these cases are nitrosamines formed by bacteria (*Escherichia coli* and certain *Proteus* species) from amines and nitrates. This process was demonstrated in vitro (Hawksworth and Hill 1971) and later in animals (Hawksworth and Hill 1974) and in humans (Radomski et al. 1978; Hicks et al. 1977). Unfortunately, nitrosamines are difficult to analyze in the urine at the present time, so there is no practical method of screening the urine for these substances.

The cocarcinogenic or "promoter" role of intravesical stones has been well documented (Clayson 1974; Harzmann et al. 1980). An indwelling catheter can likewise cause a chronic mucosal irritation that promotes the bacterial production of carcinogens. A prime risk group, then, would be patients with neurogenic bladder dysfunction who develop recurrent infections, bladder stones, and persistent catheter-induced mucosal irritation over a period of decades (Bejany et al. 1987). A similar pathogenic mechanism could account for the increased risk in patients with bladder exstrophy (Abeshouse 1943; Engel and Wilkinson 1970).

5.3.3.2 Schistosomiasis

Schistosomiasis is endemic to large portions of Africa and Arab countries. The acute stage of *Schistosoma haematobium* infection is marked by the formation of granulomatous polyps in the bladder which mimic the features of a tumor. These changes are reversible with effective treatment of the schistosomiasis. If the infection becomes chronic, processes of epithelial hyperplasia, dysplasia, and squamous cell metaplasia frequently give rise to squamous cell cancer (Morrison and Cole 1982; Hicks et al. 1977). This is believed to relate causally to an infection-induced production of nitrosamines. Examination of the urine sediment for schistosome eggs and the detection of tumor cells by urinary cytology can serve as screening methods in chronically infected patients.

5.3.3.3 Endemic Nephropathy

A relatively high incidence of urothelial cancers has been noted in certain regions of Yugoslavia, Rumania, Bulgaria, and Greece in connection with "Balkan nephropathy" (Petkovic et al. 1971). Ninety percent of the tumors arise in the upper urinary tract, and 10% are bilateral. The etiology appears to relate to a saprophytic fungus present in stored grain, which elaborates nephrotoxins and carcinogenic mucotoxins (Sattler et al. 1977).

5.3.4 Artificial Sweeteners

The artificial sweeteners saccharin and cyclamate have been increasingly used in food items for some years. In multistage experiments, rats were first given subcarcinogenic oral doses of bladder carcinogens followed by oral saccharin or cyclamate. These animals developed bladder tumors more frequently than control animals that were not fed the sweeteners. It was concluded that sweeteners act as cocarcinogens, though there was no evidence that they could initiate carcinogenesis by themselves (Hicks et al. 1978; Cohen et al. 1979).

A rise in bladder cancer has not been noted clinically since the introduction of saccharin (Armstrong and Doll 1974). It would be premature to draw a definitive conclusion, however, because saccharin has been widely used only since the 1960s. Diabetics, who consume more saccharin on average than the normal population, tend to show a lower rather than higher incidence of bladder neoplasia (Armstrong and Doll 1974). Eight of nine case control studies (IARC 1980) did not document a heightened risk in artificial sweetener users, and only Howe et al. (1977) reported on a dose-dependent risk increase of 1.6. If there is a link between artificial sweeteners and the induction of bladder cancer, it is apparently quite small. It would be more appropriate to focus on the effects of excessive caloric intake and overweight on cancer risk.

5.3.5 Coffee Consumption

Since Cole (1971) suggested that coffee was a carcinogen, other authors (Weinberg et al. 1983; Morgan and Jain 1974; Morrison et al. 1982) have stated that the carcinogenic effect of coffee is controversial, although a marked risk increase has been documented in coffee drinkers who also smoke.

5.3.6 Drugs, Radiation Therapy

Three drugs have definitely been linked to the pathogenesis of bladder cancer in humans:

- *Chlornaphazine*, used to treat polycythemia and chemically related to β-naphthylamine. It was in use until 1963.
- *Phenacetin* has been associated with the development of interstitial nephritis (phenacetin nephropathy) and an increased incidence of urothelial carcinoma mainly affecting the upper urinary tract. Five to 10% of patients with phenacetin nephropathy develop a urothelial cancer (Gonwa et al. 1980). The active carcinogenic agent is a nitrogen hydroxyl metabolite of phenacetin, which possesses the chemical structure of an aromatic amine (Rathert et al. 1975).
- *Cyclophosphamide* induces a symptomatic or asymptomatic chemical cystitis that is associated with an increased risk of bladder cancer (Pearson and Soloway 1978; Fairchild et al. 1979). Since the introduction of cystitis prophylaxis by Mesna, the risk of bladder cancer may have become negligible. Even so, it is prudent to schedule regular urinary cytologic examinations in patients who have received prolonged treatment with cyclophosphamide.
- Duncan et al. (1977) noted an increased incidence of bladder tumors in patients who received pelvic radiation for gynecologic tumors. The risk is small, however, and should not be used as an argument against appropriate radiotherapy during the planning of treatment.

5.3.7 Bracken Fern

Bracken fern (*Pteridium aquilinum*) contains a still-unidentified carcinogen that induces bladder tumors when fed to cows (Pamukcu et al. 1976). Bracken fern is rarely consumed by humans except in Japan, where a link has been established between bracken fern consumption and esophageal cancer (Hirayama 1979).

5.3.8 Endogenous Factors

Genetic dispositions have not yet been identified in the etiology of bladder cancer. However, much research has been done to detect biochemical and metabolic processes that might relate to the development of bladder tumors.

The effect of acetylation by *N*-acetyltransferase and the metabolism of the amino acid tryptophane have already been discussed. Experiments in heterotypic rat bladder transplants have additionally demonstrated a bladder tumor growth-promoting effect of normal rat urine. Most bladder tumors are highly differentiated lesions with a high rate of recurrence. Yura et al. (1989) attribute this propensity for recurrence to a growth factor present in the urine that stimulates dormant neoplastic cells. Factors isolated from fresh urine by column chromatography were tested for their induction of ornithine decarboxylase, and two of these factors administered to rat bladders caused a significantly higher incidence of *N*-methyl-*N*-nitrose urea-induced bladder tumors than in a control group treated with saline solution. Analysis of the two chromatographic fractions identified epidermal growth factor and transferrin, which promote the activation of ornithine decarboxylase and thus increase the rate of mitosis.

Ekman and Strombeck (1947) were the first to attribute bladder carcinogenicity to metabolites of the amino acid tryptophan. Dunning et al. (1950) and Cohen et al. (1979) demonstrated the promoting effect of tryptophan in the pathogenesis of bladder cancer in laboratory animals. In humans, tryptophan is almost completely metabolized to carbon dioxide and water. Only 2% is excreted as metabolites in the urine. These tryptophan metabolites may be reabsorbed through the bladder mucosa. Pyridoxine (vitamin B_6) is an essential component of the tryptophan metabolism. In patients with vitamin B_6 deficiency, the urinary excretion of tryptophan metabolites is increased to 27% (Yess et al. 1964).

Byar and Blackard (1977) investigated the clinical efficiency of vitamin B_6 therapy on bladder cancer but were unable to demonstrate a beneficial effect. An increase in the excretion of tryptophan metabolites has been detected in patients with prostatic cancer, chronic urinary tract infections, rheumatoid arthritis, lupus erythematosus, scleroderma, porphyria, and in pregnancy. Some studies indicate an elevation of urinary tryptophan metabolites in bladder cancer as well (Boyland and Williams 1956; Brown et al. 1969; Price et al. 1965), although a case control study by Friedlander and Morrison (1981) was unable to confirm this. On the whole, then, the causal significance of tryptophan metabolites in bladder carcinogenesis remains unclear, and for the present they have no clinical implications.

5.3.9 Experimental Induction of Bladder Cancer

Many chemicals have demonstrated the ability to induce bladder tumors in laboratory animals (Clayson and Cooper 1970), but only a few have proven useful in terms of developing animal models for the study of urothelial carcinogenesis, among them (*N*-[4-(5-nitro-2-furyl)-2- iazolyl]formamide (FANFT), butyl-(4-hydroxybutyl)-nitrosamine (BBN), and *N*-methyl-*N*-nitrose urea (MNU).

Experimental carcinogenesis corresponds largely to clinical observations and vividly illustrates the morphologic changes that are important in cytologic evaluations. *The urothelium is a highly specialized mucosa functioning as a barrier between the urine and interstitium* and as the lining for a hollow organ that undergoes substantial fluctuations in its volume and wall tension.

Normal cell turnover in the urothelium is slow, occurring on the order of 200 days. After injury, however, the urothelium can proliferate very rapidly. Samma et al. (1987) showed that rat bladders denuded of urothelium by detergents underwent complete reepithelialization within a few days.

Exposure to a carcinogen also incites proliferation, which can result morphologically in urothelial hyperplasia. This process is reversible if the carcinogen is withdrawn at an early stage, but even a single contact with a potent carcinogen can cause cancer to develop following an adequate latent period. Further exposure stimulates subepithelial capillary invasion and the formation of papillary structures. The process apparently is still reversible at this stage. Later the uppermost cell layer loses its regular surface structure, cellular interactions are impaired, and cytologic changes become apparent. At this stage the proliferation continues to progress even if the carcinogen is withdrawn.

Another clinically important observation is that bladder carcinogens are not urothelium-specific, but can induce adenocarcinoma even in bowel segments into which urinary discharge has been diverted. This underscores the need for a careful *endoscopic and cytologic examination in patients who have undergone cystectomy with bladder reconstruction or urinary diversion into the open bowel.*

5.3.10 Oncogenes

Many of the viruses that cause cancer in animals possess genes (oncogenes) that induce morphologic genome transformations in mouse cell cultures. When these mouse cells are reinstilled into the bladder, they form a nidus for cancer formation. It has been possible to isolate these viral oncogenes and clone them with the help of bacteria. Homologous genes have been found in the genome of healthy vertebrates (Duesberg 1983), raising questions as to the physiologic function of these genes,

their role in human carcinogenesis, and their possible identity with oncogenes.

In approximately 10% of human tumors, genes are found that, when isolated and added to mouse cell cultures, induce a malignant transformation. Their DNA code agrees with that of viral oncogenes. Human oncogenes appear to play an important role in carcinogenesis when they are no longer subject to physiologic controls.

One of these oncogenes was isolated by cloning from a human bladder tumor cell line (T24) and its DNA sequence decoded. Analysis showed a point mutation in the amino acid chain in which guanidine was replaced by thymidine (Reddy et al. 1982; Tabin et al. 1982). It is of little concern to the urologist that it was a bladder tumor cell line in which this discovery was made. Its importance lies in the introduction of carcinogenesis research at the molecular biological level. The cell transformation can be accomplished in vitro only with pathologically altered cells like the NIH 3T3 line, which probably have already undergone the initial stages of carcinogenesis. Further experiments prove that the transformation can proceed in two steps: a first step in which a kind of "immortality" is established in a normal cell culture, enabling the second oncogene to introduce the malignant transformation.

Molecular biology is a rapidly expanding science, and there is reason to hope that we will better understand the mechanism of carcinogenesis. In this regard there is growing interest in natural or synthetic substances that can reduce tumor incidence through molecular mechanisms of action. Potential tumor-preventive substances presently include vitamin A and its precursors, β-carotene, Vitamins C and E, selenium, as well as phenol antioxidants, proteinase inhibitors, and prostaglandin synthesis inhibitors.

Tumor prevention encompasses not just protection from carcinogen exposure but also the suppression of promotor activities in cells that are undergoing neoplastic change. An effective chemopreventive substance must be delivered to its site of action at the time of initiation. The primary goal of current research is the elimination of the promoting effect that sustains tumor development over a period of several years. Clinical trials will reveal the impact that the foregoing substances will have on the incidence of bladder cancer (Malone et al. 1987).

5.4 Classification

A *new version of the TNM classification* has been in effect since January 1, 1987. The essential innovation of this classification is the absence of minimum requirements in terms of diagnostic workup, which requires that additional details be given on the diagnostic measures that are performed. The basic TNM categories of *primary tumor* (T), *lymph nodes* (N), *distant metastases* (M), and *grade of malignancy of the primary tumor* (G) are preserved.

The definitions for the clinical TNM and histopathologic pTNM classifications are also retained. The clinical classification (TNM) is based on physical examination, imaging procedures, endoscopy, biopsy, surgical exploration, and other relevant findings.

The histopathologic classification (pTNM) is based on pretherapeutic findings, which are supplemented or modified by the findings at definitive operation. If the classification is in doubt, the lower category is selected. Microscopic examination is required for pathologic confirmation of lymph node involvement or distant metastases.

A special symbol for the invasion of lymphatic vessels (formerly L) is no longer used in the present system.

y Symbol: Classification during or after multimodal therapy
r Symbol: Recurrent tumor
C factor: The C (certainty) factor expresses the reliability of the classification based on the diagnostic methods employed.
C1:   Results based on standard diagnostic methods, e.g., inspection, palpation, standard X-ray views, and endoscopy.
C2:   Results based on special diagnostic measures such as imaging procedures: special X-ray views, tomography, CT, sonography, lymphography, angiography, nuclear medicine imaging, MR imaging, endoscopy, biopsy, and cytology.
C3:   Results based on surgical exploration including biopsy and cytologic examination.
C4:   Results on the extent of disease based on definitive surgery and pathologic examination of the resected specimen.
C5:   Results based on autopsy.

The clinical TNM classification corresponds to the certainty factors C1, C2, and C3, and the pathologic pTNM classification to C4.

Diagnostic measures include the history, clinical examination, excretory urography, cystoscopy, *exfoliative urinary cytology*, segmental biopsy, excisional biopsy, and bimanual palpation. These measures permit the differentiation of superficial carcinoma from carcinoma that has infiltrated the muscle. The same procedures can further differentiate superficial tumors as carcinoma in situ (Tis), noninfiltrating tumors (Ta), and tumors that have infiltrated the lamina propria (T1). Superficial tumors with a high grade of malignancy and tumors infiltrating the muscle are subdivided into metastasizing and nonmetastasizing carcinomas by checking for metastases in the liver, lung, and bone. If distant metastases have been excluded, a pelvic CT scan is indicated to exclude extensive lymph node metastases (N3). If the pelvic CT scan is negative, a staging operation is necessary to evaluate for lymph node metastasis (N1–N2). This staged diagnostic approach effectively discriminates among invasive nonmetastasizing tumors, locally advanced tumors, and tumors with lymph-node or distant metastases (Fig. 5.1).

When multiple tumors are present, the tumor is assigned to the highest T category, and "m" is added to denote multiplicity.

Primary bladder tumors are subdivided into epithelial and rare mesenchymal forms. Ninety-five percent of epithelial tumors are pure urothelial tumors with foci of glandular or squamous cell metaplasia. Pure squamous cell carcinomas or adenocarcinomas are rare.

Prognostic conclusions can be drawn from the depth of tumor infiltration and the histopathologic assessment of the degree of tumor differentiation (Bryan and Cohan 1983; Wolf et al. 1986; WHO 1973; UICC 1978).

The *prognosis is also affected by atypias* in mucosal areas adjacent to or distant from the tumor. The prognostic significance of these dysplasias also correlates with their cellular differentiation. Dysplasias are designated as:

DO: no dysplasias
D1: mild dysplasias
D2: moderate dysplasias
D3: severe dysplasias
Tis: carcinoma in situ

                                                           H. Rübben and C. Hunold

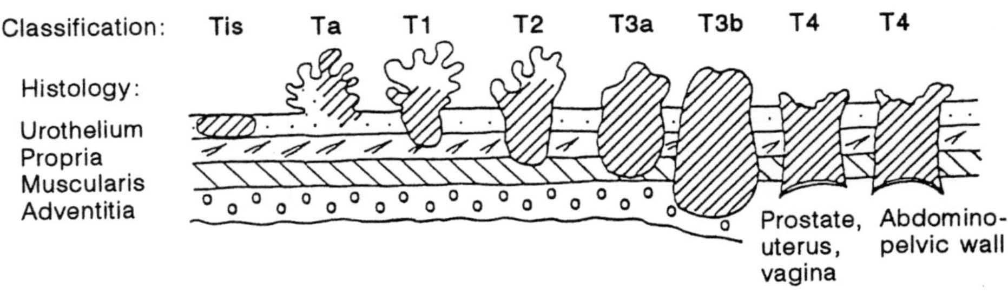

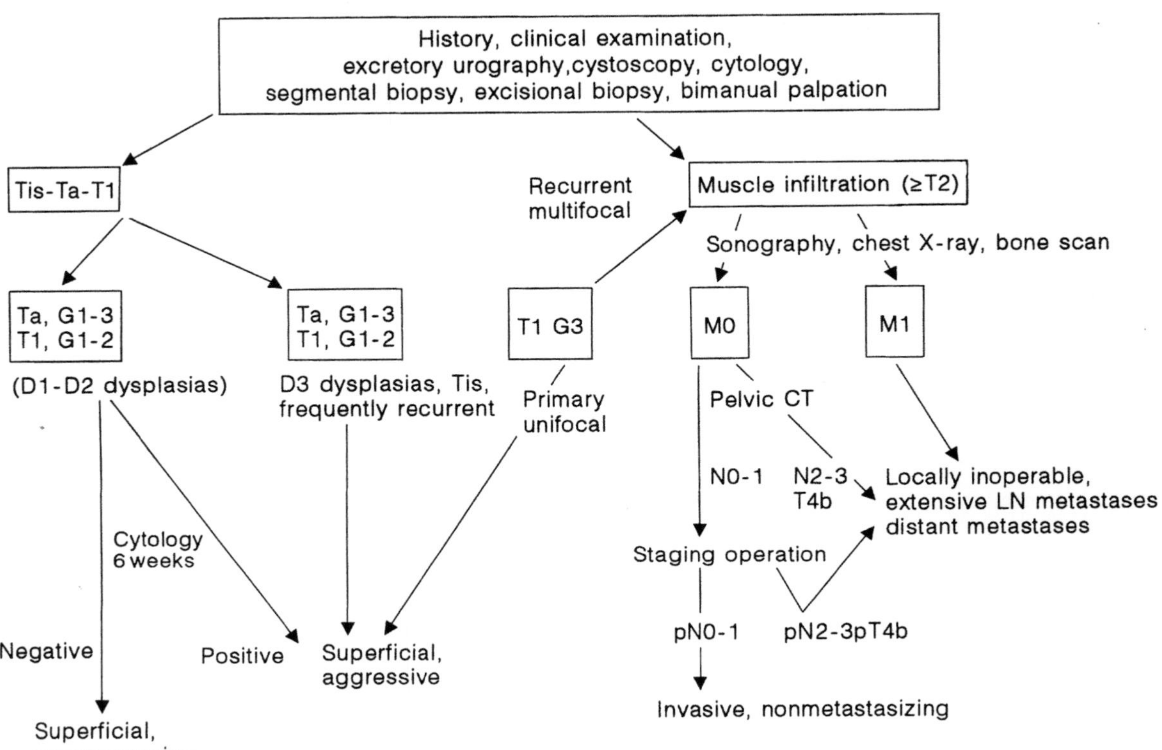

Fig. 5.1. **Flowchart for a minimal diagnostic program.** Description of the primary tumor: T category. *T0*, No evidence of a primary tumor; *Tis*, nonexophytic and noninfiltrating (carcinoma in situ); *Ta*, exophytic and noninfiltrating; *T1*, invasion of the lamina propria; *T2*, invasion of the superficial muscle; *T3a*, invasion of the deep muscle; *T3b*, invasion of the perivesical fat; *T4*, invasion of adjacent organs; *TX*, extent of infiltration cannot be assessed

Extent of lymph node metastasis (N category)
NX: regional lymph nodes cannot be evaluated
N0: no regional lymph node metastases
N1: solitary lymph node metastasis <2 cm in greatest diameter
N2: solitary or multiple lymph node metastasis 2–5 cm in greatest diameter
N3: lymph node metastasis >5 cm in greatest diameter

Extent of distant metastasis (M category)
MX: distant metastasis cannot be evaluated
M0: no distant metastases
M1: distant metastases

Determination of residual tumor
RX: residual tumor cannot be evaluated
R0: no residual tumor
R1: histologically detectable residual tumor
R2: grossly visible residual tumor

Grade of malignancy
G1: well differentiated
G2: moderately differentiated
G3: poorly differentiated
G4: undifferentiated

5.5 Summary

The rising incidence of urothelial tumors is an epidemiologic phenomenon whose etiology is based on a variety of established and probable causes. Although awareness of these causes can lead to preventive measures, the goal of providing effective treatment for urothelial cancer can be achieved only by early tumor detection in the setting of the primary diagnostic workup and follow-up. An essential part of the available diagnostic armamentarium is exfoliative urinary cytology.

References

Abeshouse BA (1943) Exstrophy of the bladder complicated by adenocarcinoma of the bladder and renal calculi. A report of a case and a review of the literature. J Urol 49: 259

Armstrong B, Doll R (1974) Bladder cancer mortality in England and Wales in relation to cigarette smoking and saccharine consumption. Br J Prev Soc Med 28: 233

Augustine A, Hebert JR, Kabat GC, Wynder EL (1988) Bladder cancer in relation to cigarette smoking. Cancer Res 48: 4405

Bejany DE, Lockardt JL, Rhamy RK (1987) Malignant vesical tumors following spinal cord injury. J Urol 138: 1390

Boyland E, Williams DC (1956) The metabolism of tryptophan. Biochem J 64: 578

Boyland E, Kinder CH, Manson D (1960) Effect of 1.4-saccharinolactone on induction of bladder cancer. Ann Rep Br Emp Cancer Campaign 38: 45

Boyland E, Wallace DM, Avis PRD, Kinder CH (1964) Attempted prophylaxis of bladder cancer with 1.4-glucosaccharolactone. Br J Urol 36: 563

Brown RR, Price JM, Satter EJ, Wear JB (1969) Tryptophan metabolism in patients with bladder cancer. Geographical differences. J NCI 43: 295

Byar D, Blackard C (1977) Comparisons of placebo, pyridoxine and topical thiotepa in preventing recurrence of stage 1 bladder cancer. Urology 10(6): 556

Cartwright RA, Glashan RW, Rogers HJ, Ahmad RA, Barham-Hall D, Higgins E, Kahn MA (1982) Role of N-acetyltransferase phenotypes in bladder carcinogenesis: a pharmacogenetic epidemiology approach to bladder cancer. Lancet 2: 842

Case RAM, Pearson JT (1954) Tumours of the urinary bladder in workmen engaged in the manufacture and use of certain dyestuff intermediates in the British chemical industry. Part II; Further consideration of the role of aniline and of the manufacture of auramine and magenta (fuchsine) as possible causative agents. Br J Ind Med 11: 213

Case RAM, Hosker ME, McDonald DB, Pearson JT (1954) Tumours of the urinary bladder in workmen engaged in the manufacture and use of certain dyestuff intermediates in the British chemical industry. Br J Ind Med 11: 75

Clayson DB (1974) Bladder carcinogenesis in rats and mice: possibility of artifacts. J Natl Cancer Inst 52: 1685

Clayson DB, Cooper EH (1970) Cancer of the urinary tract. Adv Cancer Res 13: 271

Cohen SM, Arai A, Jacobs JB, Friedell GH (1979) Promoting effects of saccharin and DL-tryptophan in urinary carcinogenesis. Cancer Res 39: 1207

Cole P (1971) Coffee-drinking and cancer of the lower urinary tract. Lancet 1: 1335

Cole P (1973) A population based study of bladder cancer. In: Doll R, Vodopija I (eds) Host environmental interactions in the aetiology of cancer in main. IARC Scientific Publication, International Agency for Research of Cancer, Lyon 7: 83

Cramer JW, Miller JA, Miller EC (1960) N-hydroxylation: A new metabolic reaction observed in the rath with the carcinogen 2-acetylaminofluorine. J Biochem Chem 235: 885

Cutler SJ, Young JL (eds) (1975) Third national cancer survey: incidence data. Natl Cancer Inst Mcnogr 41:1

Duesberg PH (1983) Retroviral transforming genes in normal cells? Nature 304: 219

Duncan RE, Ennet DW, Evans AT, Aron BS, Schellhas HF (1977) Radiation induced bladder tumours. J Urol 118: 43

Dunning WF, Curtis MR, Maun ME (1950) The effect of added dietary tryptophan on the occurence of 2-acetylaminofluorine induced liver and bladder cancer in rats. Cancer Res 10: 454

Ekman B, Strombeck JP (1947) Demonstration of tumorgenic decomposition products of 2.3-azotuolene. Acta Physiol Scand 14: 43

Engel RME, Wilkinson HA (1970) Bladder exstrophy. J Urol 104: 699 [1970]

Fairchild WV, Spence CR, Solomon HD, Gangai MP (1979) The incidence of bladder cancer after cyclophosphamide therapy. J Urol 122: 163

Friedlander E, Morrison AS (1981) Urinary tryptophan metabolites and cancer of the bladder in humans. J Natl Cancer Inst 67: 347

Gonwa TA, Corbett WT, Schey HM, Buckalew VM (1980) Analgesic associated nephropathy and transitional cell carcinoma of the urinary tract. Ann Intern Med 93: 249

Harzmann R, Gericke D, Altenahr E, Bichler K-H (1980) Induction of a transplantable urinary bladder carcinoma in dogs. Invest Urol 18: 24

Hawksworth GM, Hill MJ (1971) Bacteria and the N-nitrosation of secondary amines. Br J Cancer 25: 520

Hawksworth GM, Hill MJ (1974) The vivo formation of N-nitrosamines in the rat bladder and their subsequent absorption. Br J Cancer 29: 353

Hecht SS, Tso TC, Hoffmann D (1976) Selective reduction of tumorgenicity of tobacco smoke. IV Approaches to the reduction of N-nitrosamines and aromatic amines. Proceedings of third world conference on smoking and health. Dept Health Education and Welfare Publ (NIH) 76: 535

Hermanek P, Scheibe O, Spiessl B, Wagner G (eds) (1987) TNM-Klassifikation maligner Tumoren, 4th edn. UICC – International Union Against Cancer. Springer, Berlin Heidelberg New York

Hicks RM, Walters CL, El Sebai I, El Asser AB, El Merzabani M, Grough TA (1977) Demonstration of nitrosamines in human urine. Preliminary observations on a possible aetiology for bladder cancer in association with chronic urinary tract infection. Proc R Soc Med 70: 413

Hicks RM, Chowaniec J, Wakefield JStJ (1978) Experimental induction of bladder tumours by a two-stage system. In: Slaga TJ, Sivak A, Boutwell RK (eds) Carcinogenesis, mechanisms of tumour promotion and carcinogenesis. Raven, New York 2: 475

Hirayama T (1979) Diet and cancer. Nutr Cancer 1: 67

Hoffmann D, Masuda Y, Wynder EL (1969) Alpha-naphthylamine and beta-naphthylamine in cigarette smoke. Nature 221: 254

Howe GR, Burch JD, Miller AB (1977) Artificial sweeteners and human bladder cancer. Lancet 2: 578

Hueper WC, Wiley FH, Wolfe HD (1938) Experimental production of bladder tumours in dogs by administration of beta-naphthylamine. J Ind Hyg Toxicol 20(1): 46

IARC (1980) Monographs on the evaluation of the carcinogenic risk of chemicals to humans. Some non-nutritive sweetening agents. International Agency for Research on Cancer, Lyon, p 22

Kadlubar FF, Miller JA, Miller EC (1978) Guanyl O^6-arylamination and O^6arylation of DNA by the carcinogen N-hydroxy-1-naphthylamine. Cancer Res 38: 3628

Kaufman JM, Fam B, Jacobs SC, Gabilondo F, Yalla S, Kane JP, Rossier AB (1977) Bladder cancer and squamous metaplasia in spinal cord injury patients. J Urol 118: 967

Kunze E, Calude J, Frentzel-Beyme R (1986) Association of cancer of the lower urinary tract with consumption of alcoholic beverages – a case control study. Carcinogenesis 7: 163

Lower GM (1982) Concepts in causality: chemically induced human bladder cancer. Cancer 49: 1056

Malone WF, Kelloff GJ, Pierson H, Greenwald P (1987) Chemoprevention of bladder cancer. Cancer 60: 650

Melick WF, Naryka JJ, Kelly RE (1971) Bladder cancer due to exposure to para-aminobiphenyl: a 17 year follow up. J Urol 106: 220

Morgan RW, Jain MG (1974) Bladder cancer: smoking, beverages and artifical sweeteners. Can Med Assoc J 111: 1067

Morrison AS, Cole P (1976) Epidemiology of bladder cancer. Urol Clin North Am 3: 13

Morrison AS, Cole P (1982) Urinary tract. In: Schottenfeld D, Fraumeni FJ, Saunders WB (eds) Cancer epidemiology and prevention. Saunders, Philadelphia, p 925

Morrison AS, Buring JE, Verhoek WG, Aoki K, Leck I, Ohno Y, Obata K (1982) Coffee drinking and cancer of the lower urinary tract. J Natl Cancer Inst 68: 91

Ohkawa T, Fujinaga T, Doi J, Ebisuno S, Takamatsu M, Nakamura J, Kido R (1982) Clinical study on occupational uroepithelial cancer in Wakayama City. J Urol 128: 520

Olson CA, Vere White RW de (1979) Cancer of the bladder. In: Javadpour N (ed) Principles and management of urologic cancer. Williams & Wilkins, Baltimore, p 337

Pamukcu AM, Erturk E, Yalciner S, Bryan GT (1976) Histogenesis of urinary bladder cancer induced in rats by bracken fern. Invest Urol 14: 213

Patrianakos C, Hoffman D (1979) Chemical studies on tobacco smoke. LXIV On the analysis of aromatic amines in cigarette smoke. J Anal Toxicol 3: 150

Pearson RM, Soloway MS (1978) Does cyclophosphamide induce bladder cancer? Urology 4: 437

Petkovic S, Multavdzic M, Petronic V, Marcovic V (1971) Tumours of the renal pelvis and ureter: clinical and aetiologic studies. J Urol Nephrol (Paris) 77: 429

Pott P (1775) Cancer scroti. In: Chirurgical works. Howes Clark & Collins, London

Price JM, Brown RR, Ellis ME (1965) Quantitative studies on the urinary excretion of tryptophan metabolites by human ingesting a constant diet. J Nutr 60: 323

Radomski JL, Brill E (1970) Bladder cancer induction by aromatic amines: role of N-hydroxy metabolites. Science 167: 992

Radomski JL, Greenwald D, Hearn WL, Block NL, Woods FM (1978) Nitrosamine formation in bladder infections and its role in the etiology of bladder cancer. J Urol 120: 48

Rathert P, Melchior HJ, Lutzeyer W (1975) Phenacetin: a carcinogen for the urinary tract. J Urol 113: 653

Reddy EP, Reynolds RK, Santos E, Barbacid M (1982) A point mutation is responsible for the acquisition of transforming properties by the T 24 human bladder carcinoma oncogene. Nature 300: 149

Rehn L (1895) Blasengeschwülste bei Fuchsin-Arbeitern. Arch Klin Chir 50: 588

RUTTAC (Registry for urinary tract tumours RWTH Aachen) (1981) Arbeitssitzung und Jahresbericht des 'Register und Verbundstudie für Harnwegstumoren RWTH Aachen'. Verh Dtsch Ges Urol 33: 559

Sakashita SH, Matsuda H, Nagamori S, Sakakibara N, Maru A, Koyanagi T (1988) Papillary adenoma of the bladder in a patient with intermittent self-catheterization. Urol Int 43: 107

Samma SH, Homma Y, Oyasu R (1987) Rat urinary bladder denuded of urothelium. Am J Pathol 128: 328

Sattler TA, Dimitrov T, Hall PW (1977) Relation between endemic (Balkan) nephropathy and urinary tract tumours. Lancet 2: 278

Schulte PA, Ringen K, Hemstreet GP, Altekruse EB, Gullen WH, Tillett S, Allsbrook WC et al. (1986) Risk factors for bladder cancer in a cohort exposed to aromatic amines. Cancer 58: 2156

Tabin CJ, Bradley SM, Bargmann CI, Weinberg RA, Papageorge AG, Scolnick EM, Dhar R et al. (1982) Mechanism of activation of a human oncogene. Nature 300: 143

UICC (International Union Against Cancer) (1978) TNM classification of malignant tumors. UICC, Genf

Weinberg DM, Ross RK, Mack TM, Paganini-Hill A, Henderson BE (1983) Bladder cancer etiology. A different perspective. Cancer 51: 675

WHO (World Health Organization) (1973) International histological classification of tumors, histological typing of urinary bladder tumors. WHO, Geneva

Wolf H, Kakizoe T, Smith PH, Brosman SA, Okajima E, Rübben H, Utz DC (1986) Bladder tumors – treated natural history. In: Denis L, Niijima T, Prout G Jr, Schröder FH (eds) Developments in bladder cancer. Liss, New York, p 221–223

Wynder EL, Goldsmith R (1977) The epidemiology of bladder cancer: a second look. Cancer 40: 1246

Yess N, Price JM, Brown RR, Swann PB, Linkswiler H (1964) Vitamin B_6 depletion in man: urinary excretion of tryptophan metabolites. J Nutr 84: 229

Yura Y, Hayashi O, Kelly M, Oyasu R (1989) Identification of epidermal growth factor as a component of the rat urinary bladder tumor-enhancing urinary fractions. Cancer Res 49: 1548

6 Urothelial Atypia and Dysplasia

F. Hofstädter

CONTENTS

6.1 Introduction

Histopathology has long recognized "flat" lesions of the urothelium, a certain percentage of which are missed by gross examination. These lesions may occur in isolation or as an accompaniment to overt inflammatory or neoplastic disease. In this article we shall discuss only those lesions that bear a distinct relationship to urothelial cancer. These flat "preneoplastic" or early neoplastic changes are of major clinical and biologic significance, and they hold the key to understanding the behavior of urothelial carcinomas and their cytologic classification.

Unfortunately, inconsistencies in the *nomenclature* of the changes make it difficult to apply many of the discoveries that have been made in basic research. The different terms often carry meanings that are understood differently by different authors and tend to develop a certain life of their own. So we shall begin our discussion by reviewing the nomenclature of these lesions in an effort to standardize usage. We shall then describe associated *histopathologic criteria* with special reference to urine cytology. The significance of lesions in terms of pathogenesis and clinical manifestations derives not from their nomenclature but from their *incidence* and *clinical course*.

6.2 Problems of Nomenclature

Highly differentiated bladder carcinomas (whose basic neoplastic nature is established in some cases by invasive growth) tend to show little cellular and nuclear atypia. The morphologic spectrum includes the benign "papillomas," which in fact are covered by normal-appearing urothelium on light microscopic examination. Only the number of cell layers separates these two diagnoses from each other, and quantitative procedures have been unable to provide additional reliable criteria that would assist differentiation. All evidence suggests that these tumors, like other neoplasms derived from the superficial epithelium, develop from flat precursors. Like the tumors themselves, the precursors and intraepithelial manifestations encompass all grades of cellular anaplasia ranging from normal or nearly normal to the complete picture of *carcinoma in situ* (Fig. 6.1). While the nomenclature of the latter changes presents problems only with regard to the usefulness of grading (grade 2 or 3, analogous to overt tumors), there is still great uncertainty with respect to early changes. This problem has been addressed by a number of authors, among them Murphy and Soloway (1982).

Atypia, dysplasia, and *atypical hyperplasia* are the principal terms applied to early urothelial changes. There has been no strict diagnostic or consistent usage of these terms, nor have they been sharply differentiated from carcinoma in situ. A concept that has proven useful in practical diagnosis is described in Sect. 6.3. A major argument in favor of this concept is that it is compatible with practical techniques of exfoliative cytology.

An additional point of nomenclature should be noted: These *"flat lesions"* may be described as *primary*, and thus occurring prior to the manifestation of a visible tumor (although with low-grade lesions there may be no proof of obligate carcinogenesis), or as being *concomitant* with the tumor. Concomitant flat lesions consist of *intraepithelial extensions* in direct continuity with the tumor as well as isolated lesions (representing multifocal

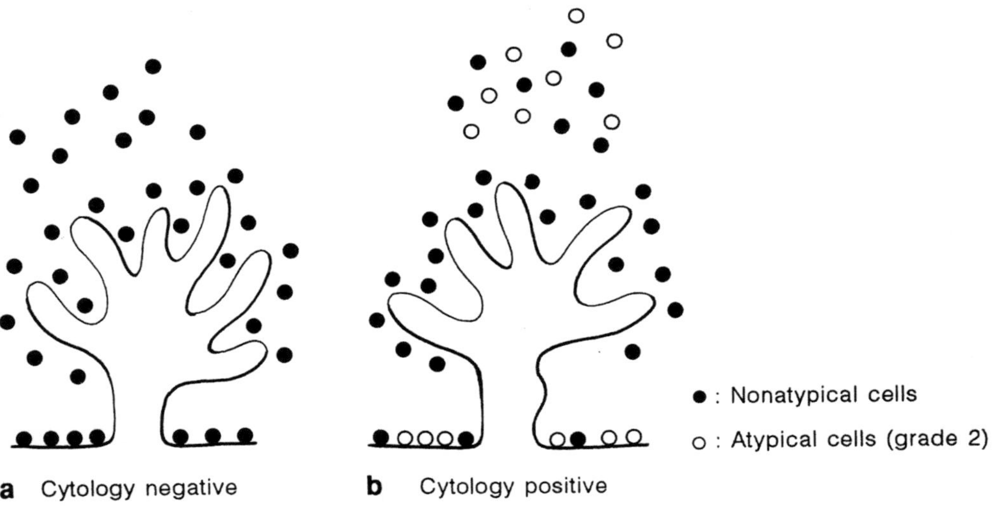

Fig. 6.1. Schematic comparison of nuclear atypias associated with an overt urothelial tumor and "flat" lesions

carcinomatous urothelial transformation). These manifestations are cytohistologically indistinguishable, even with benefit of multiple mucosal biopsies. Precise discrimination can be accomplished only by the systematic histologic analysis of cystectomy specimens (Koss et al. 1977).

6.3 Classification and Diagnosis

Some of the confusion relating to the nomenclature of "flat lesions" of the urothelium stems from an ambiguity of cytologic (mainly nuclear) and histoarchitectural criteria and from the individual clinical-biological interpretation of the examiner (and the clinical physician who subsequently utilizes the findings). It is helpful to define the morphologic parameters as unambiguously as possible (by referral to the epithelial findings associated with overt tumors) before attempting a clinical-biological evaluation. The general cytologic criteria (nuclear–cytoplasmic ratio, nuclear size, nuclear shape, chromatin structure, nucleoli) are extensively documented in this textbook and do not require further elaboration (see Chaps. 9,14). The additional information provided by histology, the second diagnostic mainstay, is based upon the relationship of the cells to one another, or the epithelial architecture.

The proliferation kinetics of the normal urothelium shows a well-ordered progression with a proliferative compartment in the basal cell layers and increasing maturation toward the superficial cell layers. A "snapshot" of this dynamic process is provided in the histologic section. The slightest disturbance of this equilibrium is manifested by an increase of proliferation with maintenance of differentiation, resulting in an increase in the number of cell layers, predominantly at the basal and intermediate levels. This change, which is not accompanied by nuclear atypias or morphologic cell changes, is termed *hyperplasia.*

In the case of the "flat lesions" discussed below, the histoarchitectural lamination of the urothelium (basal cells, intermediate cells, superficial cells) is still intact, but cytologic criteria are additionally noted. Generally the number of cell layers is increased, so hyperplasia is present. This is accompanied by nuclear atypias, which accounts for the term *"atypical hyperplasia."* Following the convention used in the pathology of cervical carcinoma, this term is usually considered to be synonymous with *dysplasia.* But while that term became well established in cervical cytology and histopathology, it has since been discarded in favor of *cervical intraepithelial neoplasia (CIN),* despite the fact that a significant percentage of CIN 1 lesions do not progress to carcinoma. A basic problem with the term "dysplasia" relates to the fact that it is also applied to malformations, so there is ample justification for objections to the term (Mostofi et al. 1988).

In any case, whether we prefer the term "atypical hyperplasia" or keep with the now-discarded term from cervical pathology, it is apparent that as the cytologic grade of malignancy increases, we encounter features like those observed in the epithelium of differentiated urothelial carcinomas. It should be emphasized, however, that this *does not prove* that these changes are an obligate precursor of carcinoma. This would be the essential condition for justifying the term *grade 1 carcinoma in situ*.

Thus, we feel that use of the term "carcinoma" is appropriate only if the criterion of irreversible growth disturbance by *destruction of the epithelial structure* is satisfied. In our experience, this can occur even in the absence of extreme nuclear pleomorphism, i.e., there are flat lesions with a very cellular epithelium, nuclear crowding, and complete loss of the normal layered cellular arrangement. The nuclei are relatively uniform, like those in the basal cell layer, but they do not show significant pleomorphism. Quantitative studies indicate, however, that the nuclei are nondiploid with a definite, uniform lineage. There are sound reasons for differentiating these two forms as *grade 2 carcinoma in situ*, exhibiting slight pleomorphism but complete loss of cellular stratification, and *grade 3 carcinoma in situ*, distinguished by pronounced nuclear pleomorphism. Maturation to superficial cells does not occur. Additionally, of course, there is classic carcinoma in situ with conspicuous nuclear irregularities including marked nuclear enlargement and chromatin clumping. One way out of the dilemma posed by diagnostic subjectivity and overemphasis on semantics (in histopathology as well as cytology) is by applying quantitative methods of analysis, not just for overt tumors (Hofstädter et al. 1984) but also for localized flat lesions (Farsund et al. 1983; Hofstädter et al. 1986). These studies have shown that moderate to severe dysplasias are often characterized by a high proportion of tetraploid nuclei. Differentiated carcinomas also display this phenomenon, underscoring the close cellular relationship of moderately severe or severe dysplasia to differentiated carcinoma and substantiating the associative impression of histomorphology.

An interesting subgroup, "denuding carcinoma," illustrates the tendency toward decreased intercellular cohesiveness, confirmed by electron microscopy, that exists in all forms of carcinoma in situ (see Chap. 4). In pronounced forms the basement membrane becomes "denuded" of cells except for an occasional urothelial cell, usually very atypical, that remains adherent to the membrane. Combining the histologic and cytologic techniques is helpful for avoiding a false-negative diagnosis, which can have dire consequences when one considers the highly malignant nature of the cellular changes.

6.4 Incidence

Much has been published on the incidence of "flat" epithelial lesions that are diagnosed concurrently with a grossly visible urothelial tumor, although there is much diversity in the histologic criteria that have been applied. Melicow first reported the finding of concomitant flat lesions in 1952. In ten cystectomy specimens he noted hyperplasia in all ten, cystitis cystica in four, carcinoma in situ in two, and squamous cell metaplasia in one. Eisenberg et al. (1960) investigated "proliferative urothelial lesions" (including Brunn's nests and papillary hyperplasia) in 161 papillary and solid urothelial carcinomas. They found in situ carcinoma in nine of 53 papillary carcinoma cases suitable for evaluation (presence of adequate non-tumor-associated mucosa). By contrast, they detected carcinoma in situ or atypical hyperplasia in 11 of 31 cases with predominantly infiltrative carcinoma. Schade and Swinney (1968) found severe proliferative lesions (atypical epithelium [42], carcinoma in situ [30], cystitis glandularis with atypical urothelium or carcinoma in situ [14]) in 86 of 100 patients with bladder carcinoma (existing tumors or tumors identified during posttreatment follow-up). A definite relationship to the visible tumor can be established only through mapping studies of resected bladders, however (Koss et al. 1977; Soto and Friedell 1977).

A strong relationship exists between the grade of malignancy of the coexisting overt tumor and the grade of the carcinoma in situ (grades 2, 3, 4 after Bergkvist et al. 1965) (Jakse et al. 1980).

The high accuracy of exfoliative cytology in the diagnosis of carcinoma in situ has been consistently demonstrated in all studies (Table 6.1).

It is interesting to observe the frequency of "flat lesions" over the course of the bladder tumor disease, i.e., to compare segmental biopsies taken from the original tumor and from various recurrences. This line of inquiry has led to evaluations of the *clinical-biological significance* of the corresponding lesions. It has been found, for example, that a marked increase in the frequency of dysplasia and carcinoma in situ is already apparent at the time of the initial recurrence (Jakse et al. 1986). Similar findings were reported by Soloway et al. (1978). Initial biopsies from normal-appearing mucosa in 42 patients with bladder cancer revealed atypia or carcinoma in situ in 33% of the patients. After 1 year (with biopsies done at 3-month intervals), this number rose to 77%. These findings indicate that there is a *definite sequence in the malig-*

Table 6. 1. **Urinary cytology in 142 patients with primary carcinoma in situ** (after Jakse et al. 1980)

Study	Patients (n)	Positive	Negative
Melamed et al.	18	18	–
Weisbrod	2	2	–
Yates-Bell	2	2	–
Barlebo et al.	6	6	–
Riddle et al.	25	19	6
Farrow et al.	69	69	–
Jakse et al.	20	19	1
Total	142	135 (95,2%)	7

nant potential not just of grossly visible tumors but also of "flat" preneoplastic and early neoplastic urothelial changes.

Less is known about the *frequency of flat lesions* without a preexisting or concomitant visible tumor. Farrow et al. (1977), in their cytologic analysis of 35 000 patients (who sought urologic consultation for dysuria or microhematuria), found 69 cases of in situ carcinoma in the absence of grossly visible bladder tumor. This study is not a true screening study, however, because the patients were all symptomatic and the population was poorly defined. It is interesting to note the studies of Koss et al. (1969), who found 13 cases of carcinoma in situ among 503 workers exposed occupationally to the carcinogen paraaminodiphenyl.

6.5 Clinical-Biological Significance

It was noted previously that the frequency of "flat" lesions relates to the grade of malignancy or the morphology (papillary or invasive) of the coexisting tumor. The presence of atypias (dysplasia) and of carcinoma in situ also has significant biological and clinical implications.

The *prognostic influence* of carcinoma in situ on the further course of the neoplastic illness is beyond question. Several specific aspects of this influence have been identified. Barlebo et al. (1972), for example, showed that the grade of malignancy has prognostic relevance. Soto and Friedell (1977) consistently detected carcinoma in situ in the presence of *multifocal* bladder tumors but never detected it in patients with unifocal growths. Prout et al. (1983) discovered another aspect of the prognostic significance of carcinoma in situ. Thirty-

eight patients found to have carcinoma in situ at the time of their initial tumor event ($n=38$) showed a significantly poorer prognosis than 32 patients who developed carcinoma in situ some time after their initial diagnosis (Table 6.2).

In any case, these and similar studies indicate that *carcinoma in situ* (with a definitely malignant cell pattern) *is a heterogeneous disease* that is determined by multiple pathogenetic factors, including the intraepithelial anatomic pattern of spread (Jakse et al. 1987).

The *biological significance of dysplasia* has been investigated by Wolf and Hojgaard (1983), who likewise demonstrated a relationship between the grade of tumor malignancy (2–4 in the classification of Bergkvist et al. 1965) and the frequency and severity of "flat lesions" (Table 6.3).

Of 53 patients, 11 developed recurrences at the site of the original tumor. Scrutiny of the remaining 42 patients revealed that seven of 27 patients (26%) developed a new urothelial carcinoma when there was no evidence of dysplasia (typically within 1 year), while 13 of 15 (87%) patients developed a new tumor when dysplasia was present. These data vividly underscore the predictive importance of dysplastic "flat lesions" detected by segmental biopsy. The authors conclude that this provides a means for assessing the biological potential of neoplastic bladder disease, quite apart from the new methodologies in immunohistochemistry and quantitative pathology (see Chap. 10).

What special significance do *urothelial atypias* separate from visible tumor have in *exfoliative cytology?* Many studies have confirmed that exfoliative cytology is relatively insensitive for detecting highly differentiated carcinomas due to their paucity of nuclear abnormalities, but is highly suc-

Table 6. 2. **Outcomes in patients with carcinoma in situ**

	Number	Debeloped muscle invasion or metastasis	Did not progress or undergo cystectomy
CIS at initial diagnosis			
CIS alone	14	5	4
CIS concomitant with overt tumor	24	8	7
Developed CIS subsequently			
CIS alone	15	1	8
CIS concomitant with tumor	17	1	11

Table 6. 3. **Relationship between the grade of malignancy of the overt tumor and the "flat lesion" of the mucosa** (after Wolf and Hojgaard 1983)

Grade of overt tumor[a]	Dysplasia			Carcinoma in situ %
	n	None %	Grade 2 %	
2	18	65	33	0
3,4	35	46	14	40

[a] Graded according to Bergkvist et al. (1965).

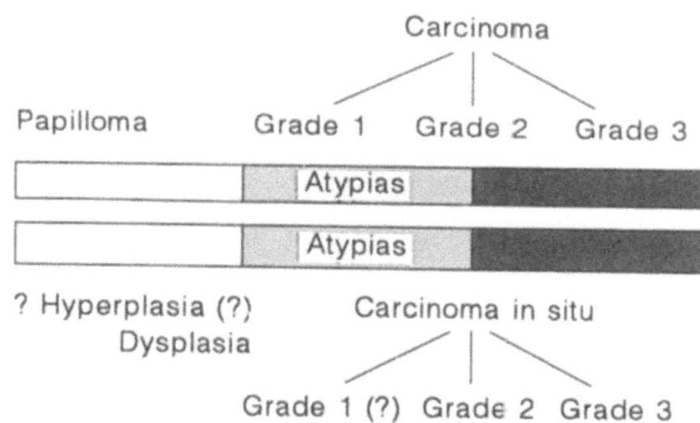

Fig. 6.2. **Schematic comparison of a low-grade tumor** *(1)* **without atypical foci in the urinary bladder mucosa with a low-grade tumor associated with flat atypical mucosal areas** *(2)*. The *upper part* of the drawing indicates what cells appear in bladder washings or urine

cessful in the diagnosis of atypical flat lesions, including carcinoma in situ. On the other hand, cytology cannot establish whether the individual cells (with atypias) originate from an overt tumor or from a "flat" atypical lesion.

It is reasonable to assume, however, that more severely atypical cells found in an exfoliative specimen from a bladder with a coexisting highly differentiated urothelial carcinoma do not originate from the tumor itself but from accompanying foci of dysplasia or even carcinoma in situ (Fig. 6.2).

Urothelial cytology thus plays a key role in the classification of the neoplastic disease. The cytologic findings must be reconciled with the tumor itself and with the result of mucosal biopsies.

I express thanks to Mrs. A. Weidner for her help in preparing the manuscript and to Mr. J. Marquardt for the artwork.

References

Barlebo H, Sorensen BL, Ohlsen AS (1972) Carcinoma in situ of the urinary bladder: flat intraepithelial neoplasia. J Urol Nephrol 6: 213–223

Bergkvist A, Ljungquist A, Moberger G (1965) Classification of bladder tumours based on the cellular pattern. Acta Chir Scand 130: 371–378

Eisenberg RB, Roth RB, Weinberg ME (1960) Bladder tumors and associated proliferative mucosal lesions. J Urol 84: 544–550

Farrow GM, Utz DC, Rife CC, Greene LF (1977) Clinical observations on sixty-nine cases of in situ carcinoma of the urinary bladder. Cancer Res 37: 2794–2798

Farsund T, Laerum OD, Hostmark J (1983) Ploidie disturbance of normal appearing bladder mucosa in patients with urothelial cancer; relationship to morphology. J Urol 130: 1076–1082

Hofstädter F, Jakse G, Lederer B, Mikuz G, Delgado R (1984) Biological behaviour and DNA-Feulgen-Cytophotometry of urothelial bladder carcinoma. Br J Urol 56: 289–295

Hofstädter F, Delgado R, Jakse G, Judmaier W (1986) Urothelial dysplasia and carcinoma in situ of the bladder. Cancer 57: 356–361

Jakse G, Hofstädter F, Leitner G, Marberger H (1980) Carcinoma in situ der Harnblase. Eine diagnostische und therapeutische Herausforderung. Urologe [A] 19: 93–99

Jakse G, Hufnagel B, Hostädter F, Rubben H (1986) Sequentielle Blasenschleimhautbiopsie beim Urothelkarzinom der Harnblase. Verh Dtsch Ges Urol 37: 200–201

Jakse G, Putz A, Hofstädter F (1987) Carcinoma in situ of the urinary bladder extending into the seminal vesicles. J Urol 137: 44–45

Koss LG, Melamed MR, Kelly RE (1969) Further cytologic and histologic studies of bladder lesions in workers exposed to para-aminodiphenyl: progress report. J Natl Cancer Inst 43: 233–243

Koss LG, Nakanishi I, Freed SZ (1977) Nonpapillary carcinoma in situ and atypical hyperplasia in cancerous bladders. Urology 9: 443–455

Melamed MR, Voutsa NG, Grabstald H (1964) Natural history and clinical behaviour of in situ carcinoma of the human urinary bladder. Cancer 17: 1533–1545

Melicow MM (1952) Histological study of vesical urothelium intervening between gross neoplasms in total cystectomy. J Urol 68: 261–279

Mostofi FK, Davies CJ Jr, Sesterhenn IA (1988) Pathology of tumors of urinary tract. In: Steinner DG, Leikovsky G (eds) Diagnosis and management of genito-urinary cancer. Saunders, Philadelphia

Murphy WM, Soloway MS (1982) Urothelial dysplasia. J Urol 127: 849–854

Prout GR, Griffin PP, Daly JJ, Heney NM (1983) Carcinoma in situ of the urinary bladder with and without associated vesical neplasms. Cancer 52: 524–532

Riddle PR, Chisholm GD, Trott PA, Pugh RC (1975) Flat carcinoma in situ of the bladder. Br J Urol 47: 829–833

Schade ROK, Swinney J (1968) Pre-cancerous changes in bladder epithelium. Lancet 2: 943–946

Soloway MS, Murphy W, Rao MK, Cox C (1978) Serial multiple biopsies in patients with bladder cancer. J Urol 120: 57–59

Soto EA, Friedell GH (1977) Bladder cancer as seen in giant histologic sections. Cancer 39: 447–455

Wolf H, Hojgaard K (1983) Urothelial dysplasia concominant with bladder tumours as a determinant factor for future new occurences. Lancet 2: 134–136

7 Cytologic Grading of Urothelial Tumors

S. ROTH, R. FRIEDRICHS, and P. RATHERT

CONTENTS

7.1 General Remarks on Clinical Cytology

The critical difference between a cytologic and histopathologic examination is that the *cytologic evaluation* disregards topographic and architectural tissue changes. While *histology* can additionally evaluate the "third dimension" of supracellular architecture and the relationship of the cells to adjacent structures, cytology is based solely on the analysis of cell morphologies. This relative information deficit of cytology compared with histology is balanced by the noninvasiveness of cytology, its simplicity, low cost, and the option of repeating the examination as often as desired.

7.1.1 Structures Evaluated in Cytologic Examinations (Fig. 7.1)

The *cell membrane* envelops the cytoplasm and possesses the important properties of permeability (see Chap. 4), selective signal transmission, and enzymatic production.

The *cytoplasm* consists of a homogeneous liquid matrix containing numerous intracellular organelles (e.g., Golgi apparatus, endoplasmic reticulum).

The *nucleus* of the cell is separated from the cytoplasm by a duplicated, porous *nuclear membrane*. The nucleoplasm is rich in nucleic acid (DNA), which joins with proteins and ribonucleic acids (RNA) to form the *chromatin*, which appears cytologically as a finely granular structure inside the nucleus.

The *nucleolus* is a round to oval-shaped intranuclear body that plays a critical regulatory function in cellular biology. The number and size of the nucleoli vary with the functional activity of the cell. For this reason intense protein biosynthesis (e.g., malignancy) is associated with multiplicity and enlargement of the nucleoli.

7.2 Cytologic Features of Normal Urothelium

The urothelium is a *multilayered epithelium* composed of superficial (luminal), intermediate, and lower (basal) cell layers (see Chap. 4). The size of the cells increases from base to surface – a circumstance useful for distinguishing the urothelial cells in cytologic specimens from squamous epithelial cells and leukocytes (Fig. 7.2).

Besides this size progression of individual urothelial cells, it is also significant that the *size relationship of the nucleus to the cytoplasm varies with the layer from which the cell originates*. This must be considered in the evaluation of malignancy, because the small basal cells normally have a

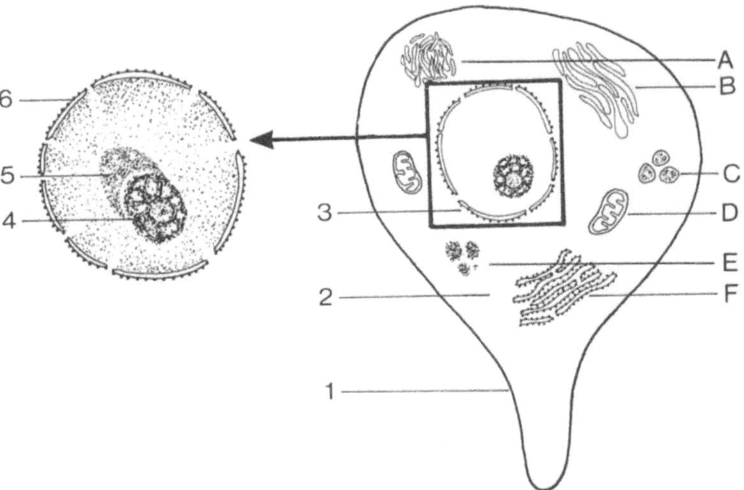

Fig. 7.1. **Schematic diagram of cellular morphology:** cell membrane (*1*), cytoplasm (*2*), nucleus (*3*), nucleolus (*4*), chromatin (*5*), nuclear membrane (*6*). The cytoplasm contains various organelles or constituents that do not contribute to the cytologic diagnosis: smooth endoplasmic reticulum (*A*), Golgi apparatus (*B*), lysosomes (*C*), mitochondria (*D*), glycogen (*E*), and endoplasmic reticulum with ribosomes (*F*)

relatively larger nucleus than the superficial umbrella cells (see Fig. 7.2).

A planimetric study by Eldidi and Patten (1982) involving two-dimensional measurements of 5000 urothelial cells in 220 healthy subjects permits a more accurate analysis of the cytologic features of normal urothelium. The basal cells of the urothelium, with an average area of 82 μ^2, are markedly smaller than the intermediate cells (229 μ^2) and umbrella cells (501 μ^2) (see Fig 7.2). Although the nuclear area of the basal cells, at 24 μ^2, is also smaller than in the intermediate cells (40 μ^2) and umbrella cells (64 μ^2), a percentage comparison indicates a *physiologic disproportion* of the nuclear–cytoplasmic ratios. Thus, while the nuclear area accounts for some 30% of the total cellular area in basal cells, it is only about 20% in the intermediate cells and 10% in the umbrella cells.

It is important to note that *multinucleation* is *not a criterion of malignancy*. Approximately 19% of the luminal umbrella cells are binucleated, and about 3% contain several nuclei (Eldidi and Patten 1982). No pathologic significance is ascribed to the abundant multinucleated cells found in washings of the upper urinary tract (see Sect. 9.6).

The *nucleoli* appear on light microscopy as dense bodies within the nucleus. Besides the chromatin, they are an essential criterion in the diagnosis of malignancy. Normal urothelial cells usually contain one or two nucleoli with a round to oval shape.

Fig. 7.2. **Relative sizes of urothelial cells** compared with other constituents of the urine sediment. The large umbrella cells (*2*) are smaller than squamous epithelial cells (*1*). The small basal cells (*4*) are slightly larger than segmented leukocytes (*5*), which in turn are larger than non-nucleated erythrocytes (*6*) and "granular" bacteria (*7*). There is a significant, physiologic difference in the nuclear–cytoplasmic ratios of the different urothelial cells (see text)

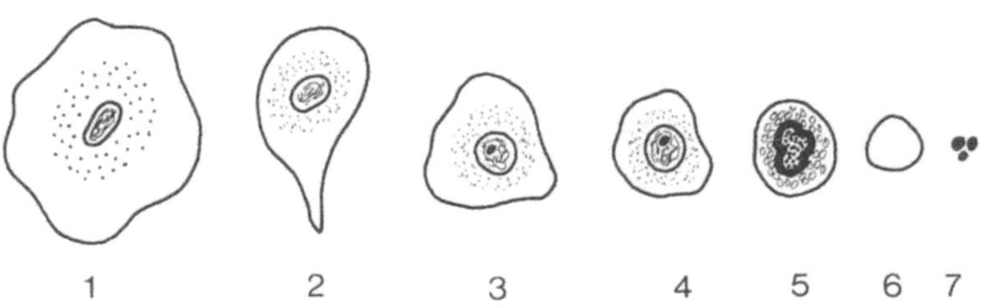

7.3 Cytologic Criteria of Malignancy

7.3.1 Changes in the Cytoplasm

The cytoplasm is important only as a reference structure for appreciating nuclear enlargement, i.e., determining the nuclear–cytoplasmic ratio. Special morphologic modifications of the cytoplasm are unimportant in terms of malignancy diagnosis. Further differentiation by means of special stains like those used in hematology is of no value in urinary cytology.

Nevertheless, there are cases in which cytoplasmic changes can furnish useful additional information. For example, a foamy, microvacuolated cytoplasm may be seen following the use of radiographic contrast medium (see Sect. 9.4, Fig. 9.23a–d), and intravesical chemotherapy commonly incites the formation of large, toxic-reactive cytoplasmic vacuoles (see Figs. 9.75 and 9.76).

7.3.2 Changes in the Nucleus

Nuclear alterations visible by light microscopy are among the *most important cytologic criteria of malignancy* (Fig. 7.3) and result from an increase in nuclear metabolism. The morphologic criteria become more pronounced as loss of differentiation progresses.

– Malignancy is associated with *enlargement of the nucleus in relation to the cytoplasm* (increased nuclear–cytoplasmic ratio).

– The *nuclear membrane* becomes irregular and shows sites of thickening, indentation, or infolding.

– As the grade of malignancy increases (from grade 1 to grade 3), *chromatin proliferation* occurs. This phenomenon, termed *hyperchromasia*, leads to a *loss of nuclear transparency*, which is interpreted as an early cytologic manifestation of urothelial carcinoma (Rübben et al. 1979).

– Alteration of the *chromatin pattern* can also occur. While normal chromatin is finely granular and evenly distributed over the nucleus, increasing loss of differentiation leads to coarsening and clumping. In extreme cases light and dark zones

Fig. 7.3. **Schematic representation of the principal morphologic criteria of malignancy. At the center is a normal urothelial cell. Changes consistent with malignant transformation are:** *A*, altered nuclear–cytoplasmic ratio; *B*, prominent and irregular nuclear membrane; *C*, hyperchromasia with loss of nuclear transparency; *D*, altered chromatin pattern (coarse granularity and clumping); *E*, multiple irregularly shaped nucleoli; *F*, irregularly shaped and pleomorphic nuclei (anisokaryosis)

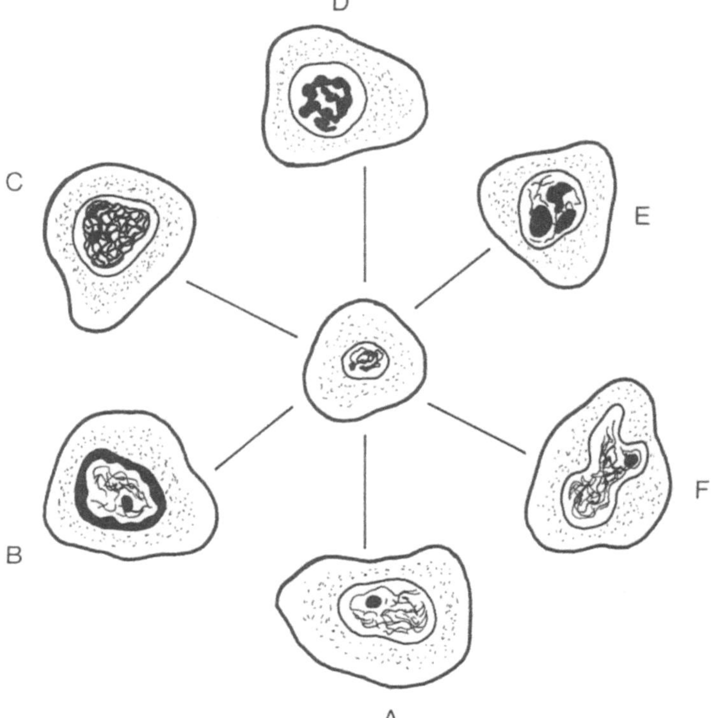

appear. From the standpoint of differential diagnosis, it is extremely important to identify remaining transparent zones in the nucleus, which persist even in cases of extreme malignant transformation. They are the only means available for ruling out overstaining, which can result from lytic cell damage, for example. Protein denaturation in these cases leads to an increased affinity for stain, which can simulate hyperchromasia. These nonviable nuclei generally appear uniformly "darkened" (see Fig. 9.24a). It is also important to note the variations in chromatin pattern and cell size associated with different staining procedures (see Fig. 9.2).

– *Nuclear pleomorphism.* The nuclei may assume bizarre shapes as loss of differentiation progresses.

– Malignancy also leads to changes in the *nucleoli.* Besides *enlargement* and *multiplicity, structural irregularities* (shape distortion, pleomorphism) may be observed.

7.3.3 Distribution of the Cells

Urothelial cells may be recovered singly or in sheets or clusters. The finding of cell clusters may reflect an increased shedding of cells by urothelial tumors. However, cell clusters in themselves are not pathognomonic for malignancy. Papillomas, defined as cytologically and histologically benign by the WHO, are associated with a relative abundance of cell clusters in cytologic samples.

Cell clusters are also more numerous following urinary tract instrumentation (e.g., cystoscopy, retrograde catheterization) and in catheter wearers. Sometimes the cells are heaped into aggregates (see Sect. 9.4, Fig. 9.22). Clusters with overlapping cells should not be interpreted, because the superposition may create a false-positive impression of hyperchromasia (Koss 1979). The patient's history can be very helpful in assessing the diagnostic significance of cell clusters.

7.4 Urinary Cytologic Grading

7.4.1 Practice of Urinary Cytologic Grading

Despite the use of specific selection criteria, conventional urinary cytology, like many other medical examinations (e.g., radiography, ultrasound), is still a subjective procedure based on empirical knowledge. Given the lack of numerical comparability in conventional cytologic studies, it is understandable that there is *ambiguity* in the grading of cytologically diagnosed cellular and nuclear atypias and malignant transformations.

The current, generally accepted grading system of the WHO subdivides urothelial tumors into three grades of malignancy:

– Highly differentiated urothelial tumors (grade 1)
– Moderately differentiated urothelial tumors (grade 2)
– Poorly differentiated urothelial tumors (grade 3)
– And sometimes undifferentiated urothelial tumors (grade 4)

The older grading systems of Papanicolaou (grades 1–5) and Bergkvist (grades 1–4) are now considered to be obsolete for urinary cytology.

Given the large variation in the cytologic appearance of the normal urothelium, *more than one criterion* must be satisfied in order to establish a *diagnosis of malignancy* (see Fig. 7.3).

This requirement is reasonable when one considers, for example, that hyperchromasia is almost always accompanied by a change in the chromatin pattern (e.g., coarse granularity) and nuclear enlargement (altered nuclear–cytoplasmic ratio). This contrasts with the nuclear swelling (apparently abnormal nuclear–cytoplasmic ratio) frequently produced by a hyperosmolar irrigant, in which the chromatin is either unchanged or "diluted" and hypochromatic (Fig. 9.21a,b). This illustrates the fact that evaluation of a single morphologic criterion would significantly increase the danger of a false-positive diagnosis (low specificity).

7.4.1.1 Cytologic Features of Highly Differentiated Urothelial Tumors (Grade 1)

A combination of several structural parameters may be seen with highly differentiated urothelial tumors:

▶ Slight hyperchromasia with no loss of nuclear transparency and no significant change in the chromatin pattern
▶ Slight prominence (thickening) of the nuclear membrane
▶ Slightly enlarged nucleoli that maintain a round to oval shape
▶ Slight increase in the nuclear–cytoplasmic ratio (more apparent in the intermediate and umbrella cells, with their more copious cytoplasm, than in the compact basal cells)
▶ All suspicious urothelial cells have a very uniform, monotonous appearance, as if "stamped from one mold" (see Sect. 9.5.1)

7.4.1.2 Cytologic Features of Moderately Differentiated Urothelial Tumors (Grade 2)

The combined occurrence of the following morphologic changes is typical:

▶ Hyperchromasia is more pronounced, causing incipient loss of nuclear transparency
▶ Altered chromatin pattern (e.g., coarsely granular)
▶ Thickened and irregular nuclear membrane
▶ Enlarged nucleoli showing incipient shape distortion
▶ Marked increase in the nuclear–cytoplasmic ratio due to the foregoing nuclear changes
▶ Pleomorphic and anisokaryotic nuclei
▶ Suspicious urothelial cells show morphologic diversity and lack the uniform appearance of grade 1 cells (see Sect. 9.5.2)

7.4.1.3 Cytologic Features of Poorly Differentiated Urothelial Tumors (Grade 3)

▶ Extreme hyperchromasia with nuclear nontransparency
▶ Coarsely granular chromatin with a tendency to form clumps
▶ Marked irregularities of the nuclear membrane
▶ Large, irregularly shaped nuclei, sometimes showing excessive multiplicity
▶ Increased mitoses
▶ Sometimes giant, very pleomorphic nuclei
▶ Highly diverse urothelial cell pattern (marked anisocytosis and anisokaryosis) (see Sect. 9.5.3)

7.4.1.4 Reactive Cell Changes

Urinary cytology is fraught with potential sources of error in cancer diagnosis. This is based on the fact that the urothelial cells, like all cells, can respond to *external or internal stimuli* with a limited number of morphologic alterations (see Fig. 7.3). Although these reactions are much more specific at the molecular and biochemical level, they cannot be differentiated by the evaluation of cytologic morphology and may *mimic the features of a tumor*. Thus, hyperchromasia that is apparently suspicious for malignancy may result from protein biosynthesis stimulated by an infectious process. The general cytologic picture may, however, offer clues to the reactive etiology of the changes, such as the inflammatory background that is associated with infections. Other practically relevant sources of error due to reactive cell changes are urolithiasis, the faulty processing of cellular material, and instrumentation of the urinary tract (see Sect. 9.4).

Certain *therapeutic procedures* also can induce urothelial cell changes that are sometimes persistent, may be characteristic, but can in some cases mimic neoplastic features (see Sect. 9.7). It is particularly important in these cases for the cytopathologist to know the patient's *history* and any therapeutic measures that have been performed (e.g., "patient received radiotherapy x months ago," or "patient is currently receiving chemotherapy for prevention of recurrence").

7.4.2 The Problem of Standardizing the Grading of Malignancy

The *problem of standardization* (ambiguity) in the grading of pathologic cell changes by exfoliative urine cytology is based on the fact that *there is no single parameter that exclusively determines the classification of cells*. Rather, the examiner must evaluate several structural parameters that are subject to different interpretations from case to case and among different examiners, according to their personal experience (Ooms et al. 1983).

Thus, some poorly differentiated tumors show extreme nuclear hyperchromasia (dense, non-transparent nucleus) with very little nuclear enlargement, while others show a greatly enlarged nucleus in which the increased chromatin can become spread out, causing little reduction of nuclear transparency.

Because conventional urinary cytology lacks the objectivity of, say, automatic image analysis systems (see Chap. 10), which generate a single machine-read parameter that can be quantified and compared with other readings, there can be substantial interindividual discrepancies of interpretation.

This disadvantage of conventional urine cytology is balanced by its modest equipment requirements and simplicity of performance. Also, *ambiguous "grading boundaries"* as a cause of interindividual discrepancies is *only a relative disadvantage* owing to the direct relationship of diagnostic accuracy to clinical relevance.

The uncertain cytologic identification of highly differentiated urothelial tumors (30%–50%) is based on the minimal or absent pathoanatomy of urothelial tumor cells (Rübben et al. 1989) and thus is a direct consequence of the blurred dividing line between normal and malignant urothelial cells. At the same time, a significant percentage of these tumors are invasive and are easily diagnosed by endoscopy, so the insufficient detection rate is of relatively minor clinical importance. The fact remains, however, that conventional urinary cytology is not a reliable procedure for replacing invasive cystourethroscopy in the primary diagnosis and follow-up of highly differentiated urothelial tumors.

On the other hand, prognostically significant lesions such as carcinoma in situ and invasive urothelial tumors almost invariably show at least a moderate loss of differentiation (grade 2). As a result, these lesions differ very markedly from normal urothelium in their cytologic presentation and can be classified as prognostically relevant dysplasias or carcinomas with reasonably high confidence, despite interindividual discrepancies (see Chap. 2).

Ultimately it does not matter whether different examiners classify a "borderline" cytologic specimen as grade 2 or grade 3. The crucial factor is the diagnosis or exclusion of malignancy (>/grade 2) or of a severe dysplasia having significant prognostic implications (see Chap. 6), for this determination forms the essential starting point for any further endoscopic and/or biopsy investigations or more sophisticated cytologic procedures (see Chaps. 10 and 11).

Despite the lack of standardization, exfoliative urinary cytology remains the first-line technique for the detection and follow-up of urothelial tumors (Badalament et al. 1987). In clinically relevant situations (e.g., where is need to assess the feasibility of radical surgery), urine cytology can be supplemented by objective, more reproducible procedures.

Here we find a promising practical realization of the requirement that the new image-analysis and immunologic techniques serve not in a competitive but in an adjunctive capacity to conventional urinary cytology.

References

Badalament RA, Hermansen DK, Kimmel M, Gay H, Herr HW, Fair WR, Whitemore WF Jr, Melamed MR (1987) The sensitivity of bladder wash flow cytometry, bladder wash cytology, and vioded cytology in the detection of bladder carcinoma. Cancer 60: 1423

Eldidi MM, Patten SF (1982) New cytologic classification of normal urothelial cells: an analytical and morphometric study. Acta Cytol 26: 725

Gompel C (1982) Atlas de cytologie clinique. Maloine, Paris

Koss LG (1979) Diagnostic cytology and its histopathologic bases, 3rd edn. Lippincott, Philadelphia

Koss LG (1989) Cytology-accuracy of diagnosis. Cancer 64 [Suppl]: 249

Koss LG, Deitch D, Ramanthan R, Sherman AB (1985) Diagnostic value of cytology of voided urine. Acta Cytol 29: 810

Leistenschneider W (1982) Zytodiagnostik. In: Hohenfellner R, Zingg EJ (Hrsg) Urologie in Klinik und Praxis, vol 1. Thieme, Stuttgart, p 326

Mostofi FK, Sorbin LH, Torloni H (1973) Histological typing of urinary bladder tumors. International Classification of tumors, vol 19. WHO, Geneva

Ooms ECM, Anderson WAD, Alons CL, Boon ME, Veldhuizen R (1983) Analysis of the performance of pathologists in the grading of bladder tumors. Hum Pathol 14: 140

Rathert P, Preiss H (1982) Urinzytologie in der urologischen Praxis. Urologe [A] 21: 67

Rübben H, Bubenzer J, Bökenkamp K, Lutzeyer W, Rathert P (1979) Grading of transitional cell tumors of the urinary tract by urinary cytology. Urol Res 7: 83

Rübben H, Rathert P, Roth St, Hofstädter F, Giani G, Terhorst B, Friedrichs R (1989) Exfoliative Urinzytologie. Harnwegstumorregister, Fort- und Weiterbildungskommission der Deutschen Urologen, Arbeitskreis Onkologie, Sektion Zytologie, 4th edn

Voogt HJ de, Rathert P, Beyer-Boon ME (1979) Praxis der Urinzytologie. Springer, Berlin Heidelberg New York

8 Working Techniques in Urinary Cytology

S. Roth

CONTENTS

8.1 General Remarks on Cytologic Technique

The success of urinary cytology depends on several components. The essential element of the cytologic examination, the cytodiagnosis itself, is dependent on the preparation of good-quality, readable specimens. In this chapter we shall describe the necessary working techniques in detail, thereby helping the physician active in cytology to make an informed choice among available devices and techniques for cytologic processing.

8.1.1 Specimen Collection

There is still controversy about many of the details involved in the collection of material for cytologic evaluation. In this chapter we shall compare irrigation specimens (washings) with the use of spontaneously voided urine (see Sect. 8.3). We shall also note certain technical points in retrograde irrigation cytology and the collection of urethral washings.

8.1.2 Specimen Processing

There are both mandatory and optional steps involved in the processing of cytologic specimens. Every clinic and office must select a procedure that is reasonable and practical for its own circumstances. The cytoprocessing routine will be influenced by the equipment that is available, the place where the cytologic evaluation will be performed (on- or off-site), and how quickly a diagnosis is required.

With regard to timing, one can choose between the options of immediate (instant) cytology and delayed or "second look" cytology.

If an immediate bedside diagnosis is needed or desired, rapid staining methods must be employed. Their exclusive use in the clinical and office setting often raises the problem of integrating the staining procedure into the clinical or office routine, making it necessary to fix or preserve the specimen for later processing.

This aspect of cytoprocessing is extremely important, because an accurate cytologic diagnosis depends on the state of preservation of the cells. No more than 2 h of contact with urine is required for proteolytic enzymes and bacterial cytolysins to induce degenerative cell changes. Thus, the cells must either be protected by the addition of a cellular preservative or fixative to the urine, or they must be recovered from the urine and transferred to a microscope slide (see Sect. 8.4 for details).

The following factors may serve as criteria for influencing the selection of a processing routine:

- Bedside diagnosis
- Integration into existing routine
- Wish or need for documentation
- Reference cytology (dispatch to an off-site lab)

Generally, the following alternatives are available for designing a cytoprocessing routine:

Immediate cytology:

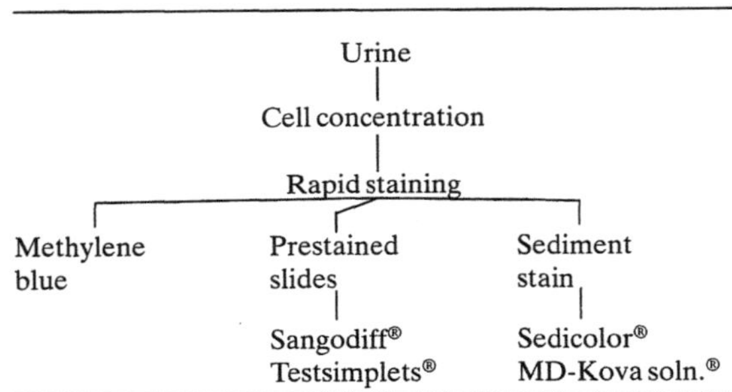

Option 1

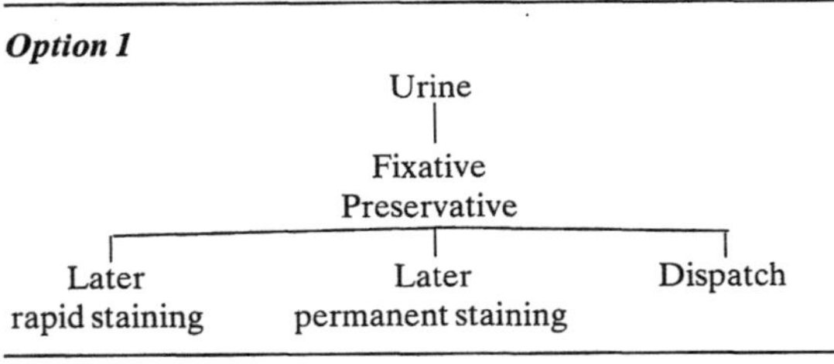

Option 2

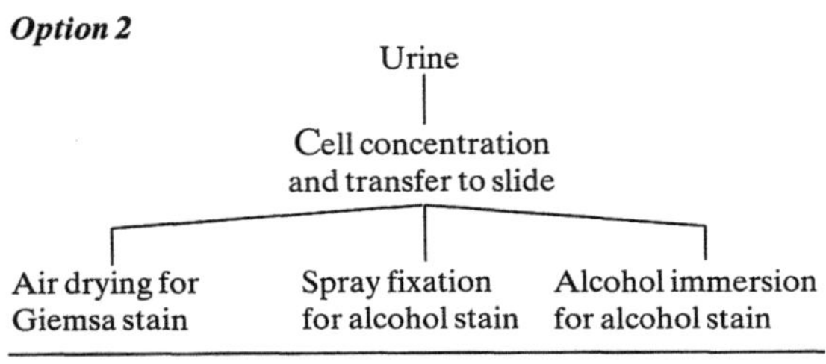

Delayed cytology: For later conduct of the cytologic evaluation, whether on-site or in a cytology laboratory, the cells may be fixed and preserved within the urine "carrier medium" (option 1), or the cells may be recovered, transferred to a slide, and fixed there (option 2). The cells may be recovered by direct centrifugation, membrane filtration, or with a cytocentrifuge (see Sect. 8.5).

When the specimen is to be *dispatched to a cytology laboratory*, the laboratory should be consulted regarding fixation and preservation of the urine to ensure that the "readability" of the specimen is not compromised by troublesome interactions. For example, air-drying must be avoided in specimens that are earmarked for alcohol staining (e.g., by the Papanicolaou method).

For cytology in the office setting, it can be useful to divide the urine sample into *two portions*. One portion can be evaluated by rapid staining right away or at the end of the visit (with more than a 2-h delay, a preservative should be added or the sediment extracted; see Sect. 8.4.1), while the second portion, with preservative added, can be saved for further processing as required (e.g., with tumor findings, suspicious or equivocal findings, or to assist follow-up of intravesical chemotherapy) (see Sect. 8.8). The second sample may be used for the preparation of a permanent or differential stain, or the cellular material may be fixed on a slide for staining in a cytology laboratory or for dispatch to a cytopathologist.

8.1.3 Cell Concentration

Urine cytology works with exfoliated cells that are collected from a relatively large volume of fluid. Given this dilution effect, the cellular material must be concentrated before it is examined so that an accurate cytologic diagnosis can be made within a reasonable period of time.

The cells can be concentrated and recovered by classic direct centrifugation to produce a sediment or "pellet," by the membrane filtration technique, or by centrifuging the cells directly onto a slide (cytocentrifugation; see Sect. 8.5). The capillary filter suction technique is a more recent method whose value has not yet been definitively proven (see Sect. 8.5.4).

Ultimately the particular clinic, institution, or office must select the method of cell concentration and recovery (see Sect. 8.5) that is most appropriate for its own circumstances.

8.1.4 Direct Microscopy and Staining Methods

The direct examination of fresh sediment by bright-field microscopy does not permit an accurate cytologic diagnosis due to the deficient contrast of the cellular structures.

A fascinatingly simple method that has become less popular with the advent of rapid staining techniques is *phase contrast microscopy* (PCM). The principle, whose discovery earned Zernicke the Nobel Prize for physics in 1953, is based on a change in the diffraction peaks of the microscopic image resulting in differences of light intensity that bring out structural details in the unstained cells (see Sect. 8.6 for details).

Interference contrast microscopy of fresh urine produces a relief-like image of the cells, as if they were being viewed under oblique illumination. This technique has no practical importance in routine oncologic cytology.

Rapid staining for *immediate cytology* may be performed with methylene blue, sediment stains, or prestained slides. Sediment stains are of limited value for oncologic urinary cytology, however.

A variety of techniques are available for the *differential and permanent staining* of specimens. After air-drying, the specimen may be stained by the Giemsa or Pappenheim (combined May-Grünwald/Giemsa stain) procedure, or the simpler Hemacolor (Merck) stain can be used.

The standard stain for urinary cytology is the Papanicolaou stain. This method yields excellent results, but the procedure is relatively complicated and time-consuming. An alternative is the rapid alcohol-based Szczepanik stain.

Other special staining techniques, like those used in hematology for the differentiation of relevant cytoplasmic structures, are not needed in urine cytology because the diagnosis of malignancy relies chiefly on the evaluation of the cell nucleus.

8.2 Working Materials

8.2.1 Slides

The glass microscope slides that hold the specimens are produced in a standard 76×26-mm format (English format) with a thickness of 1–1.2 mm. Slight deviations in slide thickness are of no importance.

The coverslip above the specimen should have the standard thickness of 0.17 mm, because microscope objectives are corrected for that thickness to eliminate "coverslip error." The coverslip thus forms part of the optical system of the microscope. Objectives are not properly corrected for coverslips of nonstandard thickness, some of which are still available commercially.

8.2.2 Centrifuge Tubes

Centrifuge tubes may be made of reusable glass (cleaned immediately after use to avoid contamination of the next sample), or disposable tubes may be used.

Sediment and cells may be lost from the tube when the hypocellular supernatant fluid is poured off (decanted) after centrifugation. This problem can be eliminated by using *conical* instead of *round-bottomed tubes.*

If, after decanting, there is too much residual urine in the sediment from a sparsely cellular sample, the concentration effect is diminished. Additional supernatant can be removed with filter strips or by inverting the tubes onto absorbent filter paper and letting them stand for 30–60 s. Again, the use of conical tubes will eliminate the danger of cell loss (Fig. 8.1).

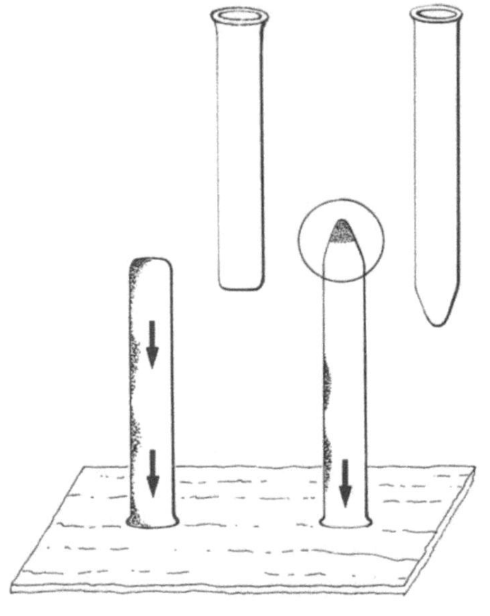

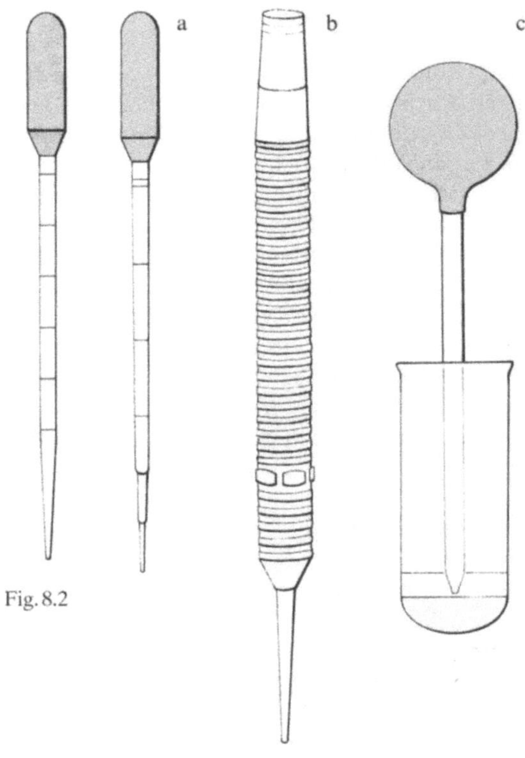

Fig. 8.2

8.2.3 Pipettes

Pipettes are used to transfer the cellular material onto the slide after centrifugation. This may be done with a disposable Pasteur pipette (a), a plunger-type pipette (Eppendorf pipette) with a replaceable plastic tip (b), or a glass pipette with a bulb syringe (c) (Fig. 8.2).

8.2.4 Fixation and Staining Utensils

For some staining procedures (see Sect. 8.7) the slides must be wet-fixed following application of the cellular material. This can be done with a *spray fixative* (Fig. 8.3) or by immersion in an *alcohol bath* (Fig. 8.4).

Glass cuvets serve as containers for fixation as well as staining. The cuvet may be manually loaded, in which case vertical guides along the side walls keep the slides from touching each other (see Fig. 8.4). If a shallower dish is used, the slides can be loaded onto a *slide carrier* for immersion (Fig. 8.5).

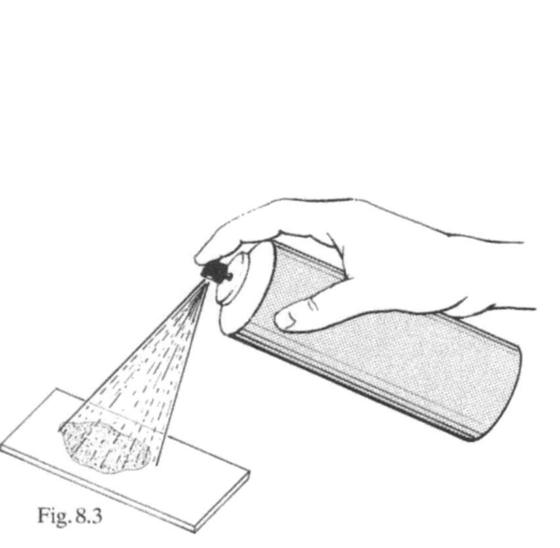

Fig. 8.3

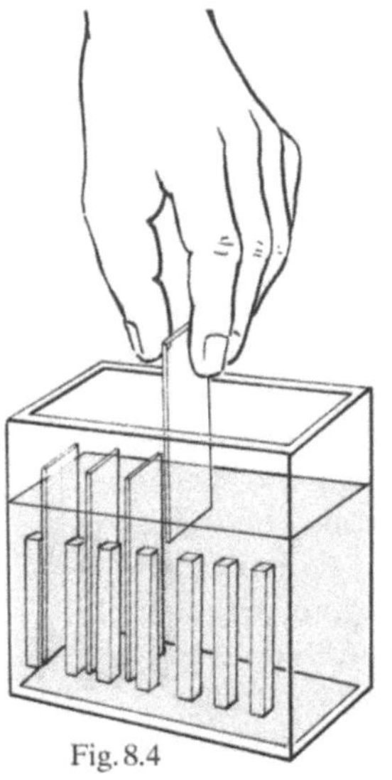

Fig. 8.4

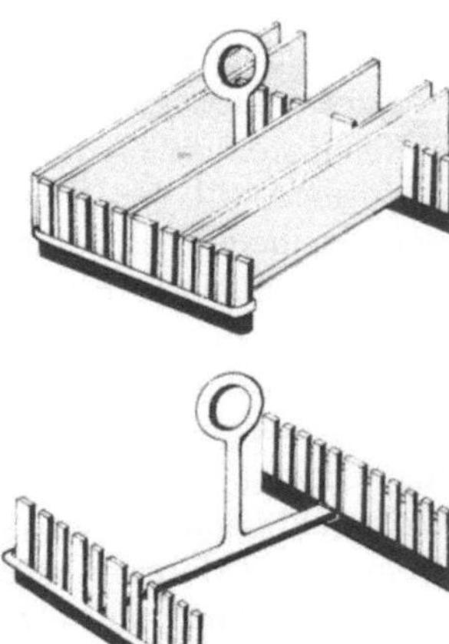

Fig. 8.5

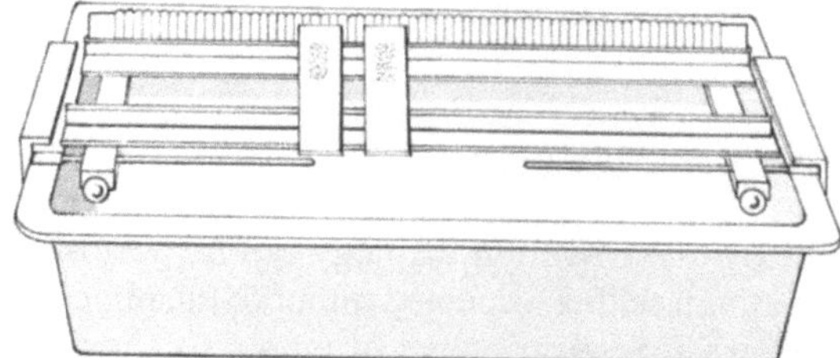

Fig. 8.6

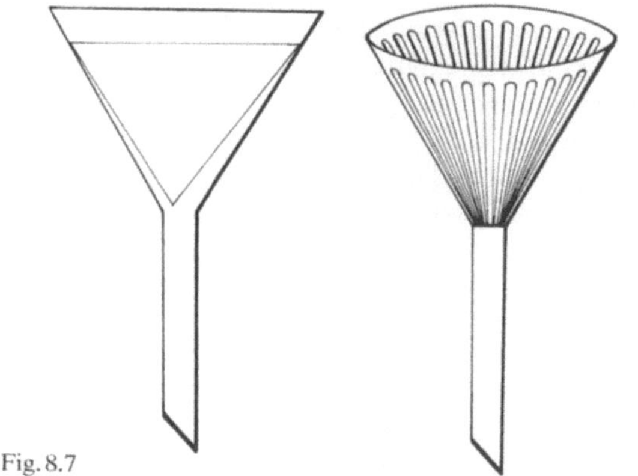

Fig. 8.7

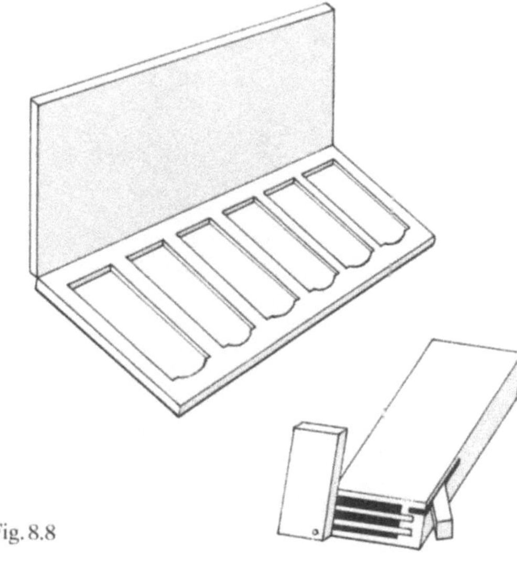

Fig. 8.8

A *staining rack* is useful for Giemsa staining (see Sect. 8.7.1.4). With this apparatus the slides can be coated with the stain while lying flat, then moved without touching to a tilted position for rinsing with water. The staining rack (Fig. 8.6) can be purchased separately from the catch basin if desired and mounted on the rim of the laboratory sink.

Some dye solutions, especially Giemsa and Papanicolaou stains, must be regularly filtered to avoid artifacts. Special *paper filters and funnels* (Fig. 8.7) can be obtained for this purpose.

8.2.5 Filing

If there is a need or desire for permanent filing of cytologic preparations, several options are available. If permanently stained specimens are mounted (see Section 8.7.3) with corbite balsam (Eukitt, Caedax, Hico-Mic), they should be thoroughly dried for 2 days before filing so that they will not stick together. Mounting is not an essential step in processing, but unmounted slides must be stored in a lightproof container.

The simplest and most economical storage method is to place the slides into their original package and label the outside of the box. Alternatively, slides may be filed in cardboard portfolios or wooden cases (Fig. 8.8), which also may be used for off-site dispatch. The use of large filing cabinets in which slides are stored on edge in narrow compartments is an effective but costly solution.

8.3 Specimen Collection

8.3.1 Timing and Technique of Urine Collection

Since micturition is a physiologic process, there is no basic difficulty in collecting material for cytologic evaluation. The use of *spontaneously voided urine* is satisfactory for routine examinations. The first morning urine should not be used, because prolonged contact of the cells with the urine leads to marked cellular degeneration by the action of proteolytic enzymes and bacterial cytolysins.

Some examiners prefer *irrigation cytology* (bladder washings) because of the improved cellular yield. After voiding, the bladder is irrigated through a catheter or cystoscope with 50–100 ml of physiologic saline solution. Isotonic physiologic saline (0.9%) or Ringer's solution is recommended to avoid osmotically-induced cell changes (see Sect. 9.4, Fig. 9.21a,b). The hypoosmolarity of dis-

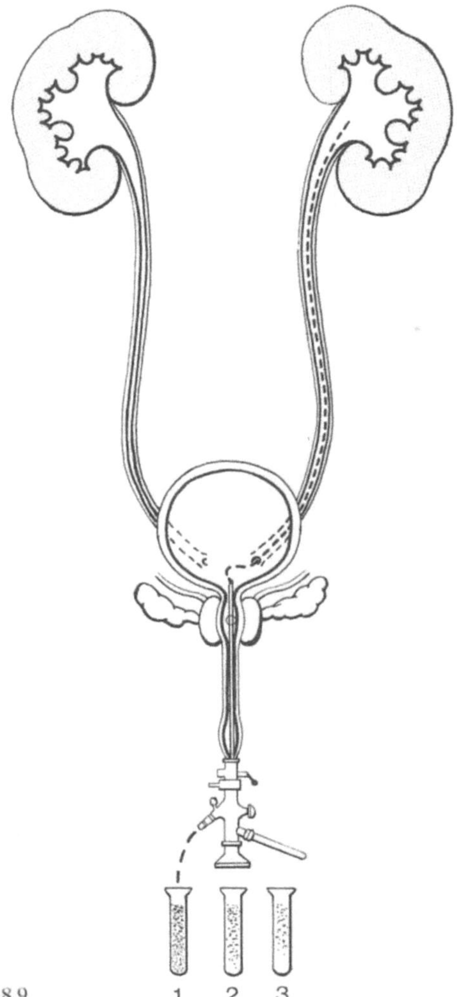

Fig. 8.9   1   2   3

8.3.2 Special Techniques

*8.3.2.1 Irrigation Cytology of
the Upper Urinary Tract*

Retrograde irrigation cytology can be performed
in patients with a suspected tumor of the upper urinary tract. Several points of emphasis should be
noted:

▶ The cytologic material should be collected *before the administration of contrast medium* to
  avoid artifactual cell changes.
▶ Experience has shown that urothelial cells re-
  trieved in washings from the upper urinary
  tract show significant reactive, manipulative
  changes that are far more pronounced than
  those seen in bladder washings. For this reason
  a *multistep procedure* has proven beneficial
  (Fig. 8.9):
  1. Drain the bladder urine (preferably no
     washings) through the cystoscope.
  2. With satisfactory diuresis, thread a catheter
     into the ureteral orifice and obtain a sample
     (several milliliters) of spontaneously pro-
     duced urine (may have to rotate UC to
     clear catheter opening of obstructing
     mucosa).
  3. Collect irrigation specimen after washing
     upper urinary tract with 10–20 ml isotonic
     saline or Ringer's solution.

If *questionably pathologic urothelial changes* are
found in the upper urinary tract washings, the bladder urine sample (1) and the spontaneous urine
sample from the catheterized ureter (2) provide a
source of additional information for evaluating a
possible *reactive etiology* of the suspicious urothelial cells in the ureteral washings (3).

Given the difficulties of the cytologic evaluation,
it is ultimately up to the individual to decide
whether to take a more passive, expectant approach or pursue a more vigorous endoscopic or
exploratory surgical examination.

The *retrograde brushing* technique, first described by Gill et al. (1973), is problematic because
passage of the Ch 7 catheter is quite painful, necessitating the use of an anesthetic, and also causes
significant artifacts. Based on the associated high
rate of false-positive findings, Koss (1989) has
completely abandoned the evaluation of cytologic
specimens obtained by the brush technique.

tilled water, for example, causes nuclear swelling
that must be taken into account when evaluating
the nuclear–cytoplasmic ratio as a criterion of malignancy.

Although irrigation cytology can increase diagnostic confidence by increasing the desquamation
of urothelial cells, it has several disadvantages:

– It sacrifices the advantage of noninvasiveness,
  which is particularly important in male patients.
– The instrumental manipulation incites reactive
  cell changes that can mimic the features of highly differentiated dysplasias or tumors ("cytologic irritation by bladder irrigation").

8.3.2.2 Irrigation Cytology of the Urethra

Because a recurrent urothelial tumor develops in the urethral remnant of 4%–18% of all cystectomized patients, urethral washings should be an essential part of the follow-up program.

While it is possible to irrigate the urethra through an olive adapted to the external meatus, this technique may not adequately irrigate the proximal part of the urethral remnant, where there is the greatest likelihood of a recurrence. Additionally, squamous epithelial cells exfoliated from the distal part of the urethra can cause a superposition effect that hampers cytologic evaluation.

It is better, therefore, to irrigate the urethra and especially the proximal remnant through a thingauge *disposable catheter* using isotonic saline or Ringer's solution.

8.3.2.3 Aspiration Cytology

The cytologic examination of aspirated material is occasionally indicated. An *anticoagulant* can be added to a sample that contains a significant amount of blood. Suitable anticoagulants are heparin (1 mg or 100 units per 10 ml aspirate), EDTA (10 mg per 10 ml aspirate), or sodium citrate (20 mg per 10 ml aspirate).

8.4 Specimen Fixation and Preservation

The fixation and preservation of cytologic material is of interest for three reasons:

1. *Dispatch to an off-site facility* for "primary" or reference cytology, or for performing a sophisticated, differential staining procedure in a cytopathology laboratory.
2. *Economy* of time and cost, since urine samples collected at different times can be processed at the same time. This also means that the collected samples can be analyzed at certain times by rapid staining or jointly subjected to differential staining.
3. *Documentation* of findings that are of clinical, scientific, or forensic interest.

Two basic methods are used in the fixation and preservation of cells before staining: The cells may be fixed and preserved while still within the urine "carrier medium," or the cells may be primarily recovered from the urine, transferred to a slide, and fixed there.

8.4.1 Preservation and Fixation of Urine

Option 1

▶ Centrifuge a urine portion.
▶ Pour off (decant) the supernatant.
▶ Add Esposti fixative to the sediment in a 1:1 ratio. Its formula is as follows:

    10 ml acetic acid 100%
    48 ml methanol
    <u>42 ml distilled water</u>
    100 ml

Note: Acetic acid can cause *hemolysis of erythrocytes.* When gross hematuria is present, this can help eliminate troublesome overlapping of urothelial cells by erythrocytes, but in studies of unexplained hematuria it can remove the valuable option of evaluating erythrocyte morphology.

Cellular suspensions fixed in this way can be kept in a refrigerator for several days. It is important to mix the sediment thoroughly with the fixative by gentle shaking.

Immediately prefixed urine samples should be stained with an alcohol stain (Papanicolaou or Szczepanik), because the presence of a fixative can degrade the quality of a Giemsa or Pappenheim stain. It is better to dry-fix the specimen on the slide when these stains are used.

Option 2

▶ Centrifuge a urine portion.

▶ Pour off (decant) the supernatant.

▶ Add alcohol to the sediment in a 1:1 ratio.
  Composition:

   90 ml ethanol (97%)

   10 ml formalin (37%)

   100 ml

Note: Formalin (even traces of formalin vapor) has such a negative effect on Giemsa and Pappenheim stains that these techniques cannot be used after ethanol-formalin fixation.

Because alcohol is a dehydrating agent, alcohol fixation causes a marked *shrinkage of the cell nuclei.* This can obscure the details of nuclear structures.

Urine samples fixed with alcohol can be stored for several months.

Option 3

▶ Mix 10–20 ml urine with approximately 50 mg of thiomersal salt. Alternatively, 2–3 ml of a thiomersal stock solution can be used. Composition: 5 g of thiomersal dissolved in 100 ml of distilled water.

Note: Thiomersal is excellent for the short-term *stabilization* of urine for *3–4 days.* The alcohol-free sample can then be processed further using a rapid stain, alcoholic stain (e.g., Papanicolaou), or Giemsa or Pappenheim stain.

Thiomersal must be stored in the dark. It contains mercury, so the supernatant left by centrifugation should be placed into a separate sealed container for *special disposal.*

Urine fixed with thiomersal should be stored in the refrigerator but may be removed for mailing, etc.

Option 4

▶ Centrifuge a urine portion.

▶ Pour off the supernatant (completely!).

▶ Resuspend approximately 1 ml in 0.9% NaCl.

Note: Degenerative cell changes can still occur with this method. It is important to remove the supernatant urine immediately and as completely as possible, retaining a highly cellular sediment. Simple decanting generally must be combined with the use of filter strips to absorb residual urine.

The sediment, resuspended in physiologic saline solution, can be stored a *maximum of 24–48 h* in the refrigerator.

Option 5

The Robapharm company offers a new "dispatch preservative" for fresh urine; 1–2 ml of the medium added to 10 ml of urine will provide adequate preservation for several days. The procedure is simple and economical, and the preservative is nontoxic.

Option 6

A urine fixation system developed in Sweden and offering the advantage of using a large urine volume with a correspondingly high cellular yield is being tested under the brand name Urotel.

Approximately 150 ml of fresh urine is poured into the labeled sediment bag (Fig. 8.10a). The chamber with the urine, clamped off at the top and bottom (b), contains a *prefixative* (3 ml of 10% trichloroacetic acid) that prevents cellular degeneration. After about 10 h of sedimentation, the upper clamp is reapplied above the sediment (c), and the lower clamp is removed. This mixes the prefixed sediment with the *fixative* (10 ml ethanol), whereupon the sediment can be further processed or dispatched at leisure (d).

One advantage of this system is that it recovers all the cells from a large urine volume, which is beneficial in terms of accurate diagnosis. Given the various alternatives that are available, it remains to be seen whether the system will become widely used.

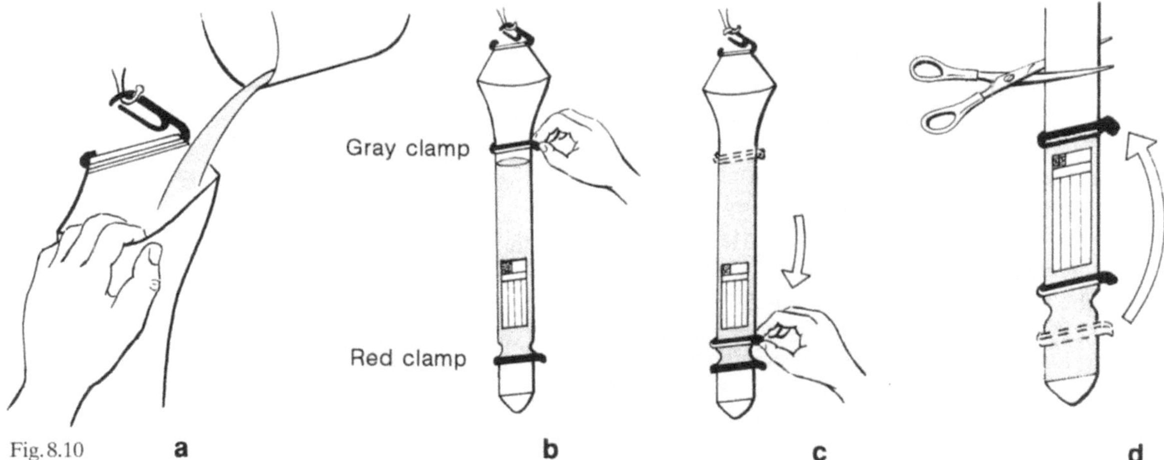

Fig. 8.10   **a**   **b**   **c**   **d**

8.4.2 Cell Fixation on Slides

8.4.2.1 Dry Fixation on Slides

The air-drying (dry fixation) of cells on the slide is
a necessary prelude to staining by the Giemsa
or combined May-Grünwald/Giemsa procedure
(Pappenheim panoptic stain) or for rapid
Hemacolor staining. After centrifugation of the
fresh or nonalcohol-preserved urine, the sediment
is spread onto the slide (Fig. 8.11) or the slide is tilted so that the sediment drop runs down it.

The specimen should be dried as quickly as possible to minimize artifacts. Drying can be hastened by
shaking the slide vigorously in the air (holding it by
the edges) or placing it briefly in a thermostatic incu-
bator until the surface sheen is gone (predrying).

Then the slide is tilted upright on its narrow edge
for definitive drying, which may require several
hours. The coated side of the slide can face down-
ward as a precaution against contamination by
dust particles.

The *fixation produced by air-drying* is adequate
for *no more than 24 h*. The fixation effect is improved and prolonged by immersing the air-dried
slide in methyl alcohol for 5–10 min, or in ethanol
or a mixture of equal parts ethanol and ether for
20–30 min. When an ether-containing fixative is
used, the fixation vessel should be tightly covered
to prevent evaporation, and the fixative solution
should be returned to the storage bottle immediately after use.

8.4.2.2 Wet Fixation on Slides

Wet fixation is necessary for alcohol staining by the
Papanicolaou or Szczepanik technique, for otherwise degenerative cell changes will occur. Either
spray fixation or alcohol immersion may be used.

Spray Fixation. Commercially available spray fixatives contain polyethylene glycol (e.g., Merckofix).
The smear or cytocentrifuge preparation is spray-
fixed while still moist. Care is taken to hold the
spray nozzle an adequate distance from the slide
(about 20 cm) so that cells are not dislodged. The
fixative should coat the slide evenly without pud-
dling (see Fig. 8.3).

The protection afforded by spray fixation permits the *collection of several specimens* for further
joint processing at a later time and is also adequate
for specimen dispatch. The alcohol contained in
the spray evaporates, leaving the residual
polyethylene glycol as a protective coating.

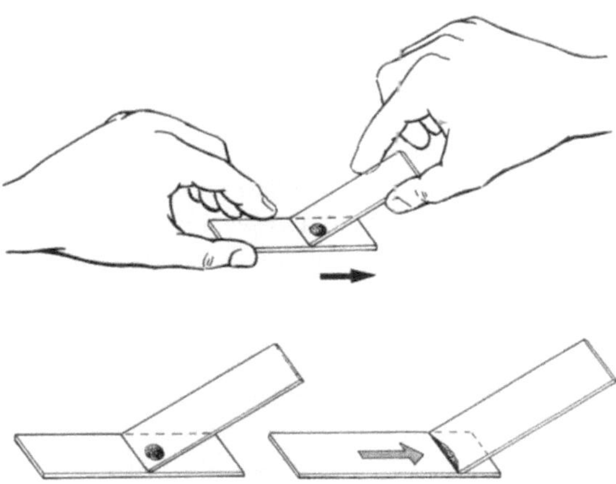

Fig. 8.11

Spray fixatives also can be applied to a large drop of material before immersing the slide into stain or alcohol. This *increases cellular adherence* to the slide, avoiding cell loss and the possible cross-contamination of slides by floaters.

Before the cell sample is stained, the protective polyethylene glycol film is dissolved by immersing the slide in diluted alcohol (30% ethanol). For Papanicolaou staining, for example, the slide is not put through the entire *initial* alcohol series (80%, 70%, 30%) before immersion in distilled water; instead, the spray-fixed specimen is dipped into 30% ethanol for only about 10 s before proceeding with the rest of the sequence (distilled water, etc.).

Wet Fixation by Alcohol Immersion. In wet fixation by alcohol immersion, the slides carrying the cellular material (applied as a drop of sediment or by another concentration technique, see Sect. 8.5) are immersed in the fixative solution for at least 15 min (see Fig. 8.4). The specimen should be fixed immediately after transfer to the slide, before drying has begun, to avoid degenerative cell changes.

When this technique is used, cells may *slough* into the alcohol bath, causing problems of cell loss and also cross-contamination of other slides that are simultaneously or subsequently placed into the cuvet. Although mucous and proteinaceous substances in the urine normally provide a natural *cellular adherence*, it is a sound precaution to pretreat the slides.

Option 1

Spread *albumin* onto the slides before applying the cellular material. Mayer's albumin is composed of equal parts of egg or serum protein and glycerine. Unfortunately, background staining artifacts may occur with this procedure.

Option 2

The clean, fat-free slides are *coated* with Poly-L-Lysin, a *synthetic protein* made from an amino acid. The slides are dipped briefly into a PLL solution (12.5 mg PLL in 100 ml distilled water) and dried in a dust-free environment. Drops of PLL should not be left on the slide, because later the protein may take up stain, depending on the staining procedure that is used. A thin PLL coating will not alter the appearance of the stained preparation.

Many slides can be coated at one time in a batch process, since the coating will retain its adhesive properties for several weeks. The slides can be used as needed during this period.

Option 3

After spreading or cytocentrifugation of the cellular material onto the slide, a small amount of spray fixative can be applied to the slide to prevent the sloughing of cells.

Various alcohols can be used as a fixative solution for alcohol immersion:

- 96% ethanol
- A mixture of equal parts of ethanol and diethyl ether (the vessel with the fixative solution should be sealable, for ether evaporates rapidly)
- Methyl alcohol
- 90% acetone
- A mixture of 5 parts 96% ethanol and 1 part glycerine
- 99% isopropyl alcohol

The addition of acetic acid to fixative solutions causes the hemolysis of erythrocytes and can be advantageous for clearing the field of red cells in grossly bloody samples. Generally, however, acetic acid should not be added because it prevents the evaluation of erythrocyte morphology as a means of identifying glomerular sources of hematuria.

If a hemolytic action is desired, 3 vol% glacial acetic acid (100% acetic acid) can be added to the fixative solution.

8.5 Methods of Cell Concentration

The *accuracy of the cytologic diagnosis* depends not just on sound preparatory and staining techniques and the experience of the cytologist but also on the *number of urothelial cells* that are available for evaluation and intercomparison. Because only small numbers of urothelial cells may exfoliate into the urine stream, cell concentration procedures must be integrated into the preparatory routine.

In contrast to ordinary centrifugation, more sophisticated techniques such as cytocentrifugation make it possible to concentrate the cells into a relatively small area on the slide. These techniques increase the likelihood of recovering all diagnostically relevant cells and improve screening efficiency. It is up to the individual to select the recovery technique that offers the most acceptable cost-benefit ratio.

8.5.1 Direct Centrifugation

Direct centrifugation is the standard technique for preparing cellular sediment from urine.

▶ Shake the urine sample to resuspend cells that have settled.
▶ Pour 10–20 ml into the centrifuge tube.
▶ Centrifuge at 2000 rpm for 5–10 min.
▶ Decant the supernatant, then proceed with further processing or fixation of the sediment according to the staining method that will be used.

8.5.2 Cytocentrifugation

Cytocentrifugation deposits the concentrated cellular material *directly onto a slide*, where it can be fixed and stained. This method offers two main advantages:

– The high rate of cell recovery concentrates many cells onto a small area of the slide where they can be easily found and interpreted (Fig. 8.12).
– Significant time saving, since the examiner does not have to screen the entire slide. The sediment is deposited at a consistent site that is easily located using the coordinates on the microscope stage.

Cytocentrifuges are offered by various manufacturers. Besides dedicated cytocentrifuges (Shandon Cytospin), whose high cost is justified

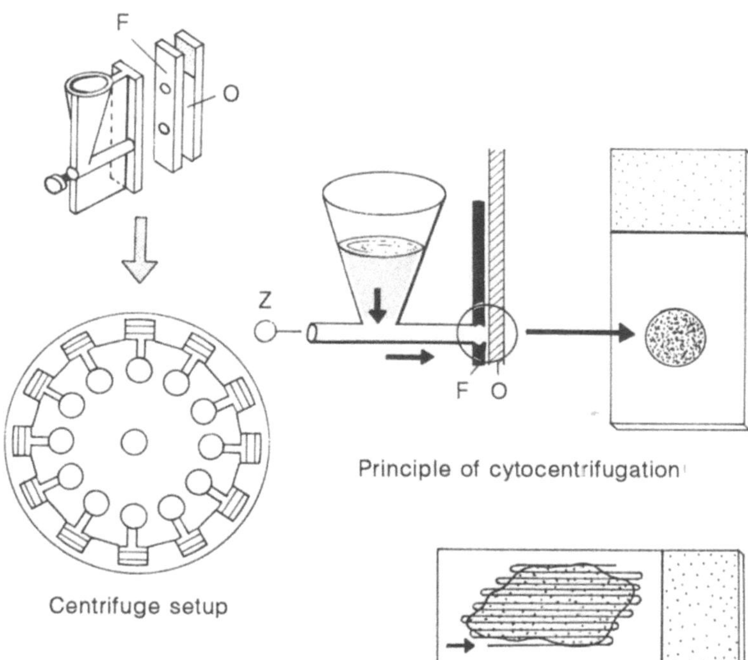

Fig. 8.12. **Several millimeters of urine sediment (the lower, settled part of the sample, not the sparsely cellular supernatant!) are pipetted into the sample chamber.** The microscope slide (*O*) and filter card (*F*) are mounted so that the slide is farthest from the centrifuge hub (*Z*). During centrifugation, the cells are concentrated into a small circular area on the slide, eliminating the need to screen an entire smear

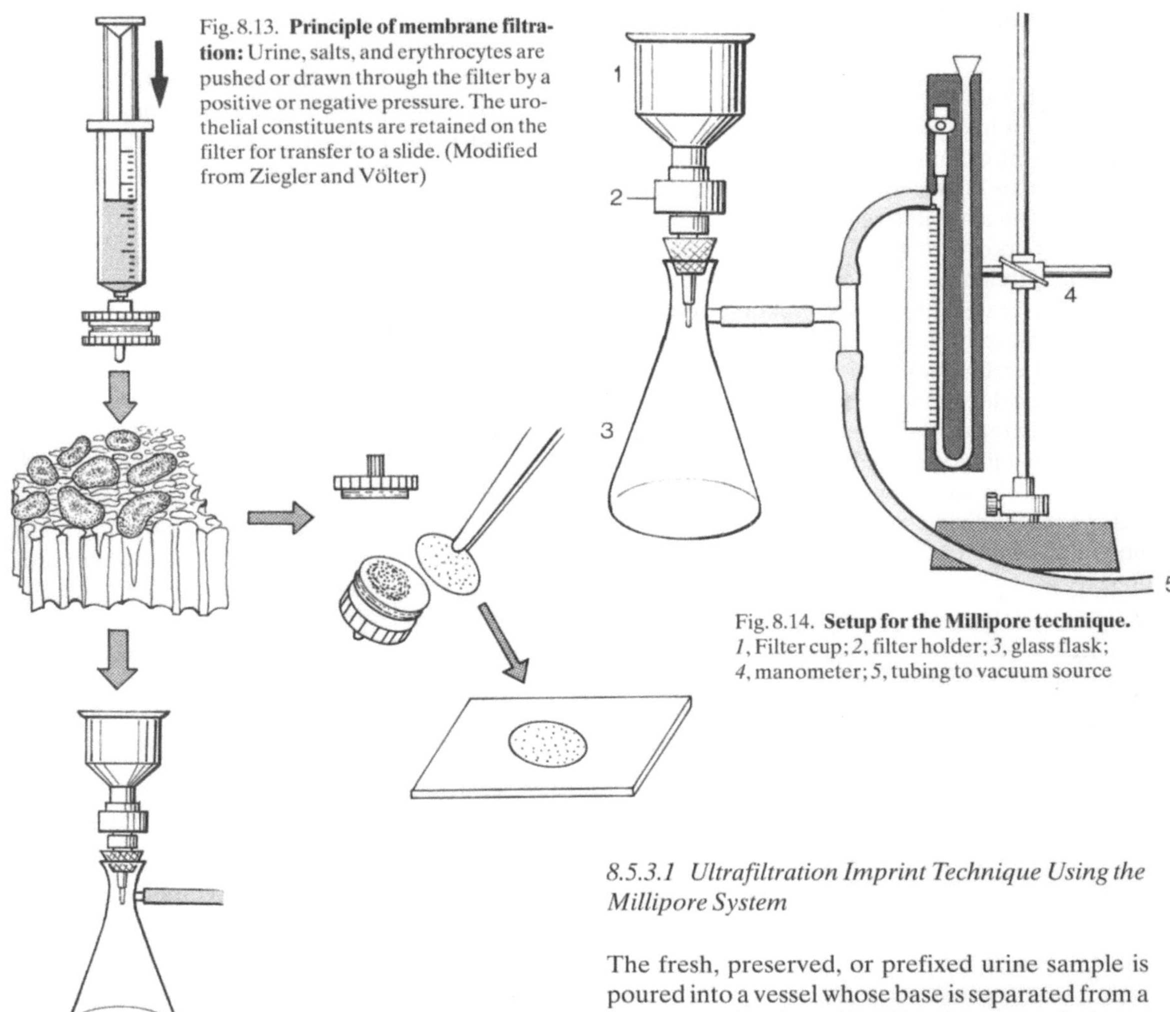

Fig. 8.13. **Principle of membrane filtration:** Urine, salts, and erythrocytes are pushed or drawn through the filter by a positive or negative pressure. The urothelial constituents are retained on the filter for transfer to a slide. (Modified from Ziegler and Völter)

Fig. 8.14. **Setup for the Millipore technique.** *1*, Filter cup; *2*, filter holder; *3*, glass flask; *4*, manometer; *5*, tubing to vacuum source

only by frequent use, some manufacturers offer devices that can be adapted for multiple tasks (Heraeus, Hettich). In these devices, special chambers for cytocentrifugation can be attached to the centrifuge rotor.

8.5.3 Membrane-Filter Techniques

In these techniques the cells are trapped by a *filter* and then transferred to a slide using a "sandwich" or imprint technique for further processing. The actual filtration processes may use a *vacuum* or *positive pressure* to draw or push the urine sample through the filter (Fig. 8.13).

8.5.3.1 Ultrafiltration Imprint Technique Using the Millipore System

The fresh, preserved, or prefixed urine sample is poured into a vessel whose base is separated from a lower container by a filter (5 or 8 μm pore size) on a coarse metal grid. A vacuum pump creates a *negative pressure* of 15–20 Torr which draws the urine through the filter pores into the lower collecting vessel (Fig. 8.14). After the urine has passed through it, the filter is removed from the apparatus with forceps and imprinted onto the slide, which is first coated with a suitable adhesive medium (e.g., Poly-L-Lysin) to improve cellular adherence (see Sect. 8.4.2).

Depending on staining requirements, the slide is subsequently air-dried or wet-fixed by spray or alcohol immersion.

8.5.3.2 Membrane-Filter Techniques (Pressure Filtration)

In the positive-pressure membrane-filter technique (Sartorius Co.), the urine is drawn up into a 10- or 20-ml disposable syringe and injected

Fig. 8.15. **Principle of positive-pressure filtration.** (Modified from Aeikens 1981)

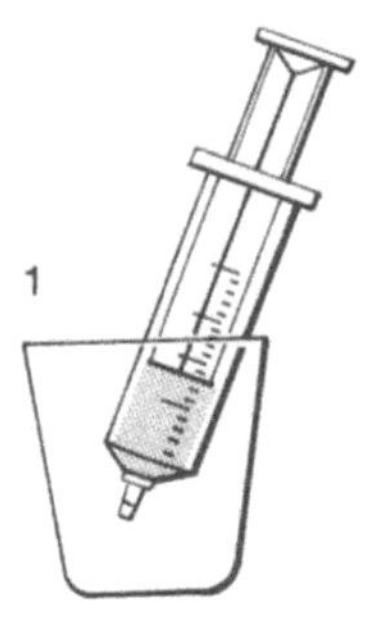
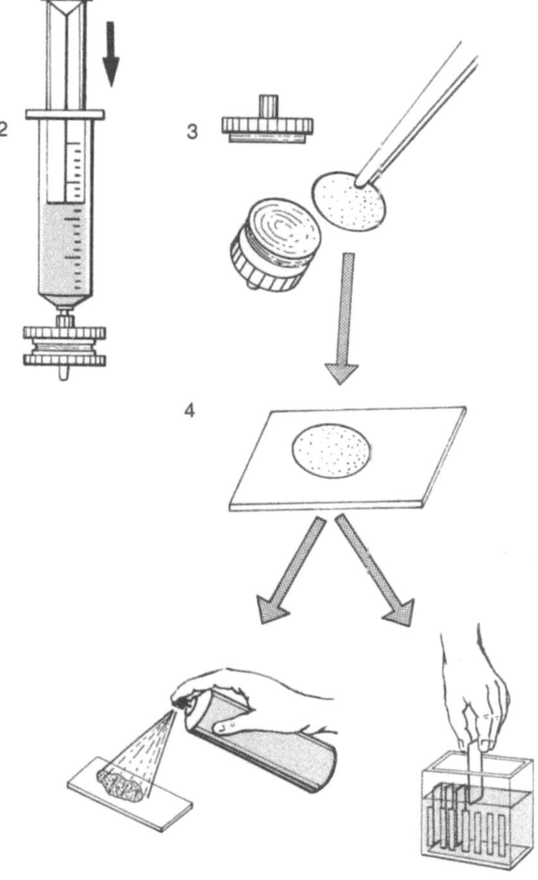

through a filter holder that screws onto the end of the syringe. The filters come in various diameters (up to 25 mm) with pore sizes of 5 or 8 µ.

The filter retains the urothelial cells while allowing smaller, undesired constituents and salts to pass through. The pores are designed so that the filter becomes clogged when it has acquired an optimum cell load. This point has been reached when no additional urine can be pressed through the filter.

After filtration is completed, the filter holder is opened, the filter is removed, and the cells on the filter are transferred to a slide previously coated with adhesive (Fig. 8.15). The specimen can then be air-dried or wet-fixed (spray or alcohol immersion) for further staining or dispatch.

Cyto-Quick System. A relatively complex apparatus for cell recovery and fixation is marketed under the brand name Cyto-Quick (Fig. 8.16). In this system *cell concentration and fixation* are accomplished in two successive filtering processes. A cap containing a filter cartridge is screwed onto a con-

Fig. 8.16. **Principle of the Cyto-Quick system** with primary filtration and fixation followed by refiltration through a secondary filter, which is then imprinted onto a slide

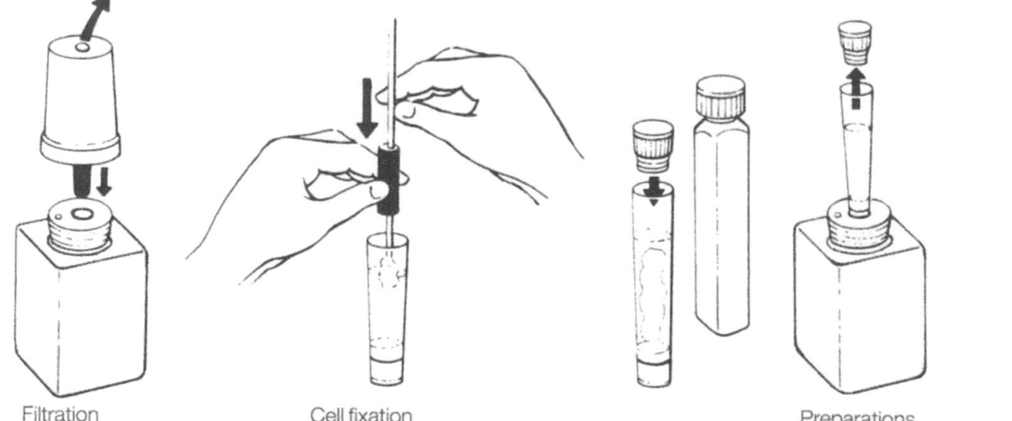

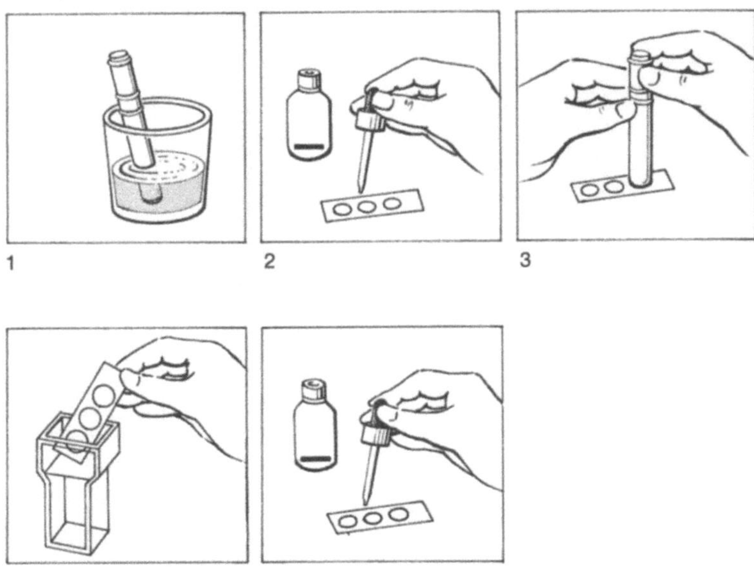

Fig. 8.17. **Principle of the Carcyt-U1 system:** The capillary filter is placed into a urine sample that has been allowed to settle (*1*). The end of the filter is then pressed onto a glass slide (*2*) previously treated with a cellular adhesive (*3*). After drying, the specimens are fixed for 15 min in alcohol (*4*) and then stained (*5*)

tainer with the urine cartridge. The container is inverted, and the cells collect in the filter filament contained in the cartridge. After 10–30 min the filter is removed from the cartridge and fixed in an alcohol medium. At this point the sample is stabilized and ready for dispatch or further examination.

In a second filtering process the urine is sent through a secondary filter, on whose surface the cells are retained. The cells are then imprinted onto a slide ("sandwich" technique) and fixed there.

8.5.4 Capillary Filter Suction Technique

A new system (Carcyt U1) currently undergoing clinical testing employs a principle similar to that of a felt-tipped pen. A small capillary tube is placed into a urine sample that has been allowed to settle, and an indicator window at the top of the tube signals when capillary suction is complete. The cells adhering to the end of the tube are then pressed onto a slide previously treated with a cellular adhesive (Fig. 8.17).

8.5.5 Sedimentation Techniques

These methods are based on the spontaneous sedimentation of cells while the fluid medium is slowly absorbed by filter paper. Gradual absorption is achieved by the use of weights (Bots' method) or brass rings (Blonk's method). The slow rate of cell

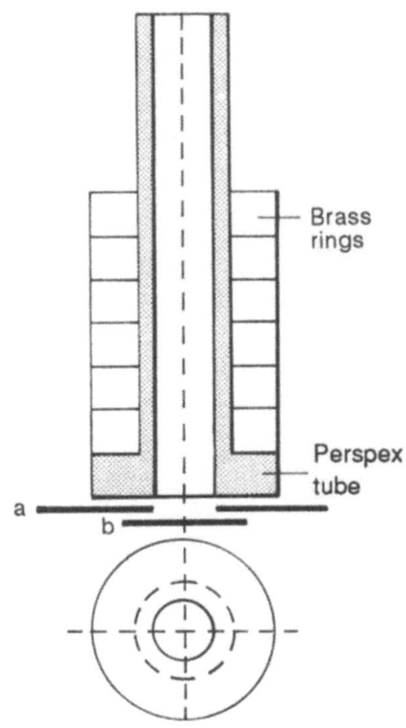

Fig. 8.18. **Principle of Blonk's sedimentation technique.** (From de Vogt et al. 1979)

sedimentation onto the slide beneath the filter paper avoids physical damage to the cells.

Procedure of Blonk's method (Fig. 8.18): A piece of hard filter paper with a circular perforation is placed over a glass slide. The size of the hole matches the inside diameter of the tube and the area in which the cells will be concentrated (13 mm). Then the tube with the brass rings is placed onto the paper so that the tube orifice lines up with the hole in the paper (see Fig. 8.18). About 1–5 ml of the urine sample is pipetted into the tube, depending on the cell concentration. The fluid should be absorbed within 30 min; if absorption takes longer, several of the brass rings are carefully removed.

8.6 Phase Contrast Microscopy

Phase contrast microscopy (PCM) is the most widely used contrast-enhancing microscopic technique for the examination of unstained specimens.

The use of an annular diaphragm (A_1) in the condenser (C) and an annular phase plate (A_2) in the objective (B) results in the separation of diffracted (I) and nondiffracted (II) light rays. The wavelength of the diffracted rays is delayed, setting

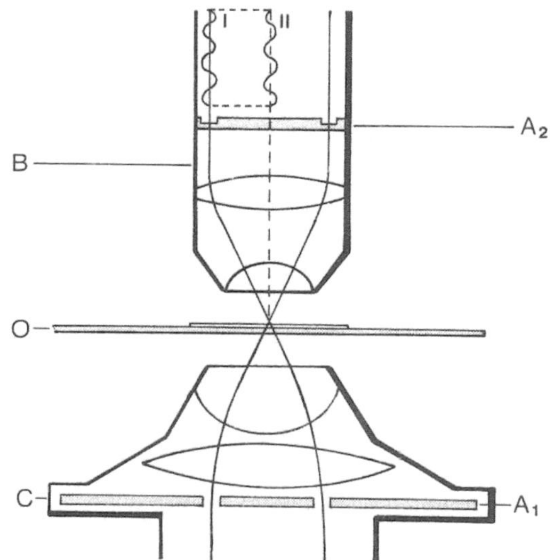

Fig. 8.19. **Principle of phase contrast microscopy** (see text). (From de Vogt et al. 1979)

up a phase difference which causes structural details within the cell to appear darker than the surrounding medium. The differences in brightness also create an optical artifact in the form of a halo around the structures of interest (Fig. 8.20).

Because PCM is *inferior to all other staining methods* in current use, it has little practical importance in oncologic urinary cytology.

A major application of PCM is in the differentiation of erythrocyte morphologies, i.e., distinguishing between a postrenal urothelial and a renal glomerular origin of hematuria (see Chap. 12).

Fig. 8.20. **Basal cells** as demonstrated by conventional light microscopy (a, c) and phase contrast microscopy (b, d). From de Vogt et al. 1979)

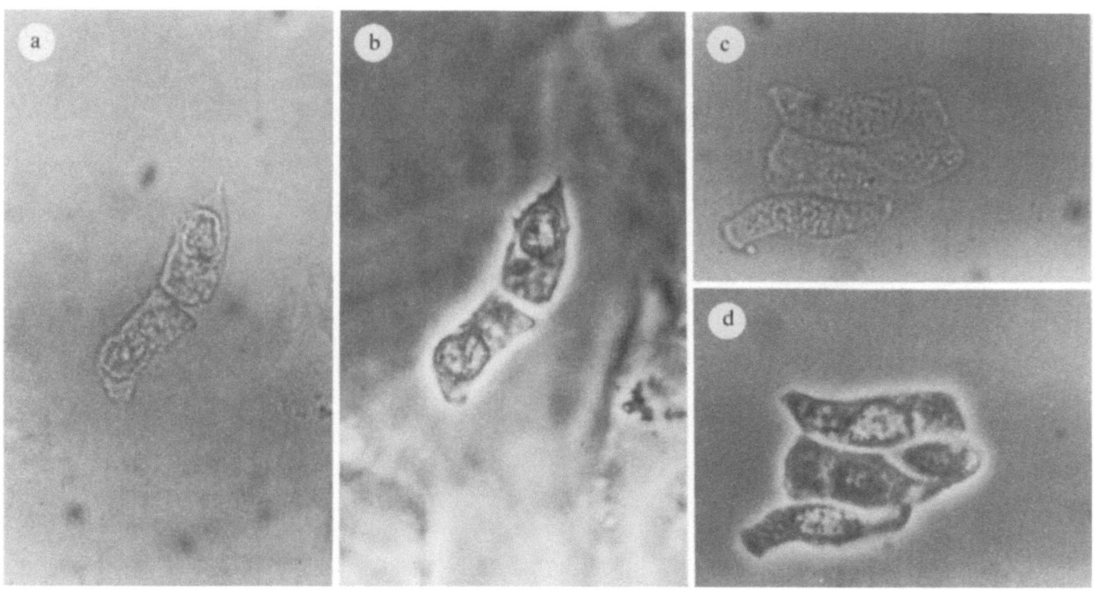

8.7 Staining Methods

8.7.1 Rapid Stains

8.7.1.1 Methylene Blue Stain

Staining with Löffler's methylene blue is a simple, rapid, and cost-effective method that imparts a blue stain to cell nuclei.

Procedure

▶ Centrifuge the fresh or preserved urine.
▶ Place a drop of sediment onto a slide.
▶ Apply the stain either next to the sediment before coverslipping or at the edge of the applied coverslip (Fig. 8.21).

Note:
– The area where the methylene blue directly contacts the sediment usually shows heavy overstaining that should not be mistaken for pathologic hyperchromasia.
– The specimen can be interpreted 5–10 min after staining.
– The specimen can be stored in a moist environment (covered Petri disk in the refrigerator) for a maximum of 24 h. A specimen stained with methylene blue cannot be mounted for permanent filing.

8.7.1.2 Sediment Stains

Unlike methylene blue, sediment stains (Sedicolor, MD-Kova solution) are mixed directly with the cellular material before the sediment is placed on the slide. Procedure:

▶ Centrifuge and decant the supernatant.
▶ Add 2 drops of stain to the sediment.
▶ Mix thoroughly.
▶ Incubate for 1 min, then apply to coverslip and slide.

Note: Sediment stains are suitable for the preliminary evaluation of urine specimens, especially for the assessment of erythrocyte morphology, but otherwise they are *qualitatively inferior* to other staining methods and are *inadequate for oncologic urinary cytology.*

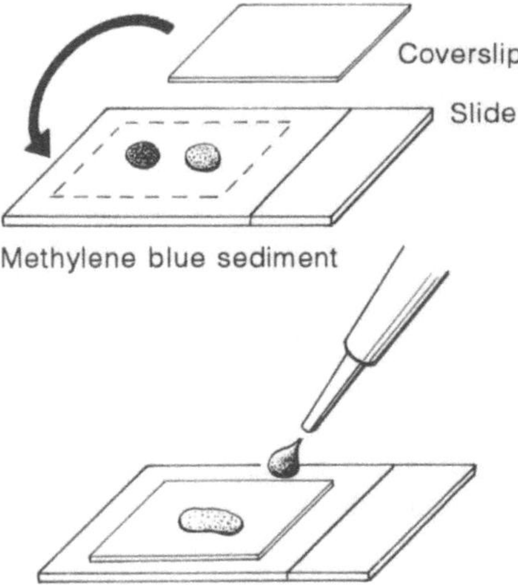

Fig. 8.21. **Staining with methylene blue**

8.7.1.3 Prestained Slides

Two products are available commercially that differ in terms of staining procedure and color quality.

Testsimplets Stain. The slides are uniformly coated with methylene blue and cresyl violet acetate, which provide a very high-quality stain with exceptional *nuclear transparency*, which is critical for the diagnosis and grading of malignancy.

Procedure
▶ Centrifuge the fresh or preserved urine.
▶ Transfer a drop of the sediment to a coverslip.
▶ Hold a slide over the sediment with the stain-coated side down; capillary forces will draw the sediment onto the slide (Fig. 8.22).

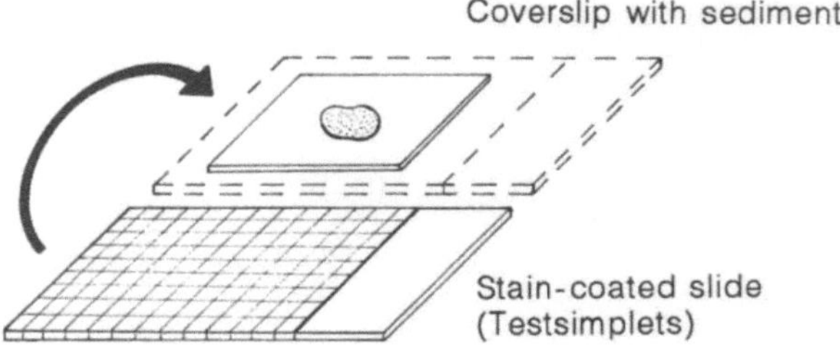

Fig. 8.22. **Staining with a prestained microscope slide** (Testsimplets)

Note:
- Primary application of the sediment to a cover-slip is necessary for uniform staining.
- Erythrocytes take little or no stain. This helps clear the field of red cells when gross hematuria is present, but it is disadvantageous for the evaluation of erythrocyte morphology.
- The specimens can be stored for up to 12 h in a moist environment (refrigerated, covered Petri dish); permanent filing is not possible.
- Staining with Testsimplets is at least equivalent in quality to Papanicolaou staining.

Sangodiff Stain. Sangodiff stain is applied *after air-drying*, so exact procedures and results will vary.

Procedure
▶ Centrifuge the fresh or preserved urine.
▶ Transfer the sediment to the slide.
▶ Air-dry the sediment or "running drop."
▶ Fix in methanol.
▶ After drying (5–10 min), the slide can be further processed at once or stored for up to 24 h for later processing.
▶ Apply the buffer solution and staining film.

Note:
- The stain does not dissolve uniformly and creates a mobile background with floating dye particles that should not be mistaken for bacteria.
- Primary air-drying creates drying artifacts in the urothelial cells; these should be taken into account when the specimen is evaluated.
- Though inferior to other rapid stains in color quality, Sangodiff does allow for the permanent *filing* of slides by removing the stain film and applying a mounting medium.

8.7.1.4 Giemsa Rapid Stain

Giemsa staining is performed on air-dried smears. The staining solution is composed of methylene azure (thionin derivative responsible for the Romanowski effect, i.e., the affinity of certain cell constituents for azure red), methylene violet, methylene blue, and eosine dissolved in methanol and glycerine. It is best to use standard solutions that are commercially available.

Procedure
▶ Fix the air-dried smear (see Sect. 8.4.2) in absolute alcohol for 30 min or in anhydrous methyl alcohol for 5–10 min, then air-dry; do not rinse.
▶ Next (within 24 h) lay the slides flat on a stain-ing rack (see Sect. 8.2) with the coated side up and stain with freshly prepared, diluted Giemsa solution (see below). The staining solution is dripped onto the slides and should cover them completely.
▶ After 20–30 min, rinse the staining solution off the slides with a squeeze bottle of neutral distilled water or phosphate buffer at pH 7.2 (do not tilt slides; direct the jet from the side).
▶ After rinsing, clean the uncoated undersurface of the slides with a dust-free cloth to avoid troublesome superimpositions during microscopic examination.
▶ Tilt the slides for air-drying.
▶ The specimen can be evaluated now or after mounting with corbite balsam (see Sect. 8.7.3). If unmounted specimens are examined with an oil-immersion system (100 x), the oil should not be removed mechanically with a cloth but by rinsing briefly in xylene. Unmounted slides can be filed by storing in a dark container.

Note:
- Like acidic water, glass surfaces that carry acidic contaminants (e.g., fingerprints) will degrade the quality of the Giemsa nuclear stain. Separate glassware should be kept specifically for Giemsa staining.
- The staining ability of the Giemsa solution is subject to variations, so the correct dilution ratio (see below) should always be checked before use.
- If the staining solution is too concentrated, if the staining time is too long, or if the smears are too thick, the nuclei will appear clumped and featureless.

Preparation of Giemsa Working Solution
- Filter the stock solution.
- Dilute 1:1 with buffered distilled water at pH 7.2 (1 drop stain to 1 ml water).
- When mixing, do not swirl the vessel any longer or more vigorously than is needed to obtain a homogeneous mixture, as this might cause precipitation of the dyes.
- The staining solution should be prepared fresh for each run and used right away. Clean the vessels by rinsing with distilled water to avoid acid residues that would affect the stain.

8.7.1.5 Hemacolor Stain

The Hemacolor staining set (Merck), consisting of a total of four solutions, is an easy-to-use, rapid staining procedure that can be used on air-dried specimens and allows for permanent filing.

Procedure
▶ Prepare an air-dried smear (see Sect. 8.4.2).
▶ Dip the smear five times for 1 s each in fixative solution one, then three times in color reagent two, and six times in color reagent 3.
▶ Rinse briefly in buffer solution (buffer tablets are included in the set).
▶ Mount if desired (Fig. 8.23).

Fig. 8.23. **Hemacolor stain:** The air-dried smear is dipped five times for 1 s in fixative solution 1, three times in color reagent 2, and six times in color reagent 3 before rinsing in buffer solution (pH 7.2)

8.7.2 Differential Stains

8.7.2.1 May-Grünwald/Giemsa (Pappenheim) Stain

The combined May-Grünwald/Giemsa stain, known also as Pappenheim's panoptic stain, is used in many laboratories owing to its excellent properties. Pappenheim staining may be performed in cuvets or on a staining rack.

Procedure (Staining Rack)
▶ Cover air-dried, unfixed smears (see Sect. 8.4.2) with filtered, concentrated May-Grünwald solution (available commercially, approx. 0.5 ml) on the staining rack for 3 min.
▶ Keeping the slides flat, add an approximately equal volume (0.5 ml) of distilled water or phosphate buffer pH 7.2 with a pipette, leave for 1–2 min.
▶ Do not rinse with water, but tilt to discard excess solution, then apply filtered Giemsa working solution to the slide and leave for 15–20 min. (Preparation of Giemsa solution, see Sect. 8.7.1.4).
▶ Rinse thoroughly with distilled water or phosphate buffer pH 7.2. Clean contaminants from the undersurface of the slide with a moist cloth, holding the slide by its edges.

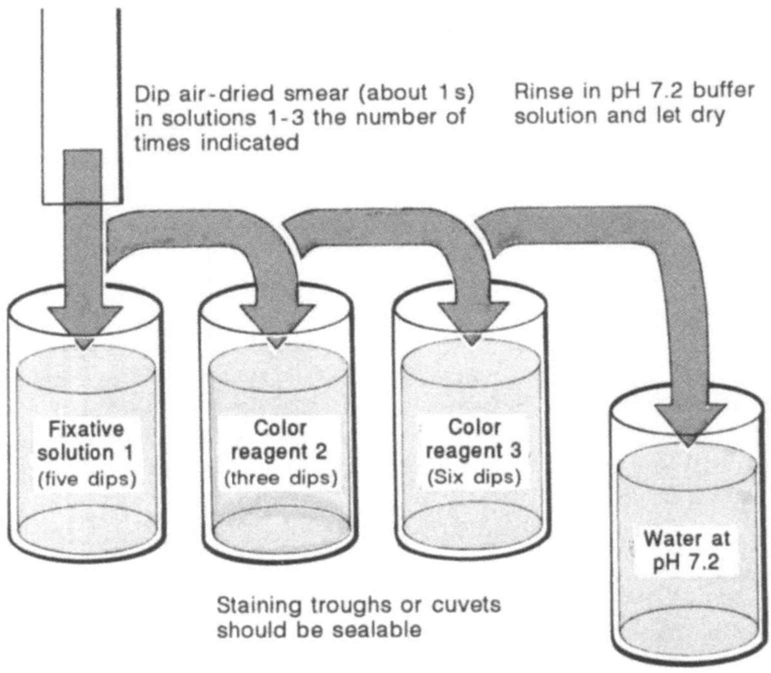

▶ Plate the tilted slide on filter or blotting paper to dry.
▶ Mount if desired.

Procedure (Cuvets):
▶ Air-dry the smears.
▶ Fix in methanol for 5 min.
▶ Place in May-Grünwald solution for 5 min.
▶ Rinse 2 min in distilled water.
▶ Place in diluted Giemsa solution (1 part Giemsa solution, 10 parts distilled water) for 15 min.
▶ Rinse with distilled water, removed dye residues from bottom of slide with a moist cloth.
▶ Dry.

8.7.2.2 Szczepanik Stain (Cytocolor)

The rapid cytologic stain developed by Szczepanik (Cytocolor, Merck) is performed with *modified Papanicolaou solutions* and approximates the appearance of a Papanicolaou stain. The alcohol-stained specimens are not susceptible to air-drying artifacts, and the total staining time is a fraction of that in the original Papanicolaou method.

Procedure
▶ Wet-fix the slide, coated with the cellular material (see Sect. 8.5), by spray fixation or alcohol immersion (propanol-2).
▶ After dipping in distilled water (10 s), place the specimen in modified hematoxylin solution for 60 s.
▶ Rinse in running water for 5 s, dip in propanol-2 for 2 s, then place into second staining solution (modified polychrome) for 60 s.
▶ After cell-preserving dealcoholization and before mounting, immerse for 5 s in distilled water and twice for 5 s in propanol-2 solution.

8.7.2.3 Papanicolaou Stain

The Papanicolaou stain has become the *standard stain* not just in gynecologic cytology but in *urinary cytology* as well. The staining procedure is technically more complex than the methods described above but yields results of outstanding quality.

Procedure
▶ Papanicolaou staining must be preceded by wet fixation (not air-drying) of the specimen (see Sect. 8.4.2). The initial step of the Pap staining procedure depends on whether the specimen has been fixed with a spray fixative or by immersion. The fixation method should be noted if the specimen is dispatched to a cytology laboratory.
▶ If the specimens are still wet in the fixative solution, put them through a descending alcohol series to distilled water; immerse for 30 s in 70%, 50%, and 30% ethanol. If the smears are already fixed and dry, place them directly in distilled water. If they are coated with polyethylene glycol (spray fixative), first dissolve the film in 30% ethanol for 30–60 s.
▶ Rinse for 30 s in distilled water.
▶ Place in Harris hematoxylin for 2 min (Pap soln. I). It is important to shorten this step (normally 3–5 min), because hematoxylin must imart the nuclear stain crucial for oncologic urine cytology without overstaining (apparent hyperchromasia).
▶ Rinse for 30 s in distilled water.
▶ For differentiation, one or two dips in 0.25% aqueous hydrochloric acid solution (concentrated HCl diluted approximately 1:400). This dissolves reddish-brown stain clouds.
▶ Rinse and blue 5 min in running tap water (e.g., overflowing cuvet). Warm water will accelerate the process.
▶ Since the next step again uses alcoholic dyes, dehydrate the specimens in ascending alcohols.
▶ Immerse in 50%, 70%, 80%, and 95% ethanol for 30 s each.
▶ Stain for 3 min in orange G solution (Pap soln. II).
▶ Rinse in 96% ethanol, immersing the slides in two different cuvets for 30 s each.
▶ Stain for 3 min in polychrome solution (Pap soln. III). Several variants of polychrome solution are available, but they do not differ significantly in their staining properties. EA 50 is widely used.
▶ Rinse again in 96% ethanol, using two different cuvets (30 s each).
▶ Place 1 min in a 1:1 mixture of ethanol (96%) and xylene.
▶ Rinse 2 min in xylene, then mount if desired. Water clouds xylene and makes it unusable, so replace xylene as needed.

8.7.3 Mounting After Staining

Mounting establishes a permanent seal between the stained and dried slide and the coverslip. It protects the cellular material from drying, mechanical damage, and wear.

Mounting is simple but *not mandatory*. Even in oil-immersion microscopy, a specimen designated for filing can be cleaned by brief immersion in xylene, obviating the need for mechanical cleaning of the slide with a cloth. Even an unmounted specimen has sufficient storage life for filing, provided it is kept in a dark container to prevent fading.

Mounting Technique

The adhesive, transparent corbite balsam mounting medium (e.g., Eukitt, Merckoglas, Hico-Mic) is spread onto the coverslip in streaks using a small wooden spatula. Immediately the coverslip is inverted and placed over the specimen on the slide. At this time the edge of the slide should be placed against blotting or filter paper and gentle pressure applied to express fine air bubbles from the sample area.

After brief drying (about 10 min), the specimen can be evaluated. It should be dried thoroughly for about 2 days before filing, however; otherwise the mounted slides may stick together.

8.8 Review of Cytopreparatory Techniques

The use of two urine samples can be helpful but is not essential. The initial processing steps may be performed separately or combined, depending on the preferences and requirements of the user.

Various options exist with regard to processing of the second sample

Option 1

Option 2

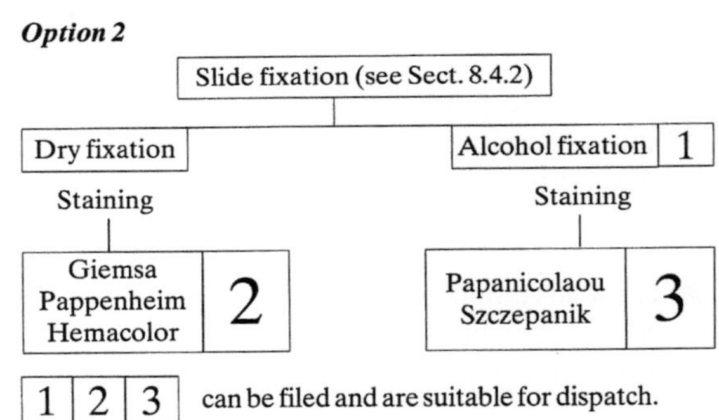

Option 3

Compact system with centrifuge-independent cell recovery and simple, permanent staining (Carcyt U1 capillary suction system). The individual components can be used separately and "mixed and matched" as desired. The system has not yet been clinically tested.

References

Aeikens B, Wittekind D (1981) Vereinfachte Methode zur Anfertigung urinzytologischer Präparate mit Hilfe von Membranfiltern und einer Papanicolaou-Ersatzfärbung. Aktuel Urol 12: 23–25

Gill WB, Lu CT, Thomson S (1973) Retrograde brushing: a new technique for obtaining histologic and cytologic material from ureteral, renal pelvic and caliceal lesions. J Urol 109: 573–574

Graham RM (1972) The cytologic diagnosis of cancer, 3rd edn. Saunders, Philadelphia

Koss LG (1989) Zytologie der Harnwege – Histologie und Ultrastruktur. In: Schenck U, Soost H-J (eds) 10. Fortbildungsveranstaltung für klinische Zytologie, 3–9 december 1989, Munich (Abstracts), pp 49–56

Näslund I (1988) Reliability and accuracy of urotel test, a device for urinary cytology. Eur Urol 15: 54–58

Rathert P, Preiss H (1982) Urinzytologie in der urologischen Praxis. Urologe [A] 21: 67–72

Romeis B (ed Böck P) (1989) Mikroskopische Technik, 17th edn. Urban & Schwarzenberg, Munich

Roth St, Rathert P (1988) Praxis der Urinzytologie, parts I–III. Urologe [B] 28: 75–78; 141–143; 201–203

Rübben H, Terhorst B, Deutz F-J, Lagrange W, Giani G (1986) Unterschiedliche Färbeverfahren der exfoliativen Urinzytologie. Urologe [A] 25: 302–304

Rübben H, Rathert P, Roth St, Hofstädter F, Giani G, Terhorst B, Friedrichs R (1989) Exfoliative Urinzytologie, 4th edn. Harnwegstumorregister. Publication of the Fort- und Weiterbildungskommission der Deutschen Urologen, Arbeitskreis Onkologie, Sektion Urinzytologie, Aachen

Soost HJ (1976) Lehrbuch der klinischen Zytodiagnostik. Thieme, Stuttgart

Suppan A, Suppan C, Höltl W (1989) Urinzytologie mit Hilfe von Membranfiltern unter Berücksichtigung der Tumorzelldiagnostik. Akt Urol 20: 239–243

Völter D, Keller AJ, Schubert GE (1987) Diagnostik der Harnblase. Schattauer, Stuttgart

Voogt HJ de, Rathert P, Beyer-Boom M (1979) Praxis der Urinzytologie. Springer, Berlin Heidelberg New York

Wegener H, Böhm E, Ebert B (1989) Ein neues standardisiertes System zur Optimierung urinzytologischer Untersuchungen. Urologe [B] 29: 343–347

9 Atlas Section

S. Roth and P. Rathert

CONTENTS

9.1 General Introductory Remarks

9.1.1 What Magnification Should Be Used?

Screening at $\times 100$ magnification ($\times 10$ eyepiece, $\times 10$ objective) is appropriate for the identification of highly cellular areas on the microscope slide. Often this is unnecessary, however, if a recovery technique has been used that concentrates the cellular material onto a well-defined area of the slide that is easily located using the coordinates on the microscope stage (see Sect. 8.5).

A *magnification of $\times 400$* ($\times 10$ eyepiece, $\times 40$ objective) is generally adequate for a reasonably prompt and effective urinary cytologic analysis. An oil-immersion objective of $\times 40$ magnification is advantageous for bringing out details but is not essential. One disadvantage of this method is that specimens must be subsequently cleaned with cloths or xylene for permanent filing.

A magnification of $\times 630$ ($\times 10$ eyepiece, $\times 63$ objective) or $\times 1000^*$ ($\times 10$ eyepiece, $\times 100$ objective) can be useful in cases requiring the comparison of specific cells. This must be done using the oil-immersion technique.

9.1.2 Magnifications Used in the Atlas Figures

In the great majority of figures contained in the atlas section, the relevant cell structures are shown *at magnifications of both × 400* and x 1000**. This is done first because of the reduction effect of reproducing the × 400 photomicrographs on the printed page (the microscope image is perceived as larger due to the projection factor) and second because we wished to provide a high-power view that would detail the essential structural features of the cellular changes.

9.1.3 Selection of Stains

Most of the specimens shown in the atlas were stained by the *Papanicolaou* method, which is the worldwide standard for urinary cytology and is excellent for permanent filing. Use of the Papanicolaou stain enabled us to select from more than 25 000 urinary cytologic examinations performed over a period of more than 5 years. Alternative staining methods, especially rapid stains for immediate analysis (see Chap. 8), have been integrated into the illustrative materials.

It is important to note that *alcoholic stains* (Papanicolaou, Cytocolor), rapid stains with methylene blue or Testsimplets, and Carcyt are *qualitatively different from stains used on air-dried specimens* (Giemsa, Pappenheim, May-Grünwald-Giemsa, Hemacolor, Sangodiff). It is best, therefore, to use stains from just one of these two categories (with or without air-drying; see Chap. 8) so that the eye will not have to continually readjust to different staining properties during cytologic evaluations. For completeness, the *principal differences* between air-dried and non-air-dried specimens are described in Sect. 9.2.

---

*For typographic reasons, the photomicrographs in this book were reduced by 15% relative to the magnification factor indicated.

9.1.4 Organization of the Atlas

9.1.4.1 Importance of Comparisons

Urinary cytology is a descriptive method based on the comparison of observed cytomorphologic features. For optimum clarity, therefore, we have not only organized the atlas by *clinically relevant groups of indications* but have also included *comparative findings* that will be particularly useful in the sections on normal findings and neoplastic changes in the urothelium. Thus, the section on the Normal Urothelium includes examples of urothelial carcinomas, while the sections on Urothelial Tumors include normal findings, dysplasias, as well as carcinoma of higher or lower grades of differentiation. The repetition of some illustrations will aid the atlas user in "reading the specimens" without having to leaf back and forth through the text.

9.1.4.2 Presentation of Findings According to Clinical Relevance

In accordance with the clinical role of urinary cytology as discussed in Chap. 2, we have compiled the atlas in such a way that:

– Little attention is given to rare findings.
– Common findings are given extensive coverage.
– The figures are presented in clinically relevant cytologic diagnostic groups:
● Normal urinary cytologic findings
● Suspicious findings requiring further investigation or close follow-up
● Positive urinary cytologic findings

The physician active in urinary cytology should bear in mind that the *urothelial cell does not react to internal and external stimuli with a morphologically differentiated response but with a limited and undifferentiated number of morphologic changes.*

Reactive changes caused by endogenous factors such as stones or infection and by external manipulations (e.g., instillations and retrograde irrigations) can *mimic neoplastic changes*. Because there are no special stains or immunocytologic procedures that can reliably distinguish among these changes, allowance must be made for this lack of differentiated cell response. Although this naturally reduces the percentage of "unequivocal" normal and pathologic findings, there remains a *clinically relevant yield of positive findings* that will influence further diagnostic and therapeutic procedures.

We therefore selected illustrative specimens that would enable the cytologist to assign findings to the clinically relevant diagnostic groups (normal/suspicious/ positive).

9.2 Comparison of Different Stains

In this section we shall present examples showing how different stains can affect the appearance of cytologic specimens. (Specific tumor characteristics are discussed in later sections, and details on cytopreparatory techniques are given in Chap. 8.) All the specimens were obtained from patients with a moderately differentiated (grade 2) urothelial carcinoma.

9.2.1 Effect of Air-Drying

Category I: stains used on non-air-dried specimens:
Papanicolaou, Cytocolor, Testsimplets, methylene blue, Carcyt-U1.

Category II: stains used on air-dried specimens:
Giemsa, May-Grünwald-Giemsa, Pappenheim, Hemacolor, Sangodiff.

With the alcoholic Papanicolaou stain, the dehydrating effect of the alcohol causes a *cell shrinkage* that also affects the nucleus, which of course is critical for the diagnosis and grading of malignancy. This is also true of Cytocolor stain but not of the other rapid stains in category I, which, though used on non-air-dried specimens, do not employ alcohol.

With all the stains in category II, *air drying causes significant nuclear enlargement*, and an examiner accustomed to the Papanicolaou technique will describe these nuclei as "swollen" or "expanded." The cells appear spread out like a "fried egg" (Lopez-Cardozo 1976). This accounts for the clear definition of the chromatin pattern even in small cells. While the remaining cytodiagnostic criteria are the same for both categories of stain (see Table 9.1), there is a significant difference in the appearance of the chromatin pattern.

Table 9.1. **Differences in the appearance of chromatin stained with an alcoholic stain (e. g., Papanicolaou) and after air-drying (e. g., Giemsa).** Other essential malignant criteria (shape of the nucleoli, size and shape of the nuclei, ratio of nucleolar to nuclear size) appear besically the same after Papanicolaou and Giemsa staining (modified from Beyer-Boon 1979)

Chromatin appearance in atypia (dysplasia)	
Papanicolaou stain	Giemsa stain (etc.)
Slightly prominent nuclear membrane	Clear areas in dark nuclei
Slight chromatin condensation	No loose chromatin pattern

Chromatin appearance in malignancy	
Papanicolaou stain	Giemsa stain (etc.)
Very prominent nuclear membrane	Loose chromatin pattern
Chromatin condensation	Broad, irregular dark bands
Irregular distribution of chromatin	Granulated chromatin

Table 9. 2. **Different appearances of nuclear structures of comparable malignancy following Papanicolaou and MGG staining** (after Beyer-Boon 1979)

A Chromatin pattern of Papanicolaou-stained abnormal urothelial cells. *a* Nuclei of atypical cells. Slightly prominent nuclear membrane and slightly coarse chromatin. *b* Malignant cells with very prominent nuclear membrane, clumping and irregular distribution of chromatin. *c* Malignant cells with chromatin particles of markedly varying size. *d* Small malignant nuclei with very prominent nuclear membrane and unfavorable nucleolar-nuclear ratio. *e* Giant nucleus with very xoarse, irregular chromatin. Pronounced hyperchromasia. B Chromatin pattern of MGG-stained abnormal urothelial cells. *a* Nuclei of atypical cells. Lighter areas in the dark nucleus (left). Completely dark nucleus (right). *b* Loose or "open" chromatin pattern of malignant cells. *c* Granulated chromatin of a malignant cell. *d* Broad, irregular dark chromatin bands in malignant cells

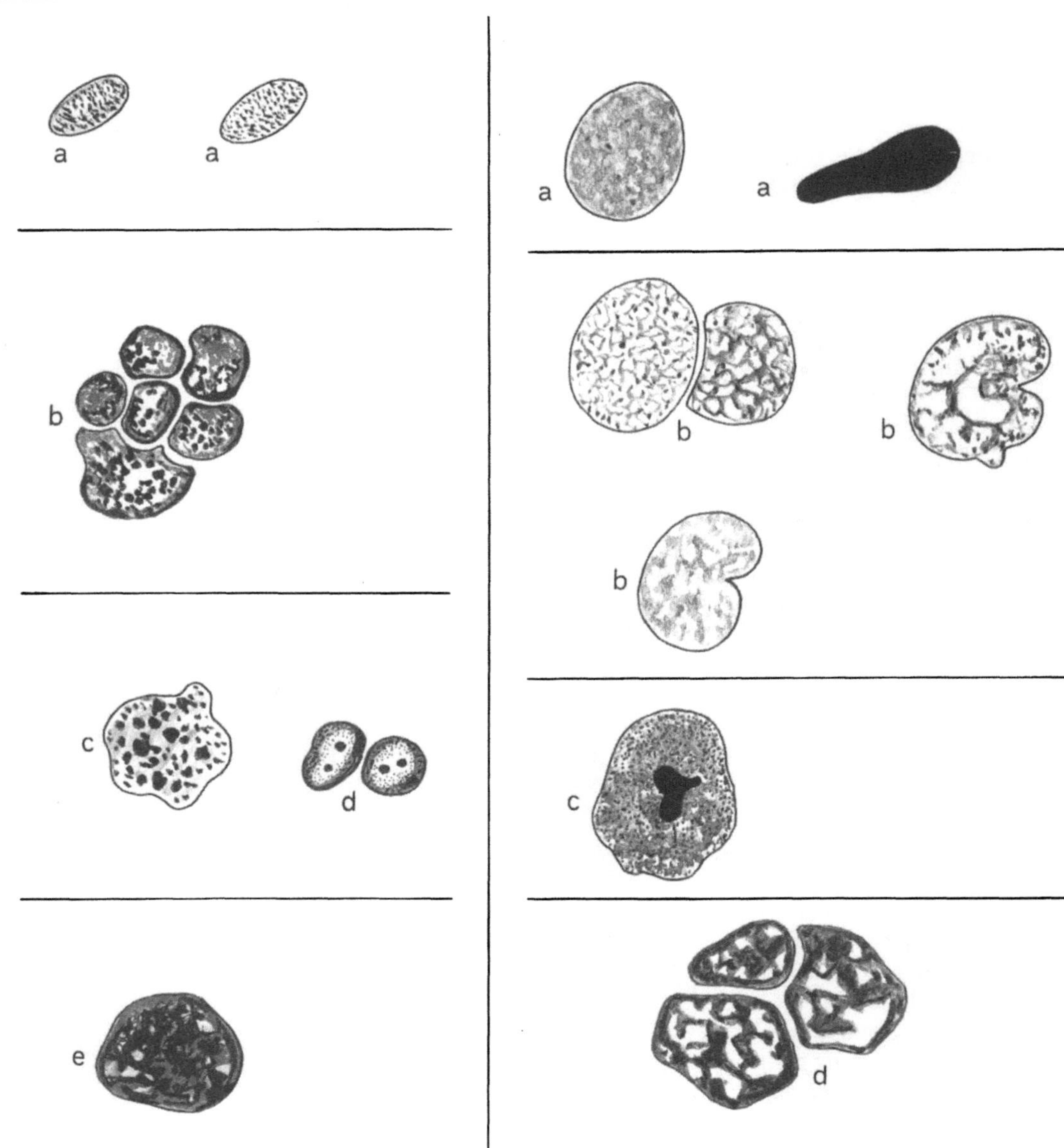

A Papanicolaou   B MGG

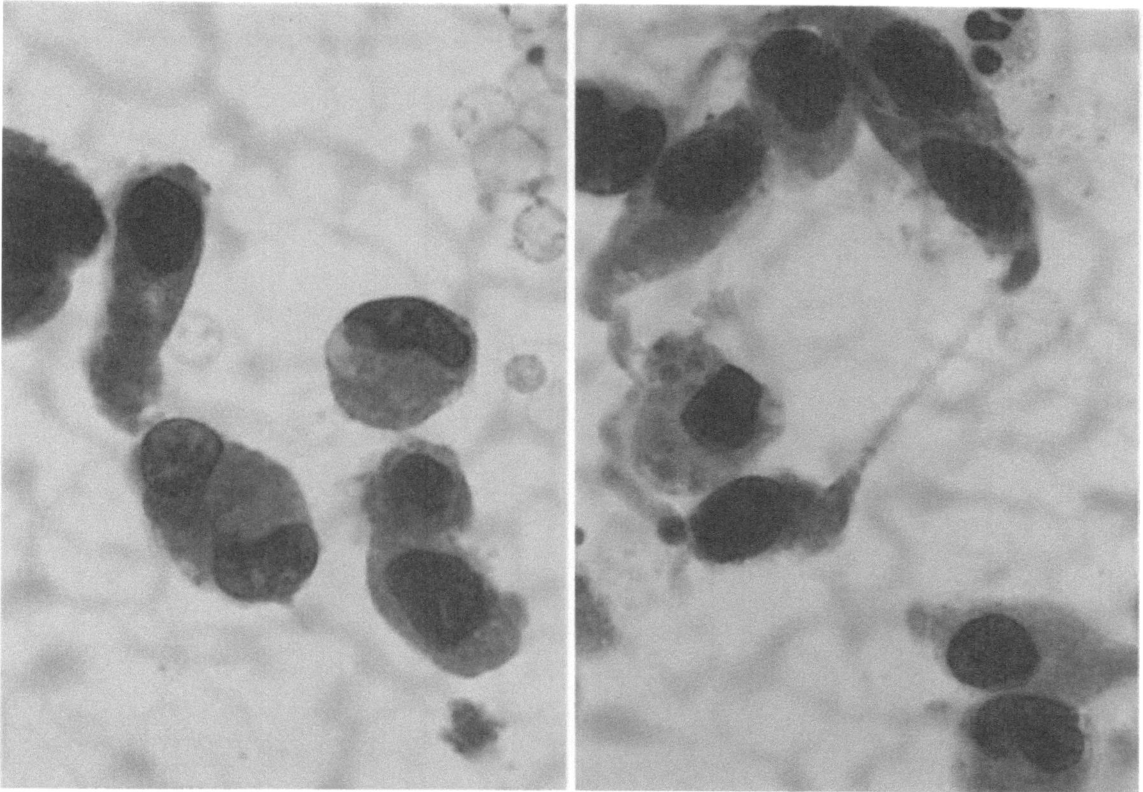

a   b

Fig. 9.1a,b. **Testsimplets stain** gives excellent transparency of the nucleus, which is critical for the assessment of malignancy. The reticulated background is caused by the stain coating on the slide and does not hinder interpretation (×850)

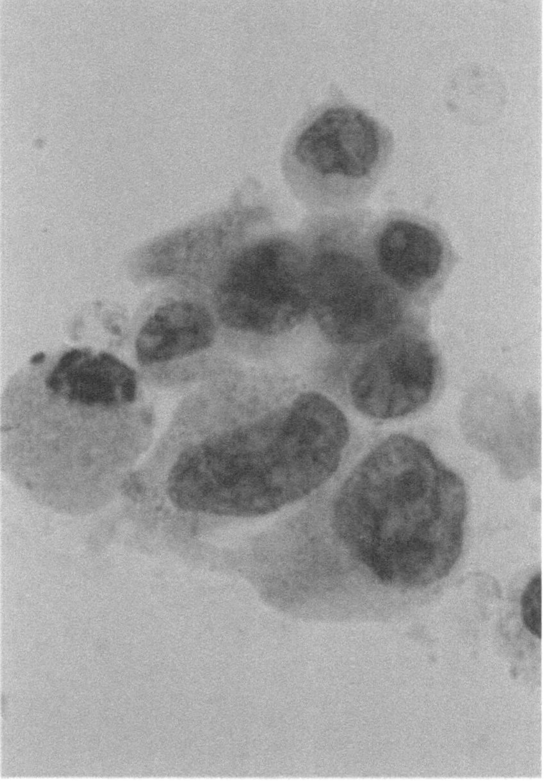

Fig. 9.2. **Methylene blue stain.** Noting that cells are often ▷ overstained at sites where the dye is in direct contact with the sediment, the examiner should interpret adjacent areas that are correctly stained (×850)

                                                                S. Roth and P. Rathert

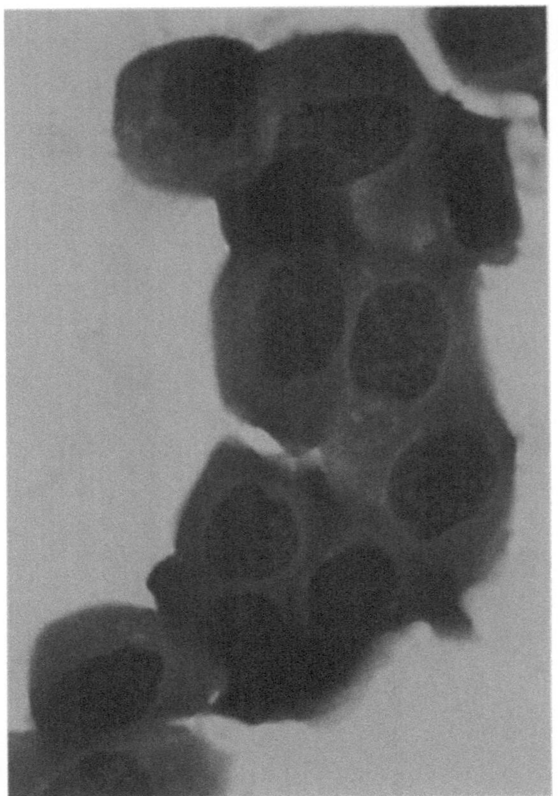

Fig. 9.3. **Hemacolor stain** is an easy-to-use stain for air-dried specimens; it allows for storage and permanent filing (×850)

Fig. 9.4a, b. **Carcyt system.** Cellular material that has been concentrated by the capillary filter suction technique can be stained with the rapid stain supplied with the system or with any commercially available stain (×850)
▽

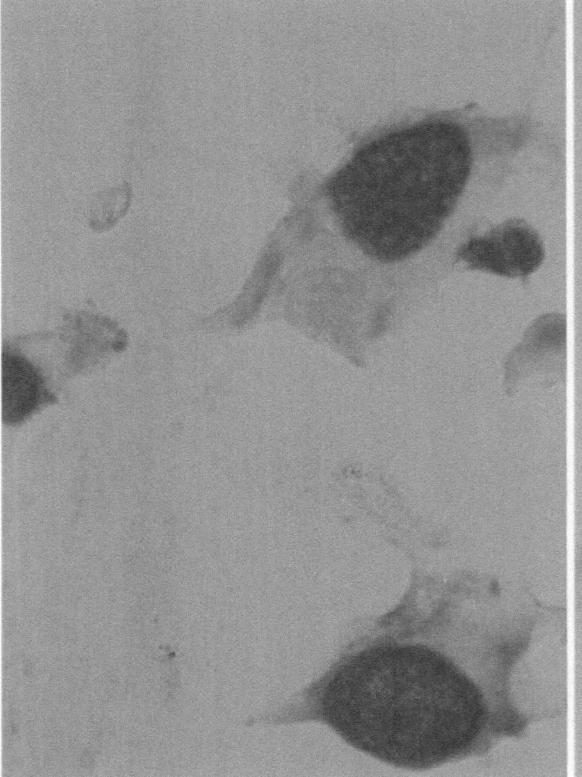

a

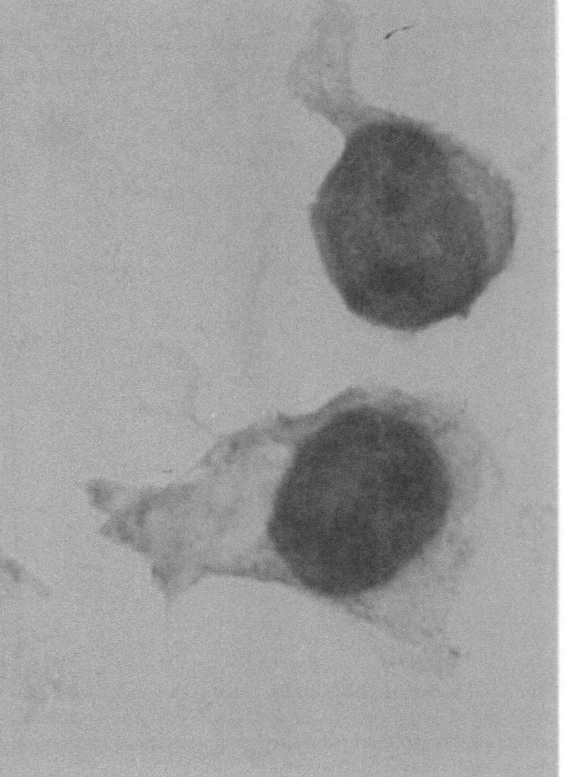

b

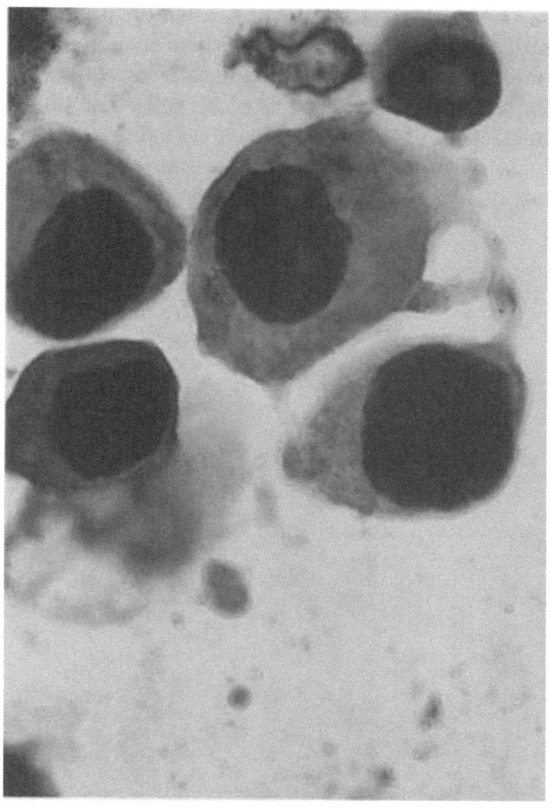

9.5

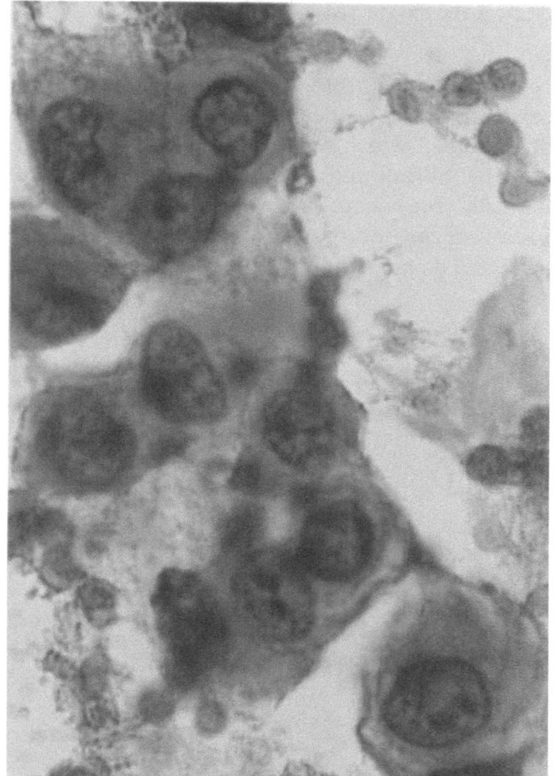

9.6

Fig. 9.5. **Sangodiff stain** is a rapid stain applied as a film to
the air-dried specimen, which then can be permanently filed
(×850)

Fig. 9.6. **Papanicolaou stain**, the standard stain for urinary
cytology, yields excellent results with high definition of nuclear detail (×850)

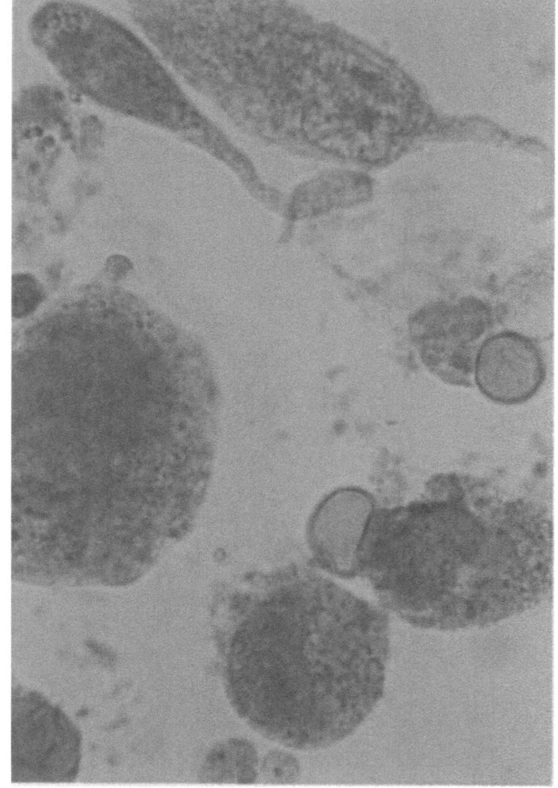

Fig. 9.7. **Sediment stain** (Sedicolor). Commercially avail- ▷
able sediment stains are unsatisfactory for oncologic urinary
cytology, as illustrated by this specimen from a patient with a
grade 2 urothelial carcinoma. They are acceptable for other
types of microscopic urinalysis (infection, erythrocyte morphology, quantitative evaluation of leukocyturia and hematuria) (×850)

9.3 The Normal Urothelium

Normal human urothelial tissue is composed of seven layers. The deepest layer consists of the "basal cells," which are firmly attached to the basement membrane and are the smallest of the urothelial cells (see Chap. 7). Because of their size, the *basal cells can be difficult to distinguish from leukocytes* that do not exhibit marked segmentation.

The most superficial layer consists of large, often multinucleated cells, each of which stretches over several underlying intermediate cells. They are therefore called umbrella or cap cells. The multinucleation of these superficial cells illustrates the fact that *multinuclearity is not a criterion for malignancy* in oncologic urinary cytology.

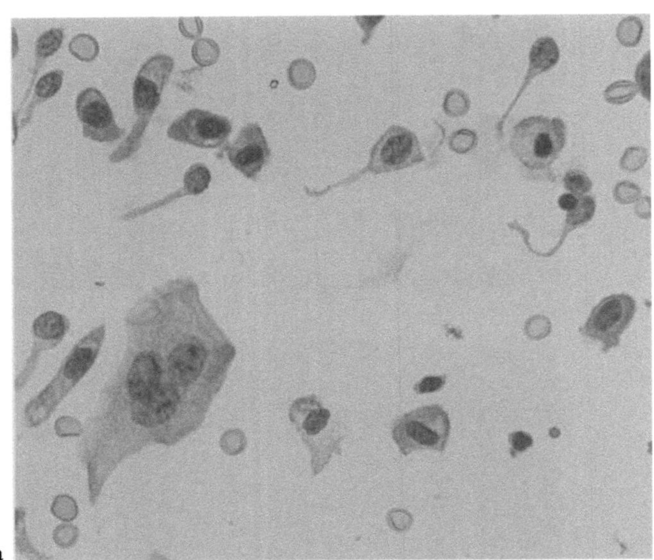

a

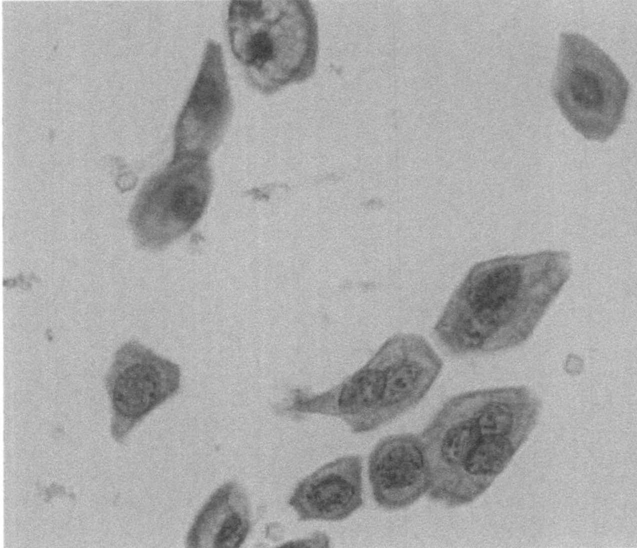

b

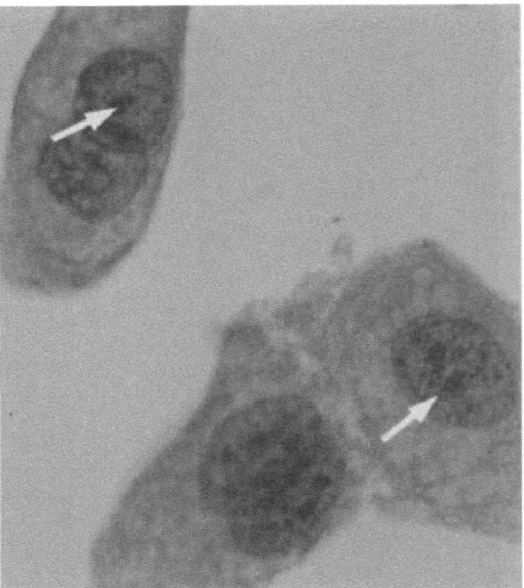

c

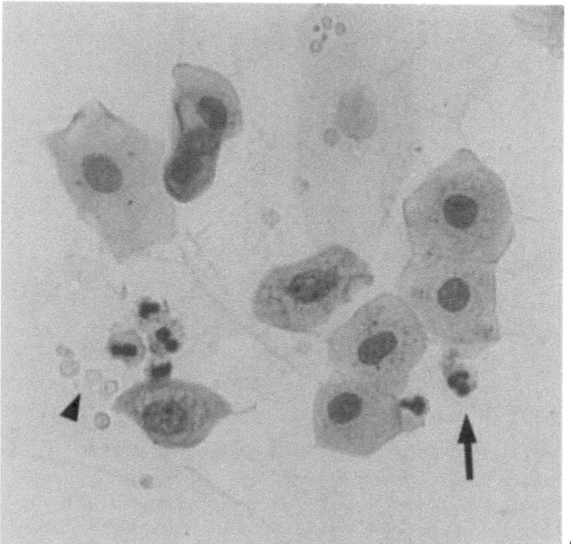

a

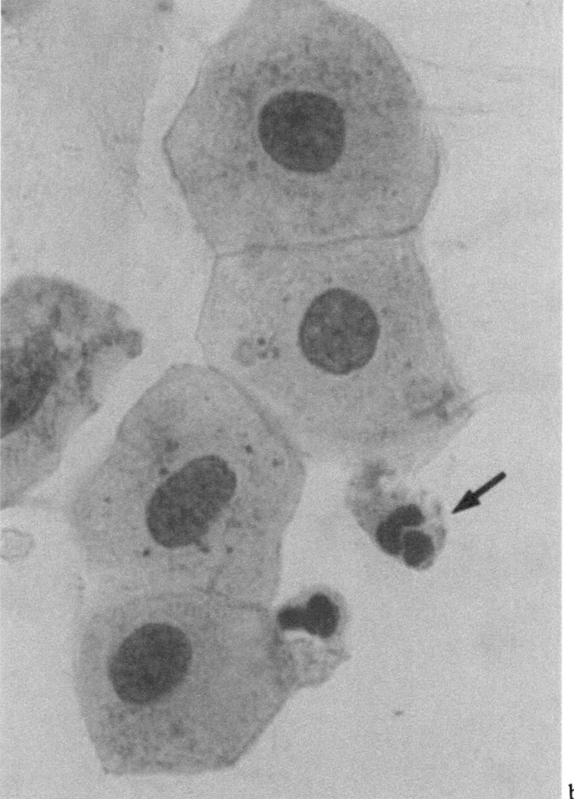

b

◁ Fig. 9.8 a,b. **Morphologic diversity** of normal urothelial cells collected in urinary tract washings. This field contains larger, multinucleated umbrella cells along with many smaller cells originating from deeper urothelial layers. These cells frequently have an oblong shape caused by their sin-gle-point attachment to the basement membrane (× 340; Papanicolaou stain). c High-power view demonstrates the finely granular nuclear chromatin pattern and normal-size nucleoli (← ←) (×850; Papanicolaou stain)

Fig. 9.9 a, b. **Normal urothelial cells** together with segmented leukocytes (← ←) and erythrocytes (◀). All the urothelial cells have finely granular nuclear chromatin with good nuclear transparency (i.e., it is easy to see through the nucleus, with or without focusing). Decreased nuclear transparency signifies a pathologic chromatin increase, which is an important criterion for the diagnosis of malignancy (a × 340; b × 850; Papanicolaou stain)

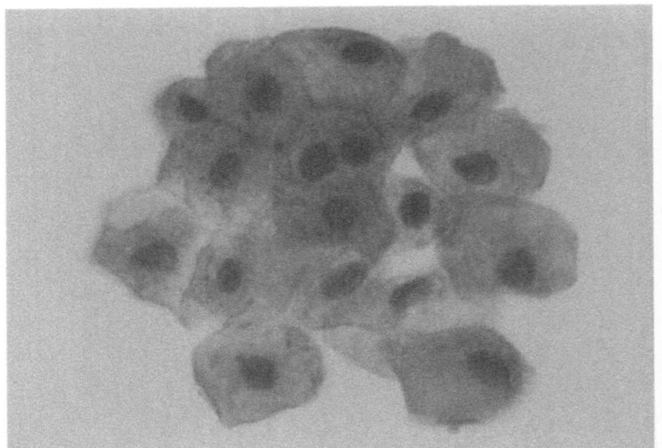

a

9.10

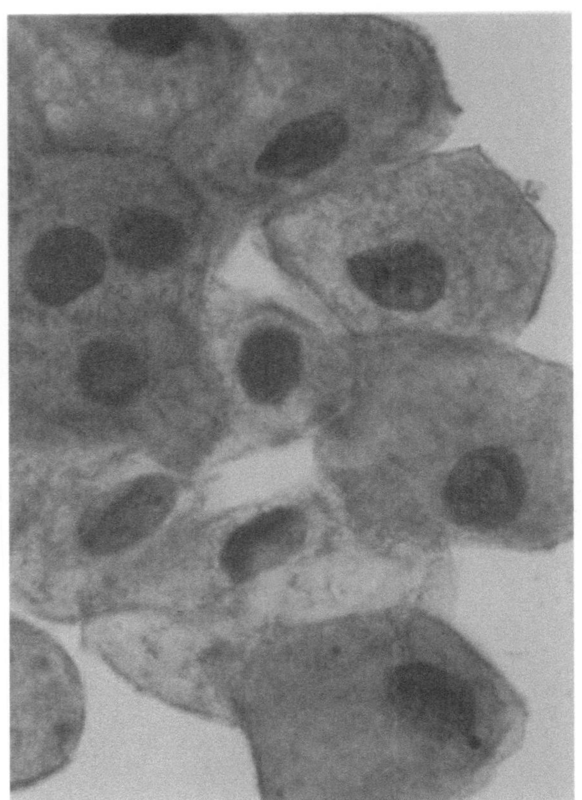

b

Figs. 9.10 a,b and 9.11a,b. **Normal urothelial cells** with normal-appearing nuclei. An important criterion is the regular shape of the nuclei, i.e., they are uniformly round or follow the external shape of the cell. Also, there is normal nuclear transparency (9.10 a and 9.11 a, × 340; 9.10 b and 9.11b, × 850; Papanicolaou stain)

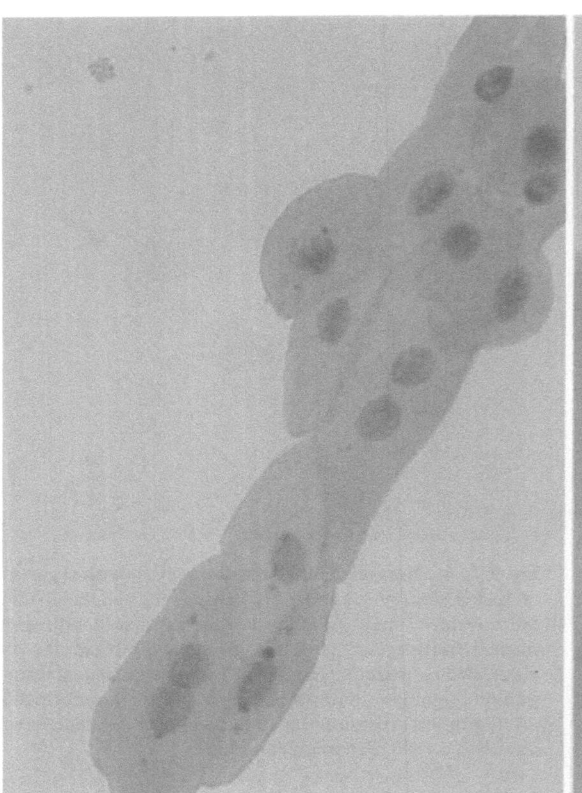

a

9.11

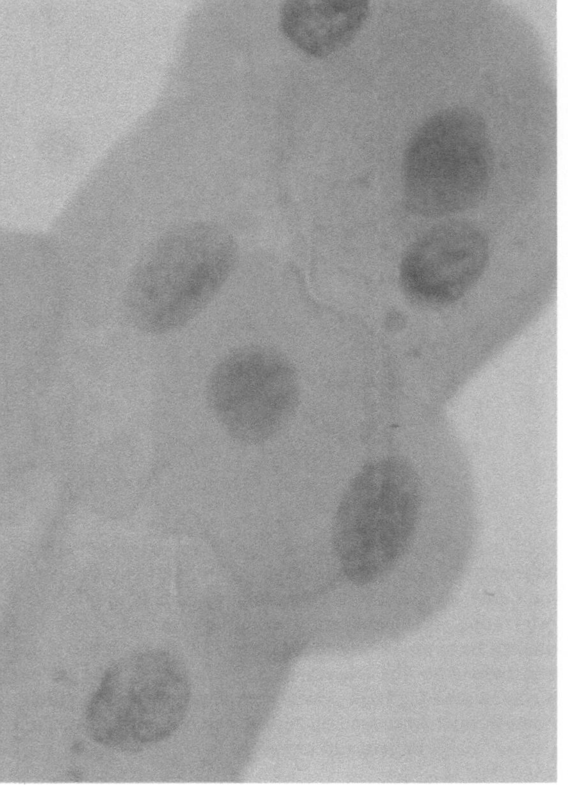

b

Comparison with Abnormal Findings

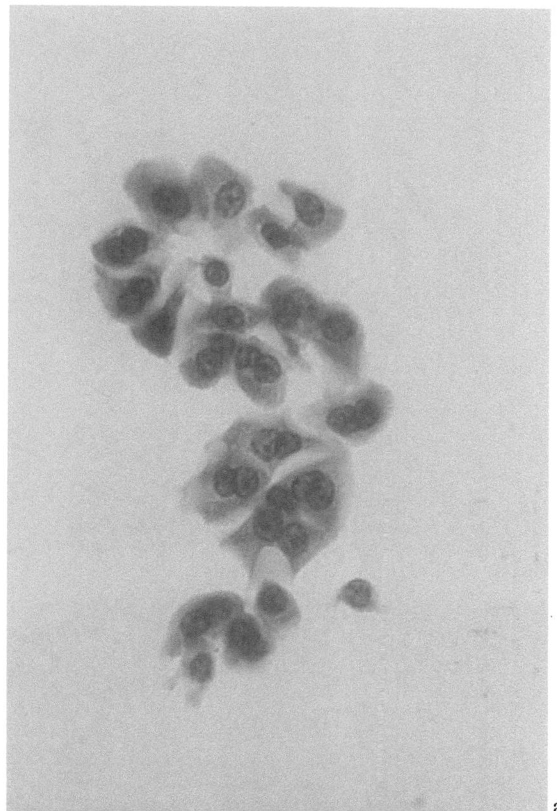

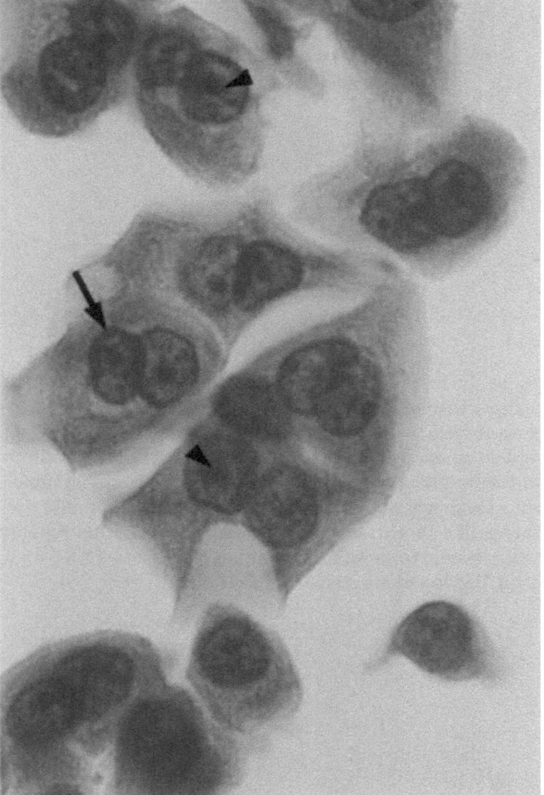

Fig. 9.12 a, b. **Cells from a well-differentiated urothelial carcinoma (grade 1)** contrast with normal cells by their thickened nuclear membrane (←) and prominent nucleoli (◀ ◀). Thickening of the nuclear membrane is caused by a chromatin increase. The presence of multiple nuclei is not a criterion of malignancy. (a, ×340; b, ×850; Papanicolaou stain)

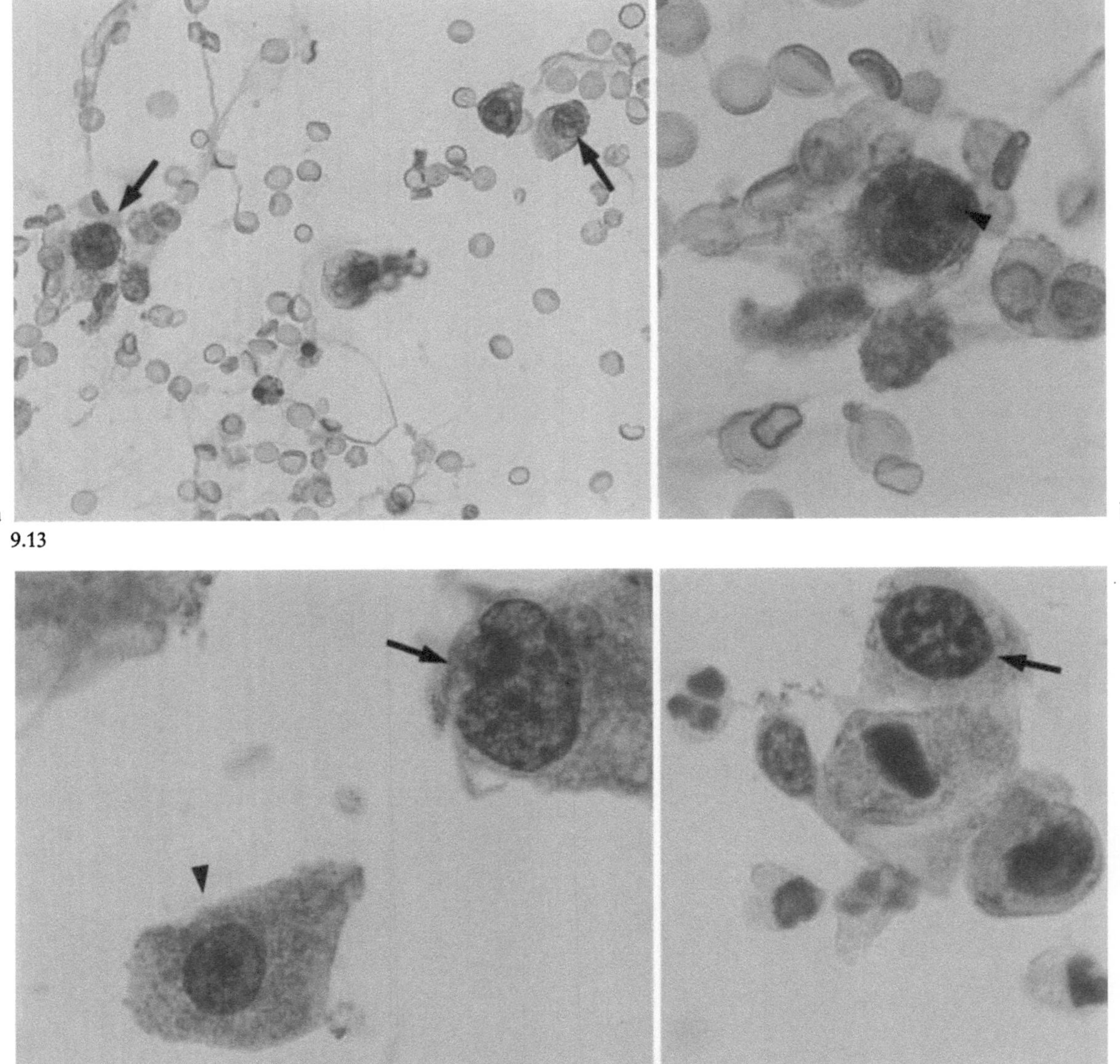

9.13

9.14

Fig. 9.13 a, b. **Dysplasia secondary to urolithiasis.** Cytologic changes with nuclear enlargement (increased nuclear–cytoplasmic ratio; ← ←) and a slightly prominent nucleolus (◀) raised suspicion of grade 1 urothelial carcinoma. Cystoscopy and uroradiology showed no evidence of a urothelial tumor, but a stone was detected. This case illustrates that mild dysplasias like those caused by urolithiasis cannot be reliably distinguished from well-differentiated neoplasia (a, × 340; b, ×850; Papanicolaou stain)

Fig. 9.14 a, b. **Moderately differentiated urothelial carcinoma (grade 2)** is marked by an enlarged nucleus with pronounced hyperchromasia and chromatin clumping (← ←) alongside normal urothelial cells (◀ ◀). Note segmented leukocytes in b (× 850; Papanicolaou stain)

9.4 Important Cytologic Differential Diagnoses and Pitfalls

In this section we shall look at important differential diagnoses and pitfalls that may be encountered in urinary cytology, both separate from and in connection with oncologic investigations.

It is important for the cytopathologist to *read the specimens with a maximum information yield*, especially when supplementary clinical findings are unavailable. In this way the cytopathologist can sometimes diagnose conditions that are missed by routine clinical evaluation. This is illustrated by urinary tract fungal infections which do not grow on conventional culture media but are easily diagnosed by cytologic examination.

Whenever there is cytologic suspicion of carcinoma, the possibility of a *simple reactive cell change* should be considered. The differential diagnosis should include, for example, urinary tract infections with a typical "infectious" cell pattern as well as calculi, which may be accompanied by crystalluria.

Some specimens may show evidence of artifacts caused by a *faulty examination technique* (e.g., hypoosmolar irrigating solution) or *improper preservative measures* (e.g., cytolysis) that likewise must be considered during interpretation.

Normal Urothelial Cells

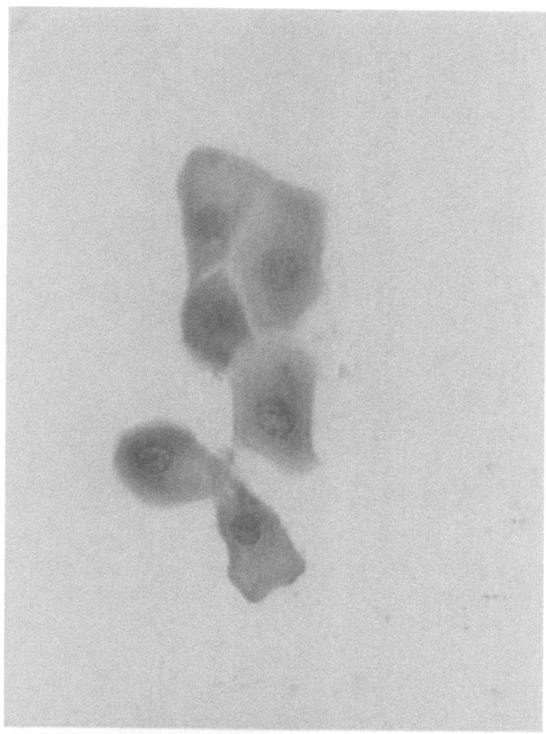

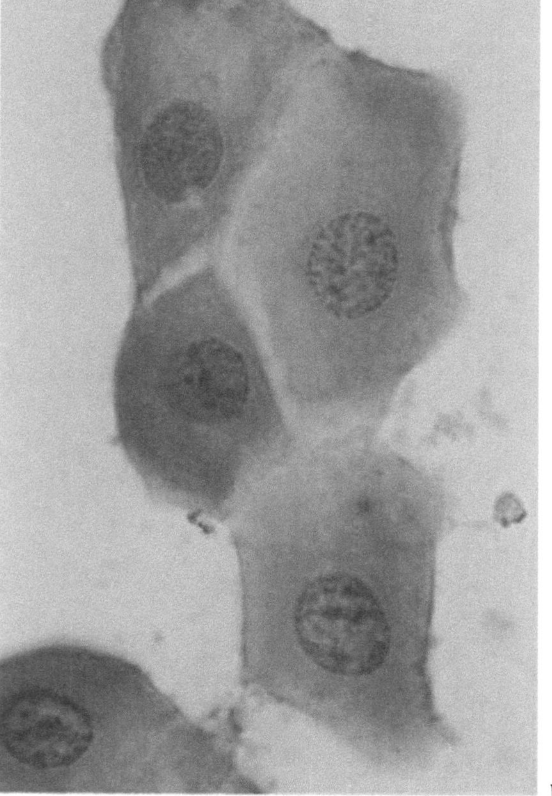

Fig. 9.15 a, b. **Normal urothelial cells.** a, × 340; b, × 850; Papanicolaou stain

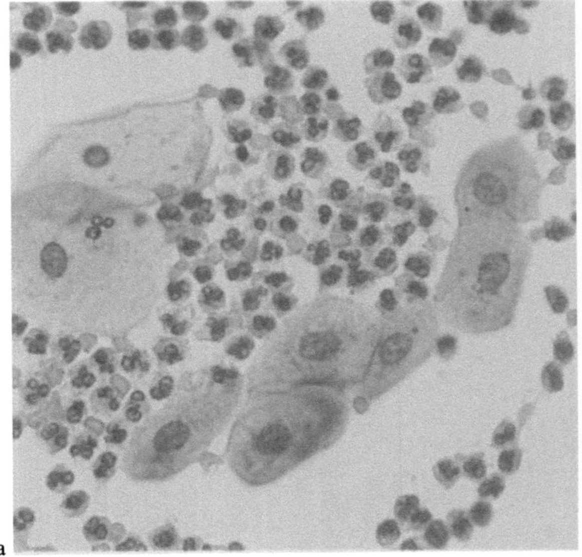

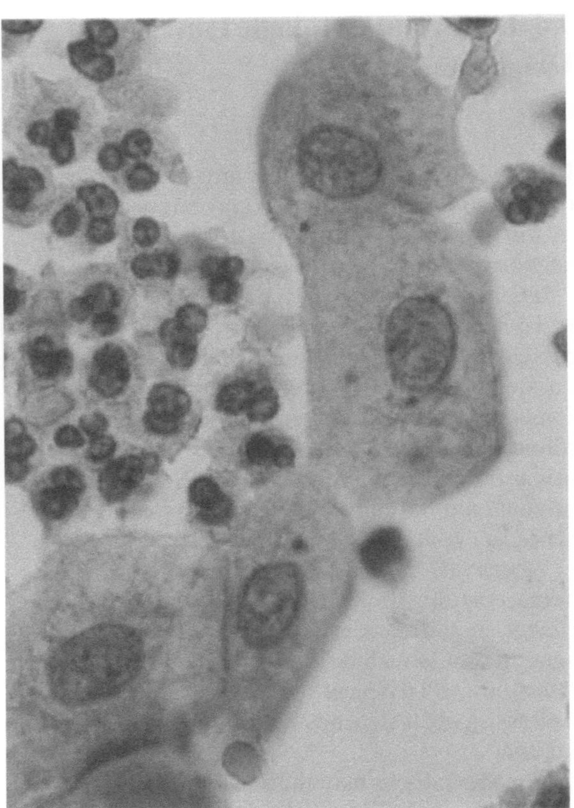

Fig. 9.16 a, b. **"Sterile" leukocyturia** 4 weeks after TUR with normal urothelial cells (a, × 340; b, × 850; Papanicolaou stain)

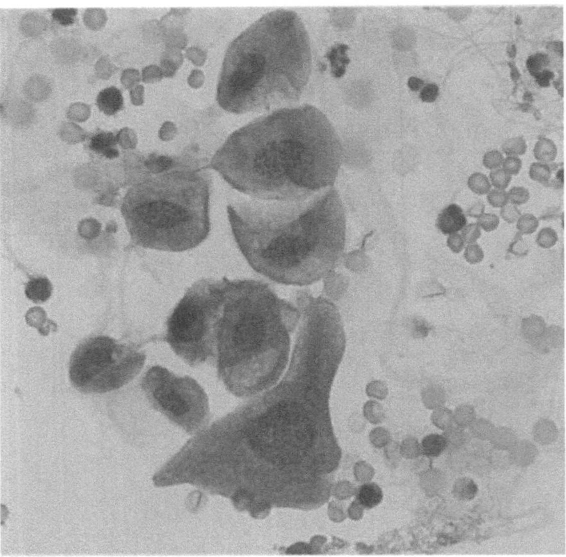

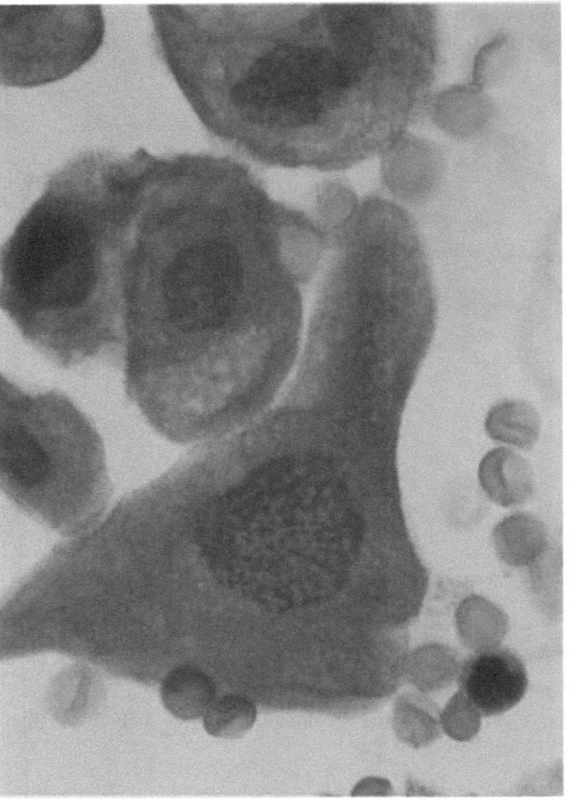

Fig. 9.17 a, b. **Mild dysplasia** with slight prominence of the nuclear chromatin. A reactive etiology is evidenced cytologically by erythrocyturia and leukocyturia. Clinically the patient had a ureteral stone (a, × 340; b, × 850; Papanicolaou stain)

Bacterial and Fungal Infections

Normally the urine is devoid of microorganisms. Their presence in a microscopic specimen may represent a true pathologic finding or may result from iatrogenic contamination (faulty collection technique, contaminated transport vessels). The absence of typical inflammatory cells (leukocytes) implies an iatrogenic cause.

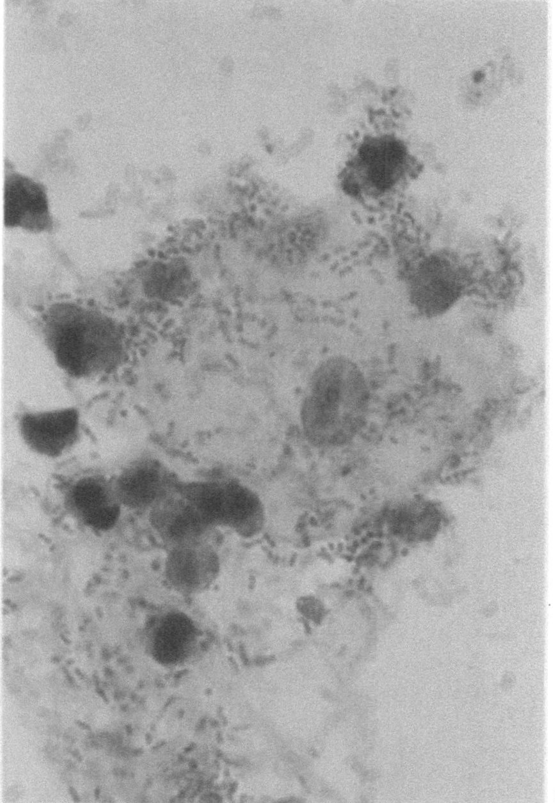

a

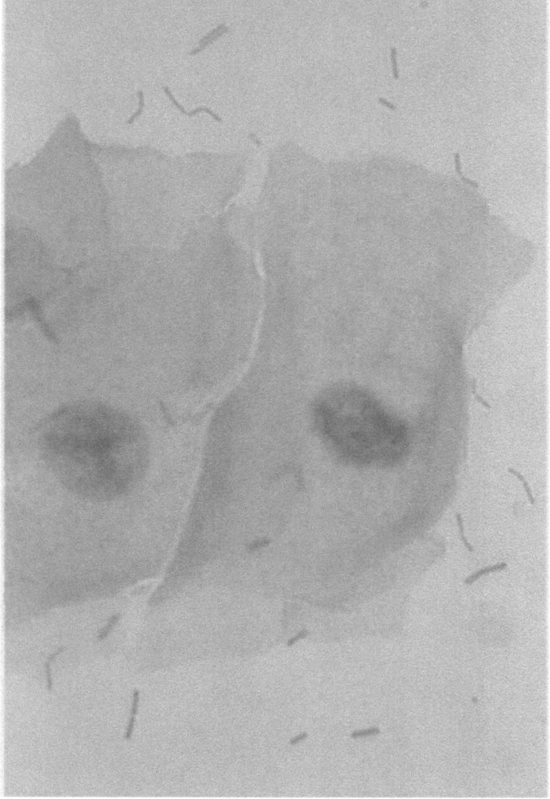

b

Fig. 9.18 a, b. **Typical infectious pattern** (a) and **physiologic Döderlein contamination** (b) in a female. Döderlein's bacilli are finer and more elongated than the smaller and plumper coliforms. They are usually accompanied by numerous squamous epithelial cells, and typical infectious signs such as leukocytic infiltration are absent (a, × 340; b, × 850; Papanicolaou stain)

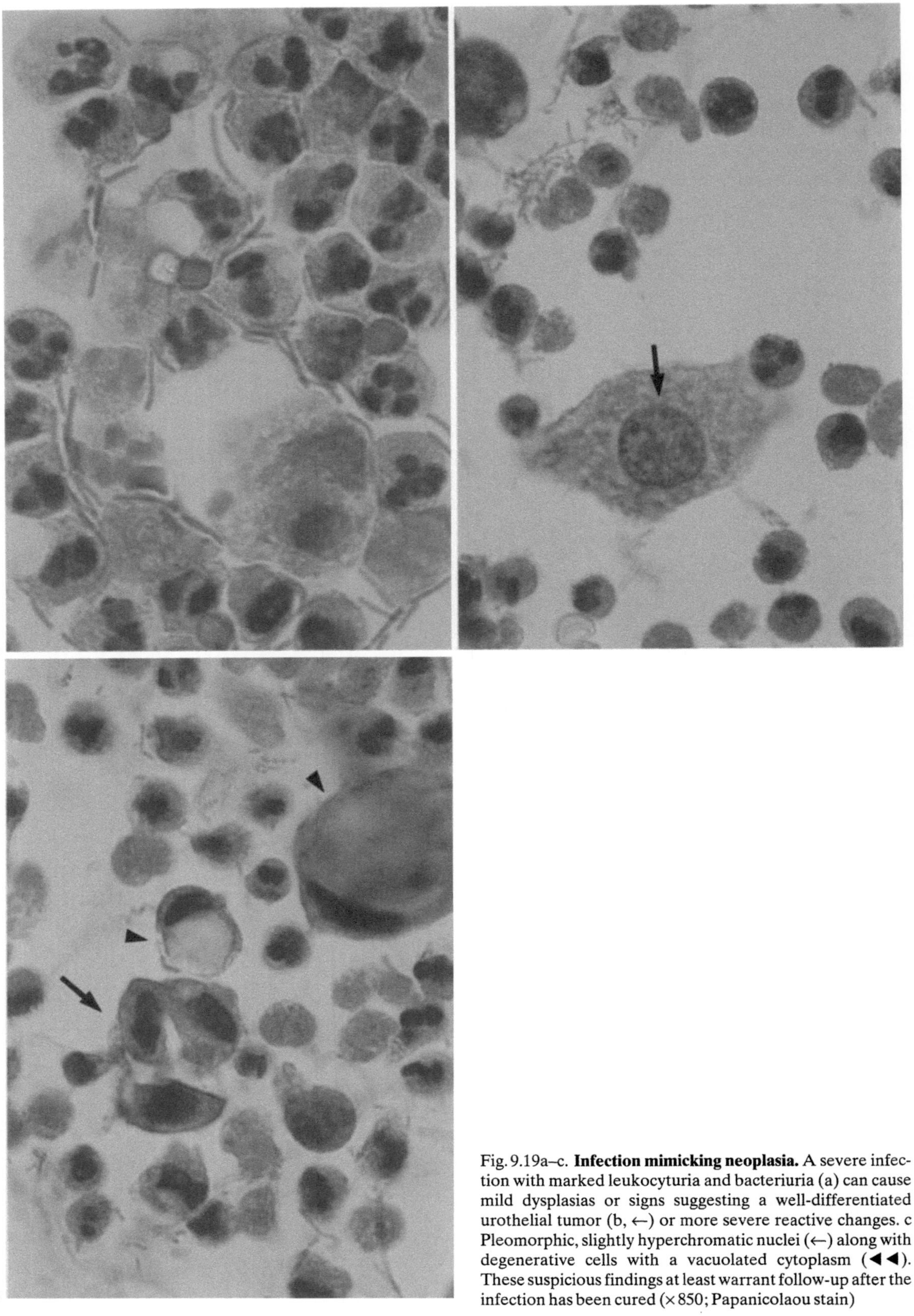

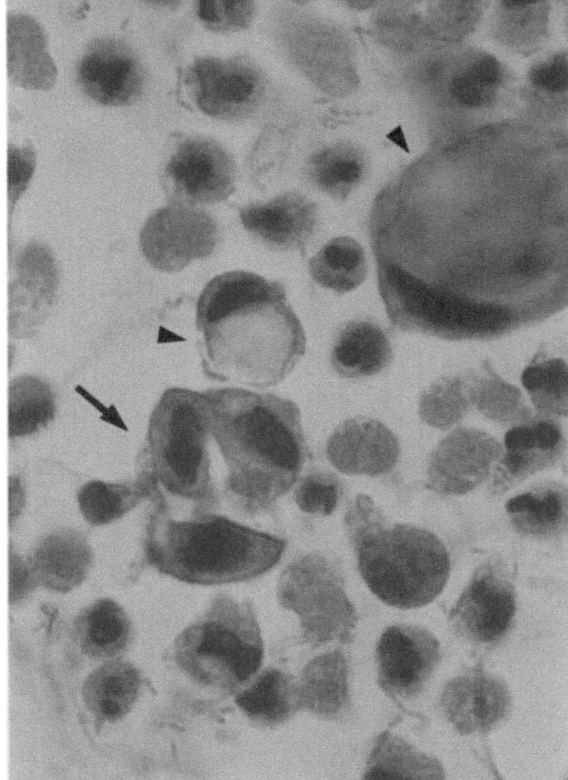

Fig. 9.19a–c. **Infection mimicking neoplasia.** A severe infection with marked leukocyturia and bacteriuria (a) can cause mild dysplasias or signs suggesting a well-differentiated urothelial tumor (b, ←) or more severe reactive changes. c Pleomorphic, slightly hyperchromatic nuclei (←) along with degenerative cells with a vacuolated cytoplasm (◀◀). These suspicious findings at least warrant follow-up after the infection has been cured (×850; Papanicolaou stain)

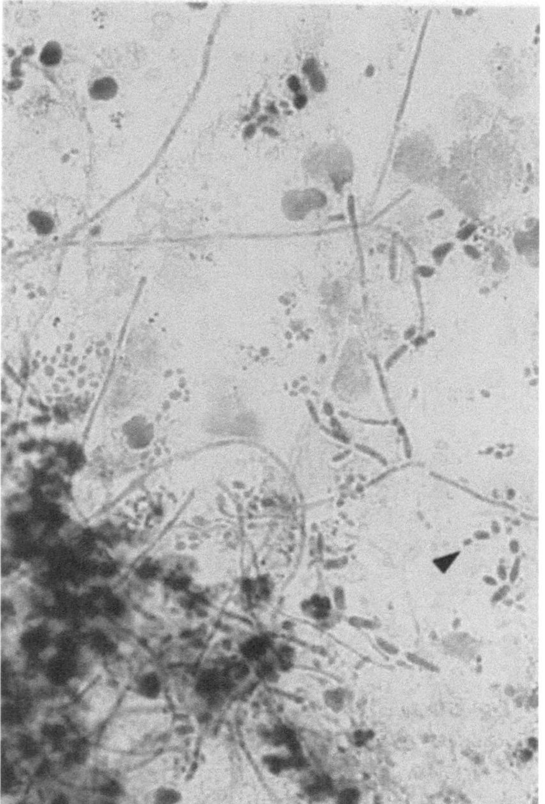

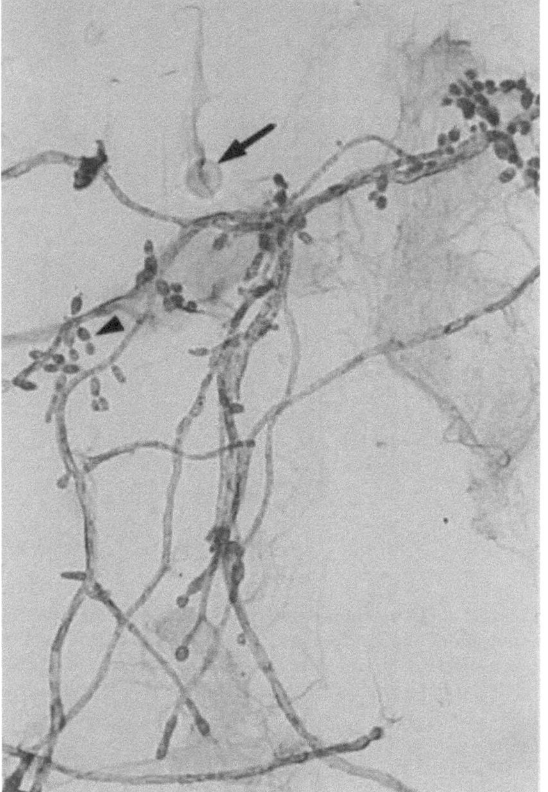

Fig. 9.20 a, b. **Fungal infection.** Fungi are easier to identify morphologically than bacteria. Because of their rapid division, they commonly appear in characteristic aggregates as yeast structures with filamentous hyphae. The individual buds or spores (◄ ◄) are smaller and more elliptical than erythrocytes (←) and have thicker walls (× 340; Papanicolaou stain)

Cellular Changes Caused by External Manipulations

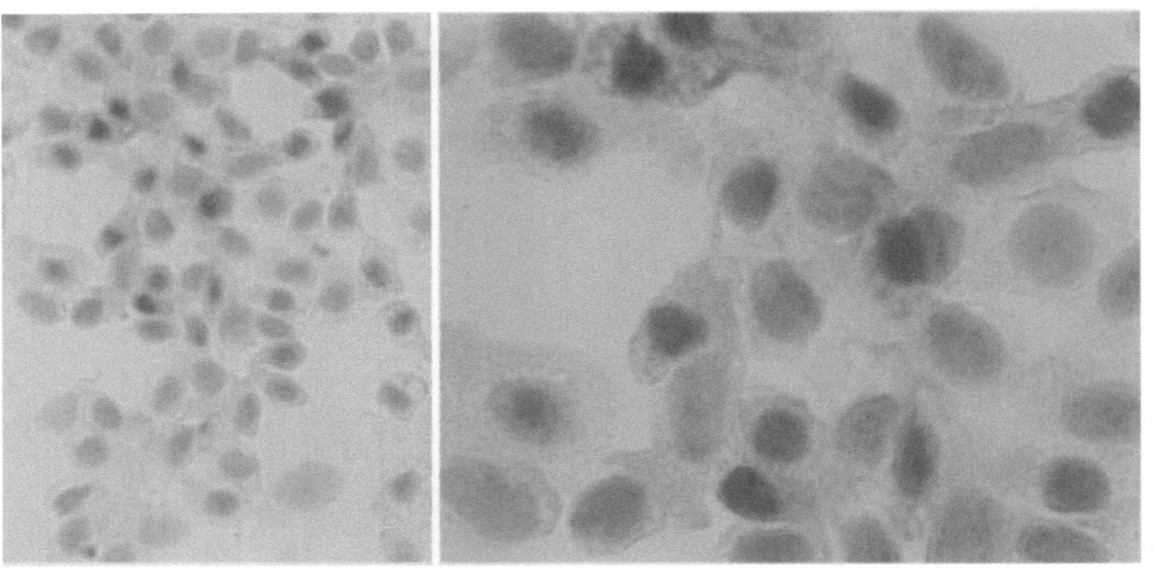

a   b

Fig. 9.21 a, b. **Hyperosmolar irrigating solutions.** As a result of osmotic processes, the use of hypoosmolar fluids (e.g., cystoscopy with distilled water) causes a nuclear swelling that must not be mistaken for a pathologically altered nucle-ar–cytoplasmic ratio. The swollen nuclei are distinguished by their uniformity and indistinct margins. The nuclear chromatin has a "diluted" hypochromatic appearance (a, × 340; b, × 850; Papanicolaou stain)

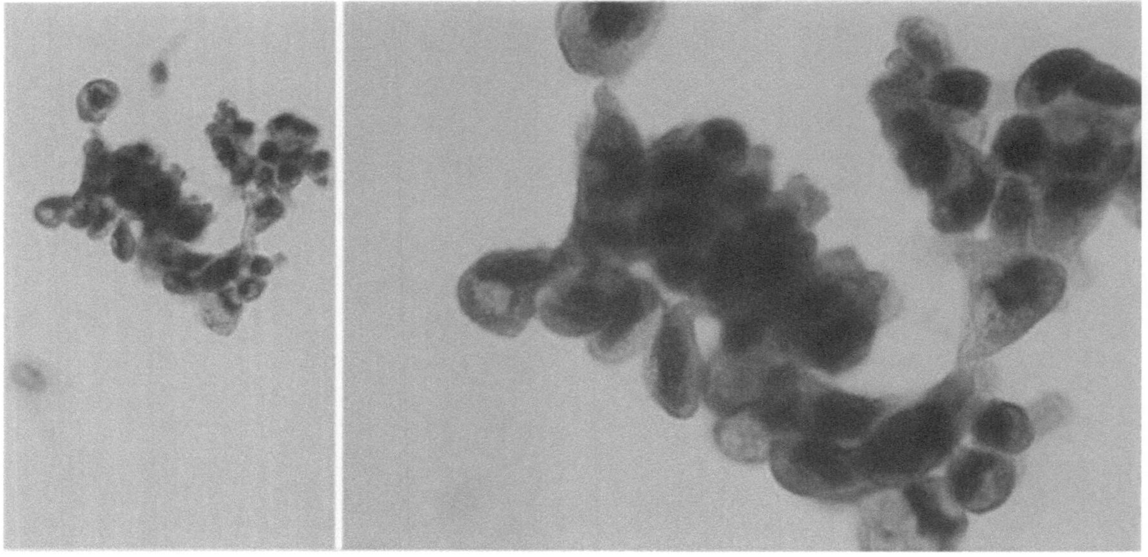

a   b

Fig. 9.22 a, b. **Overlapping cells mimicking hyperchromasia.** A cystoscopic or catheterized urine sample may contain clusters of overlapping cells, causing a loss of transparency that resembles an abnormal chromatin increase (hyperchro-masia). Generally these cell clusters occur in isolation with an otherwise normal-appearing general cytologic pattern (a, × 340; b, × 850; Papanicolaou stain)

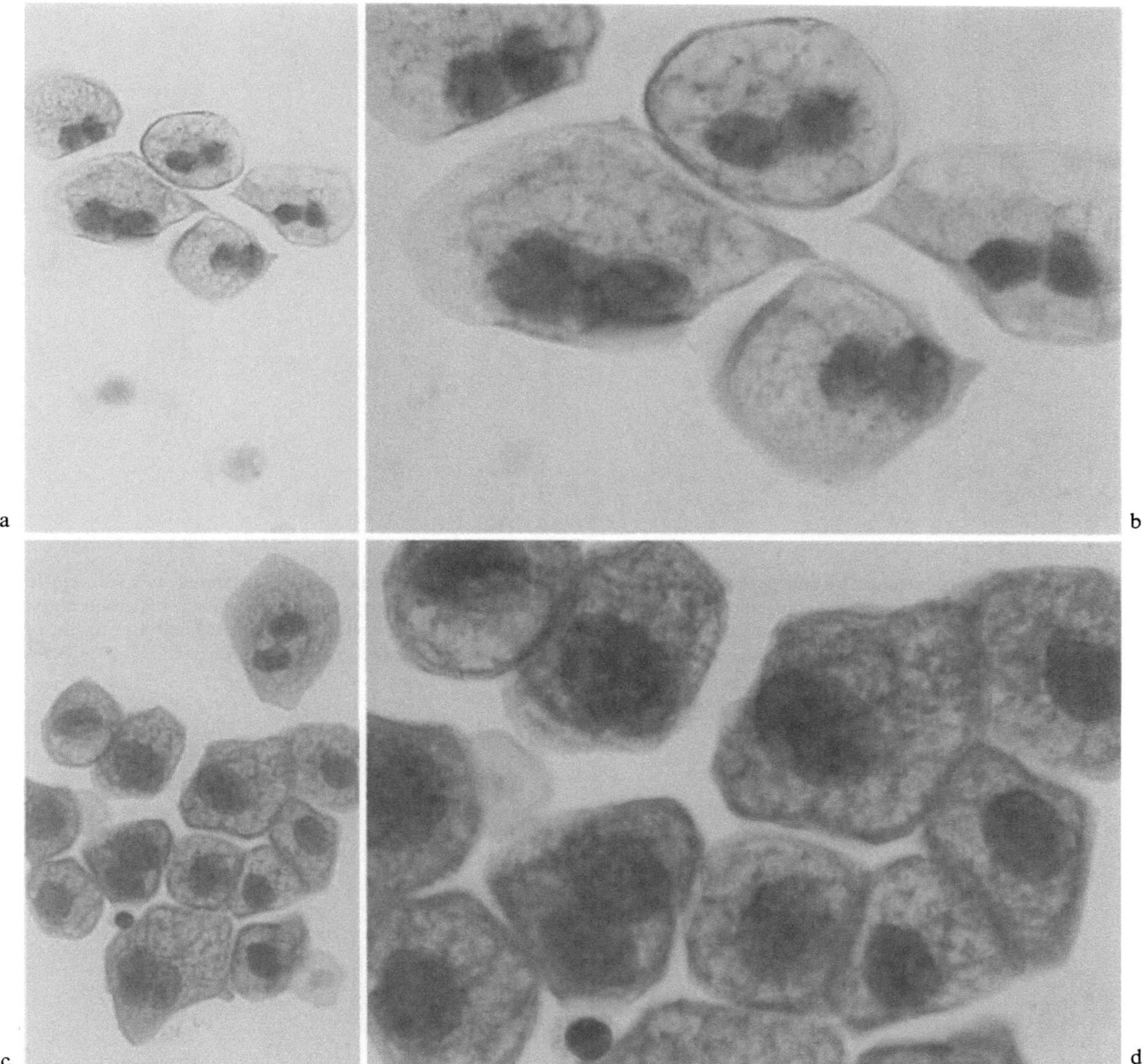

Fig. 9.23a–d. **Cellular changes caused by contrast medium.**
When urine cytology is performed immediately after intra-
venous urography (a, b) or retrograde pyelography (c, d), it
is common to see cytoplasmic vacuolation with the formation of "foam cells" (a and c, × 340; b and d, × 850;
Papanicolaou stain)

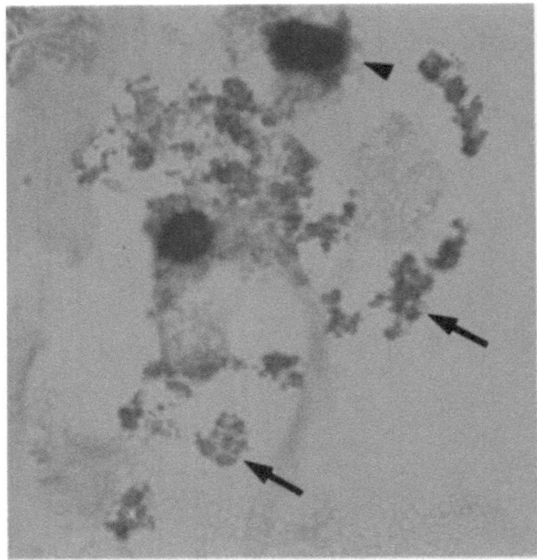
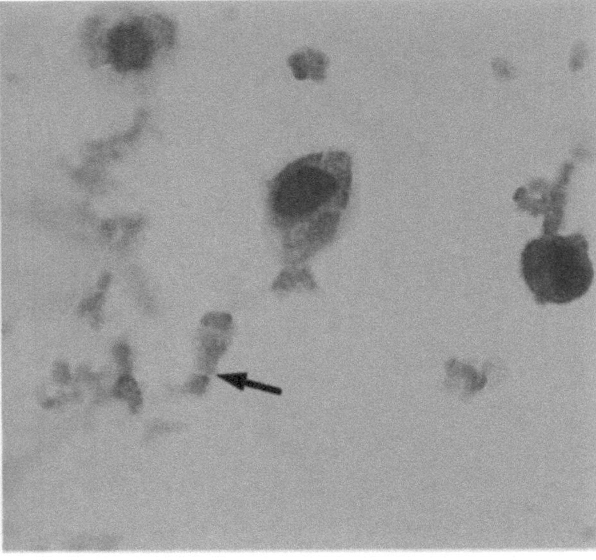

a   b

Fig. 9.24 a, b. **Cytolysis caused by improper preservation or fixation.** The erythrocytes are a reliable indicator of the adequate preservation or fixation of a urine sample, for they are the most sensitive of the urinary corpuscular elements ("internal quality control"). Their destruction ($\leftarrow\leftarrow$) will be accompanied by degenerative cytoplasmic changes in the urothelial cells. Increased nuclear staining due to protein denaturation ($\blacktriangleleft$) can simulate pathologic hyperchromasia. Conversely, intact erythrocytes and degenerative urothelial cells indicate satisfactory preservation ($\times 850$; Papanicolaou stain)

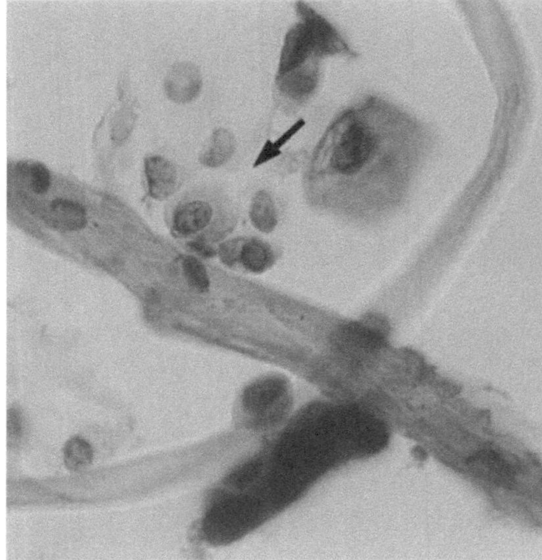
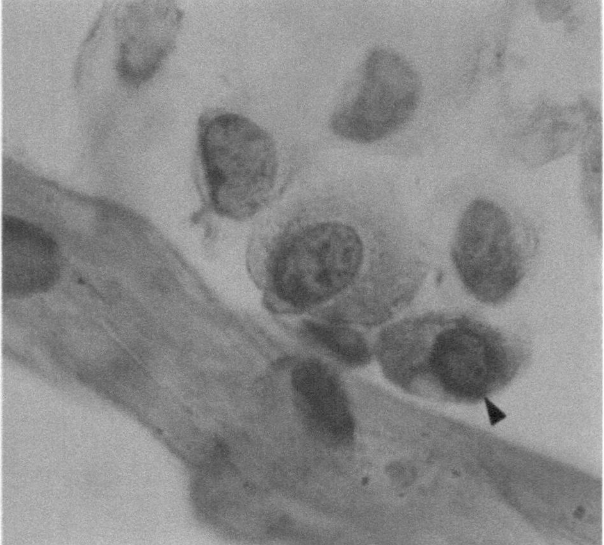

a   b

Fig. 9.25 a, b. **Findings in a patient with external urinary diversion.** Any instrumental manipulation of the urinary tract (e.g., suprapubic/transurethral catheterization, internal splintage) will cause profound morphologic reactions that can make cytologic evaluation impossible. This specimen contains urothelial cells ($\leftarrow$) that exhibit prominent nucleoli, a prominent nuclear envelope ($\blacktriangleleft$), and slight nuclear enlargement (altered nuclear–cytoplasmic ratio), suggestive of well-differentiated carcinoma, just 1 day after a suprapubic diversion (a, $\times 340$; b, $\times 850$; Papanicolaou stain)

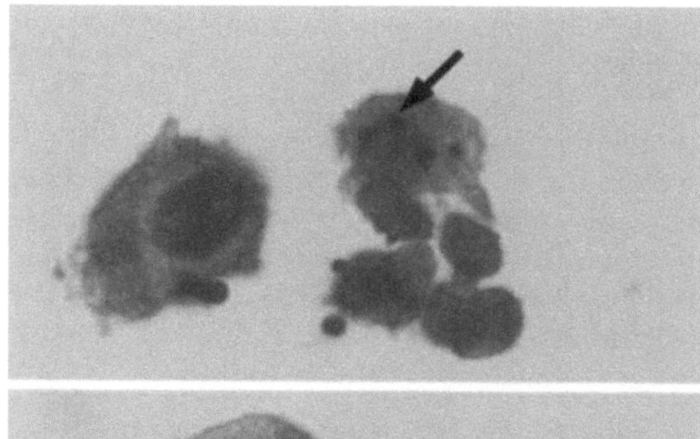

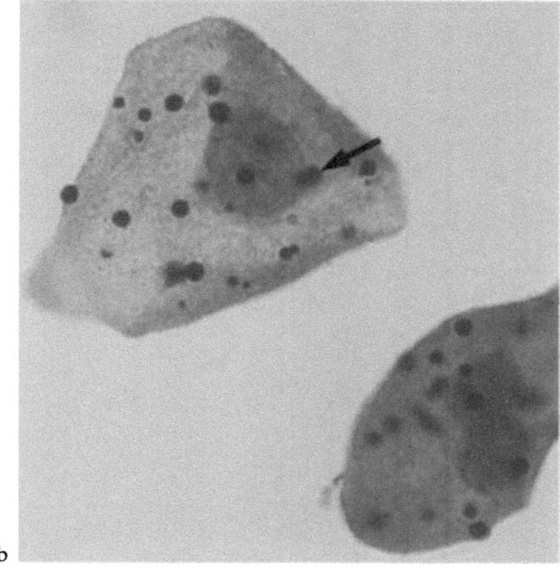

Fig. 9.26 a, b. **Staining artifacts.** Dye particles, so-called "keratohyaline bodies" (← ←), projected onto the nucleus must be differentiated from nucleoli. Usually these bodies are scattered over an area transcending the nuclear boundaries (a, × 340; b, × 850; Papanicolaou stain)

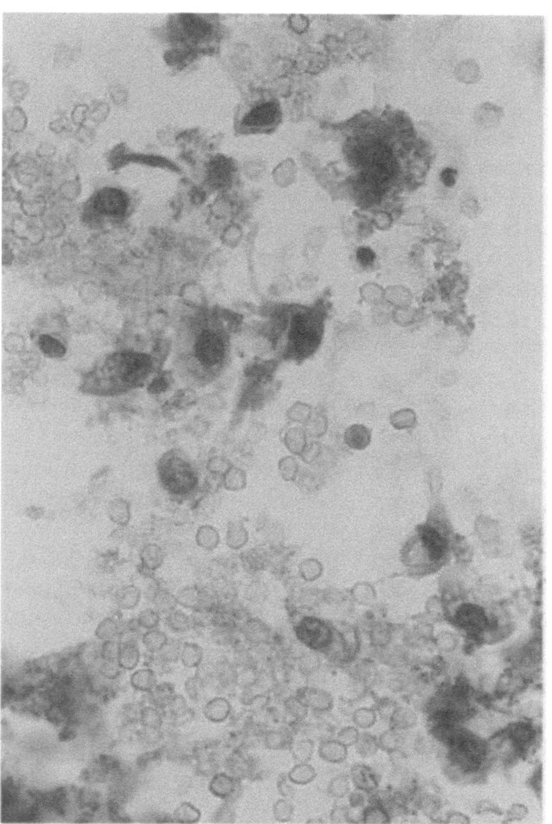

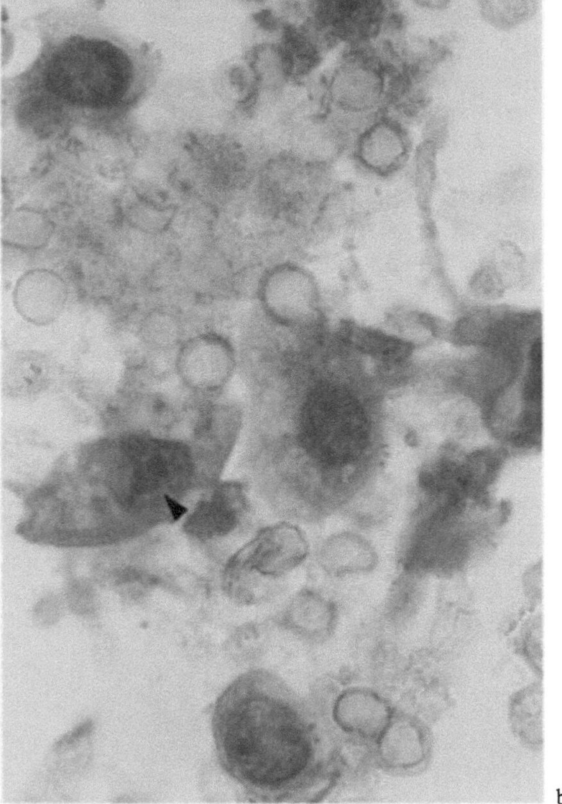

Fig. 9.27 a, b. **Lithiasis simulating neoplasia.** Stones can alter ▷ the appearance of single cells or whole cell clusters, making them indistinguishable from well-differentiated tumor cells. This figure shows enlarged nuclei with some structural diversity (a) and prominent nucleoli (b, ◀) (a, × 340; b, × 850; Papanicolaou stain)

9.5 Urothelial Tumors

The cells that are shed (exfoliated) from *well-differentiated urothelial tumors* exhibit few if any of the recognizable morphologic criteria of malignancy. This phenomenon, which has been referred to as the "lack of pathoanatomy in the individual cell" (Rübben et al. 1989), is the reason for the *low, approximately 50% accuracy rate* of urinary cytology in the detection of these neoplasms (Esposti et al. 1978; Murphy et al. 1986; Koss et al. 1985).

At the same time, well-differentiated urothelial carcinomas are rarely invasive in their growth and are easily diagnosed by endoscopy, so the relatively poor sensitivity of conventional urinary cytology *does not have major clinical significance* (Rübben et al. 1989). It is *significant*, however, that *moderately differentiated (grade 2) carcinomas* can be detected with a *sensitivity of 65%–80%* while *poorly differentiated (grade 3) carcinomas* are detected with *85%–95% sensitivity* (Jakse et al. 1980; Murphy et al. 1986; Esposti et al. 1978; Rübben et al. 1989).

Thus, if a urinary cytologic specimen is declared normal, there is about a 50% chance that a well-differentiated carcinoma has been missed. It is more *clinically significant*, however, that a moderately differentiated carcinoma can be excluded with approximately 80% confidence and a poorly differentiated carcinoma with greater than 90% confidence (see Chap. 2).

It is also noteworthy that dysplasia without pathologic significance can simulate the changes associated with a well-differentiated carcinoma (Jakse et al. 1986). Cancer can also be mimicked by reactive changes induced by infection (often recognizable by an accompanying general inflammatory pattern) or possibly by lithiasis.

The *pathologic significance* of moderately and poorly differentiated *dysplasias*, whose cytologic appearance is indistinguishable from that of moderately and poorly differentiated malignancies, is discussed in a separate chapter (see Chap. 6). Since the corresponding cytologic picture is always an indication for further investigation by endoscopy and biopsy, which ultimately will determine the choice of treatment, the lack of differentiating criteria for dysplasias and carcinomas is of only minor importance.

9.5.1 Well-Differentiated Urothelial Tumors (Grade 1)

The micrographs presented in this section will illustrate that the difference between *normal findings*, *well-differentiated carcinomas*, *dysplasia*s, and *reactive changes* is one of *degree only*. They also illustrate, through pictorial comparisons, the conspicuous features that differentiate these changes from higher-grade carcinomas.

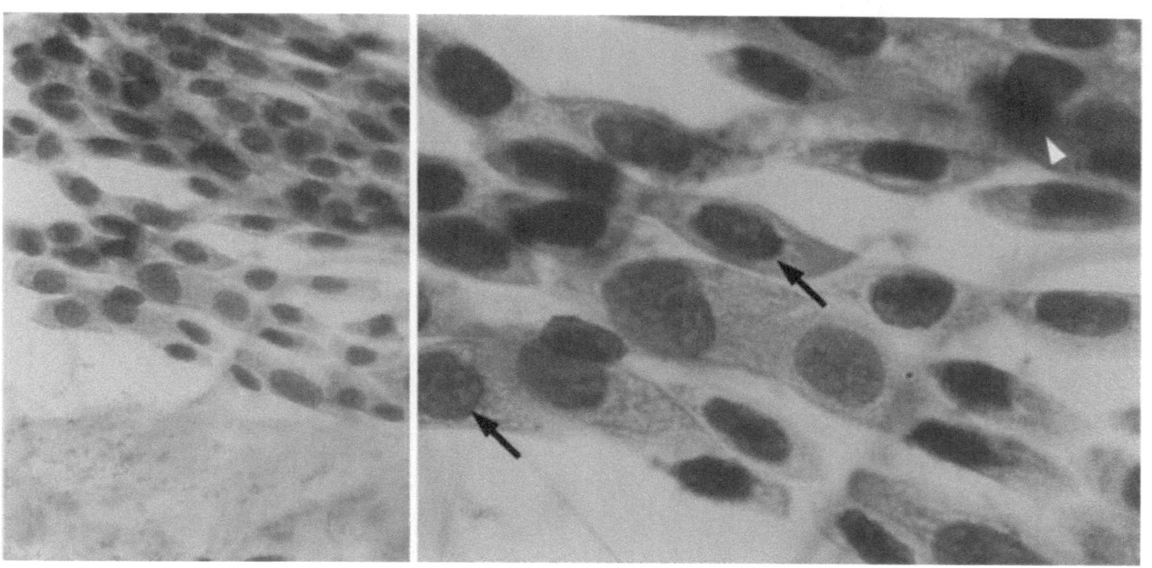

Fig. 9.28 a, b. **Well-differentiated urothelial carcinoma (grade 1).** The cells exhibit a fine chromatin pattern with minimal thickening of the nuclear membrane (← ←). The most conspicuous criterion of this grade 1 tumor is the in-creased nuclear–cytoplasmic ratio. True hyperchromasia is absent but is simulated by overlapping nuclei (◄) or imperfect focusing (a, × 340; b, × 850; Papanicolaou stain)

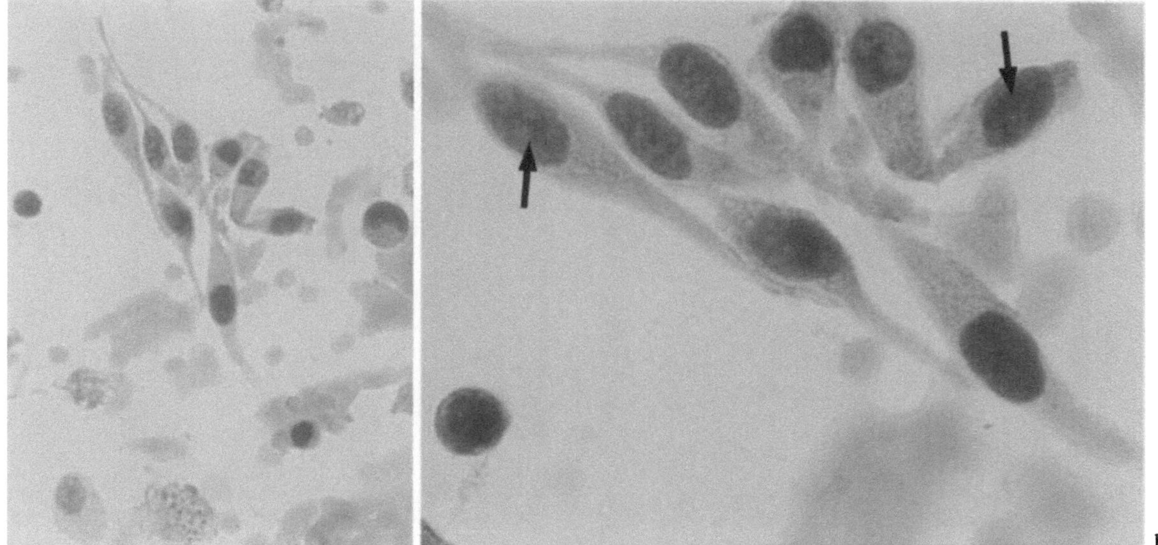

Fig. 9.29 a, b. **Well-differentiated urothelial carcinoma (grade 1)** with slightly enlarged nuclei and slight clumping of the nuclear chromatin (← ←) (a, × 340; b, × 850; Papanicolaou stain)

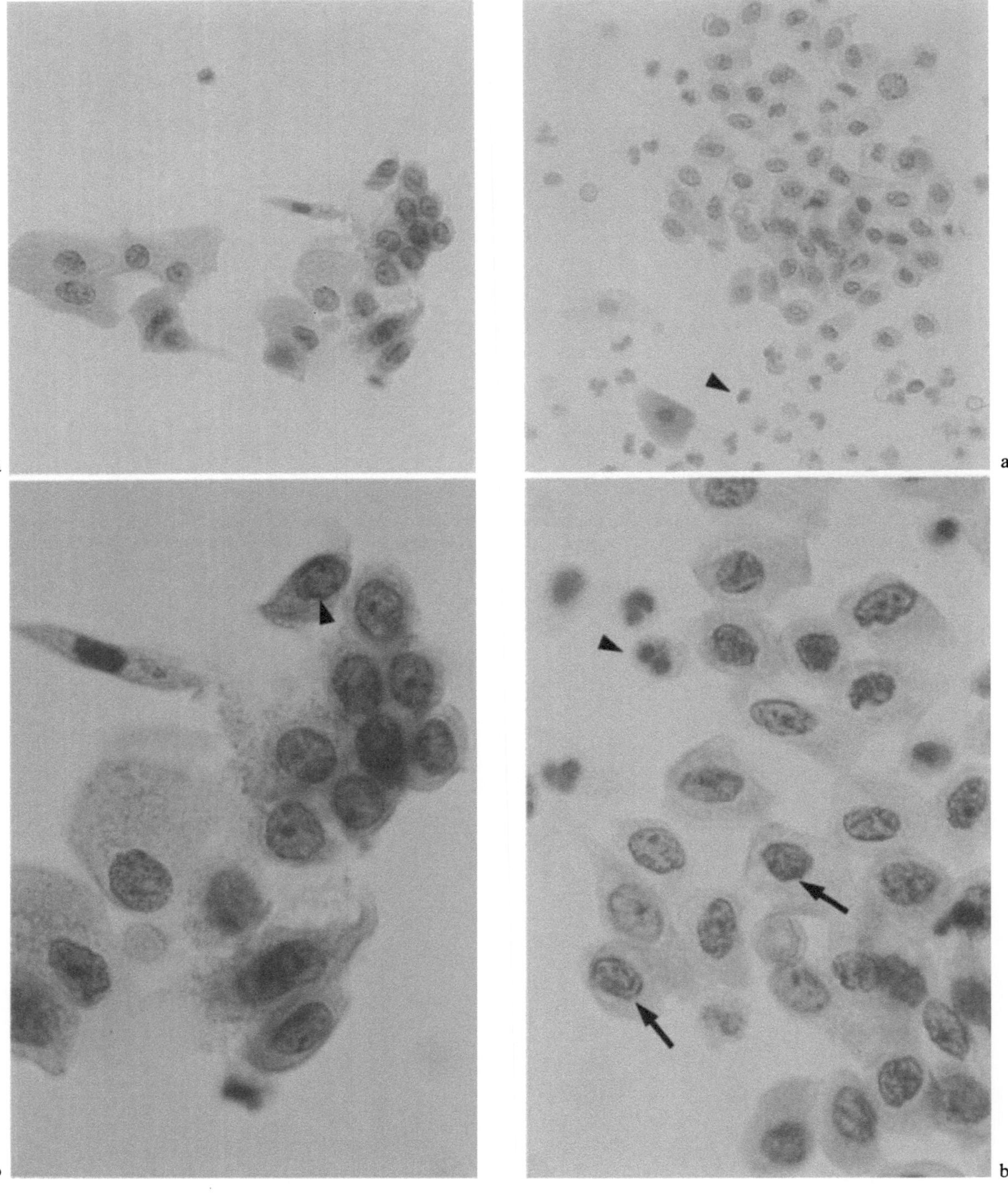

Fig. 9.30 a, b. **Well-differentiated urothelial carcinoma (grade 1)** with minimally enlarged nuclei and slight prominence of the nuclear membrane (◄). The very subtle morphologic changes illustrate the uncertainty inherent in the cytodiagnosis of grade 1 tumors (a, × 340; b, × 850; Papanicolaou stain)

Fig. 9.31 a, b. **Well-differentiated urothelial carcinoma (grade 1)** with slightly coarse chromatin and slight prominence of the nuclear membrane (← ←). Leukocytic infiltration (◄ ◄) without bacteriuria ("sterile leukocyturia") may be seen with urothelial tumors, lithiasis, or after TUR, so a reactive etiology of the cellular changes should be considered (a, × 340; b, × 850; Papanicolaou stain)

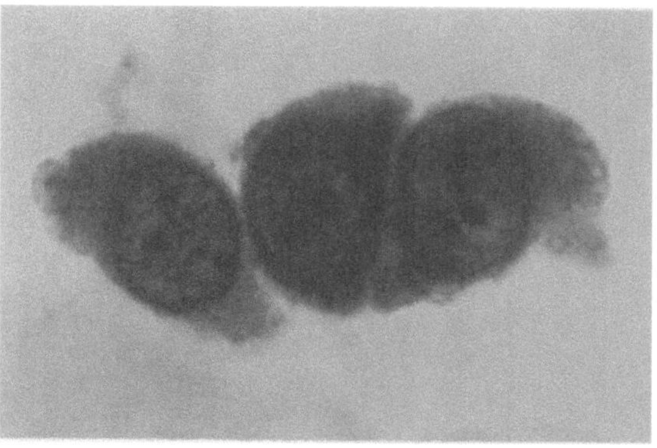

9.32

Figs. 9.32 and 9.33. **Well-differentiated urothelial carcinomas (grade 1)** demonstrated with rapid stains (Fig. 9.32 *methylene blue*; Fig. 9.22 *Testsimplets*). Both specimens show marked prominence of the nuclear membrane. The nuclei are significantly larger than in Papanicolaou-stained specimens (e.g., Figs. 9.28–9.31) due to lack of alcohol-induced dehydration and cell shrinkage (×850)

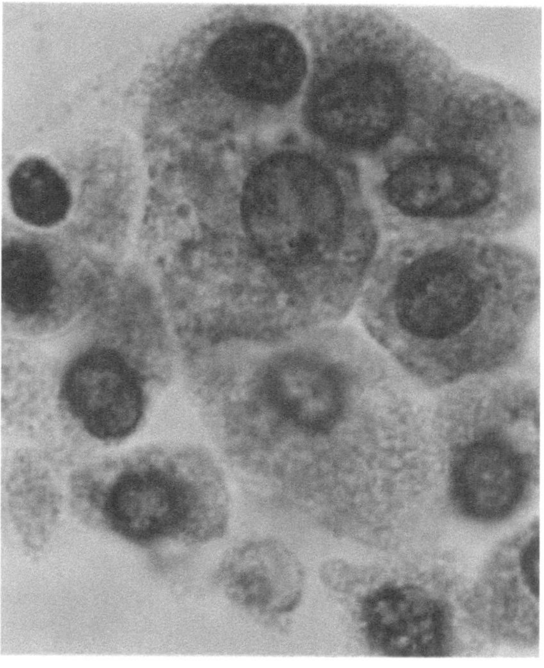

9.33

Comparison with Other Findings

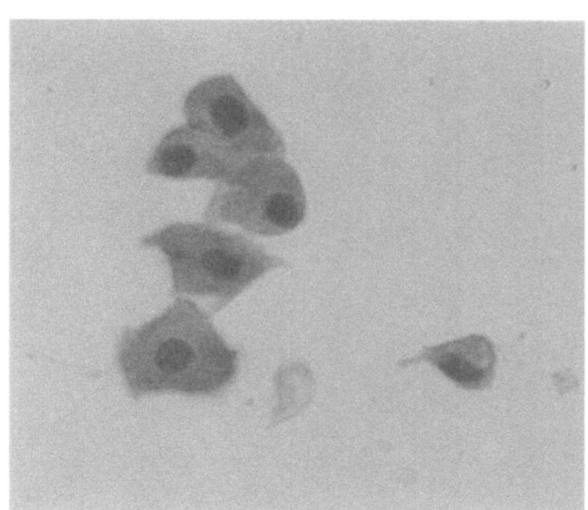

a

Fig. 9.34 a, b. **Normal urothelial cells** (a, × 340; b, × 850; Papanicolaou stain)

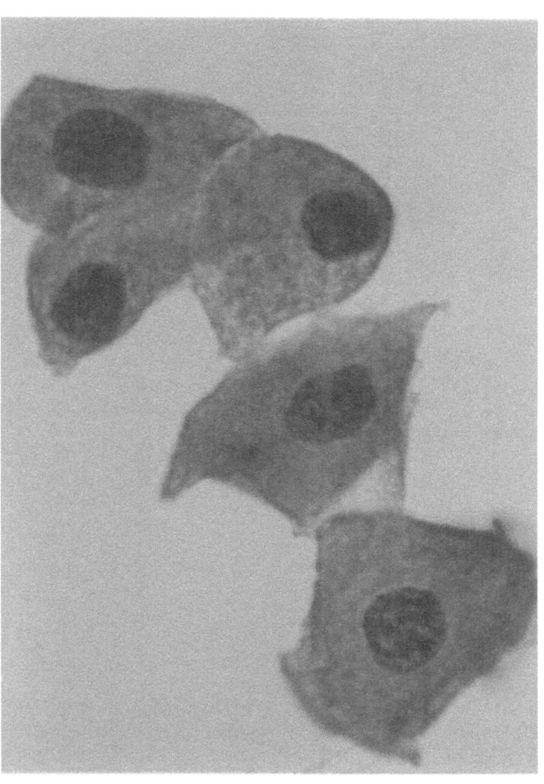

b

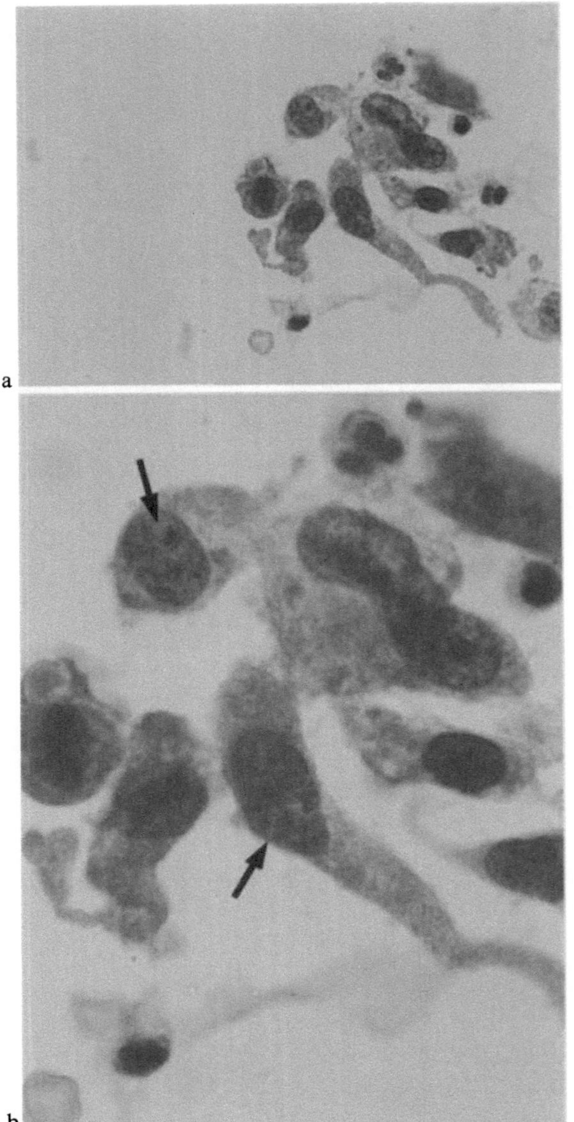

Fig. 9.35 a, b. **Moderately differentiated urothelial carcinoma (grade 2)** with a markedly altered nuclear–cytoplasmic ratio, slight coarseness and clumping of the nuclear chromatin (← ←), and a prominent nuclear membrane (a, × 340; b, × 850; Papanicolaou stain)

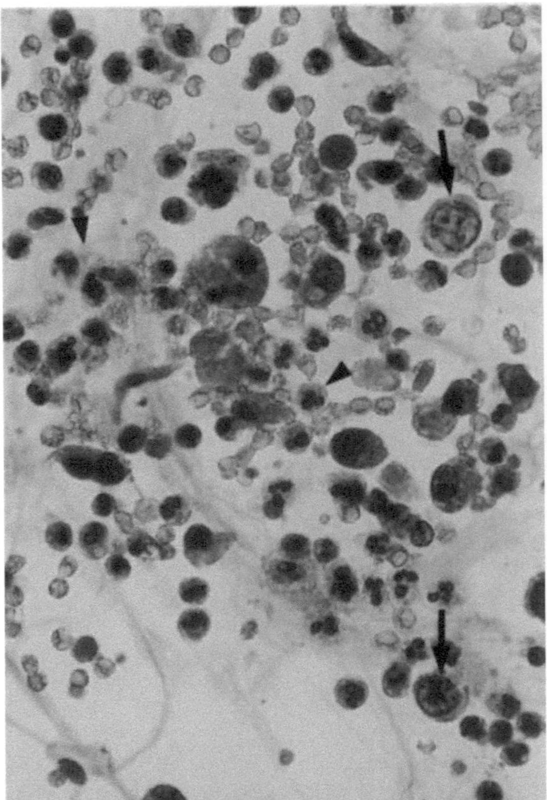

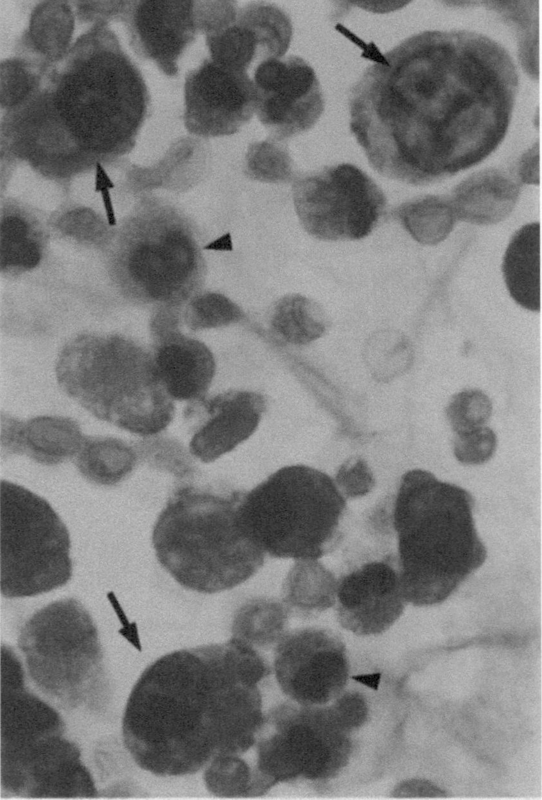

Fig. 9.36 a, b. **Poorly differentiated urothelial carcinoma ▷ (CIS, grade 3)** with massive leukocytic infiltration (◄ ◄) and pathologic urothelial cells (← ←) (a, × 340; b, × 850; Papanicolaou stain)

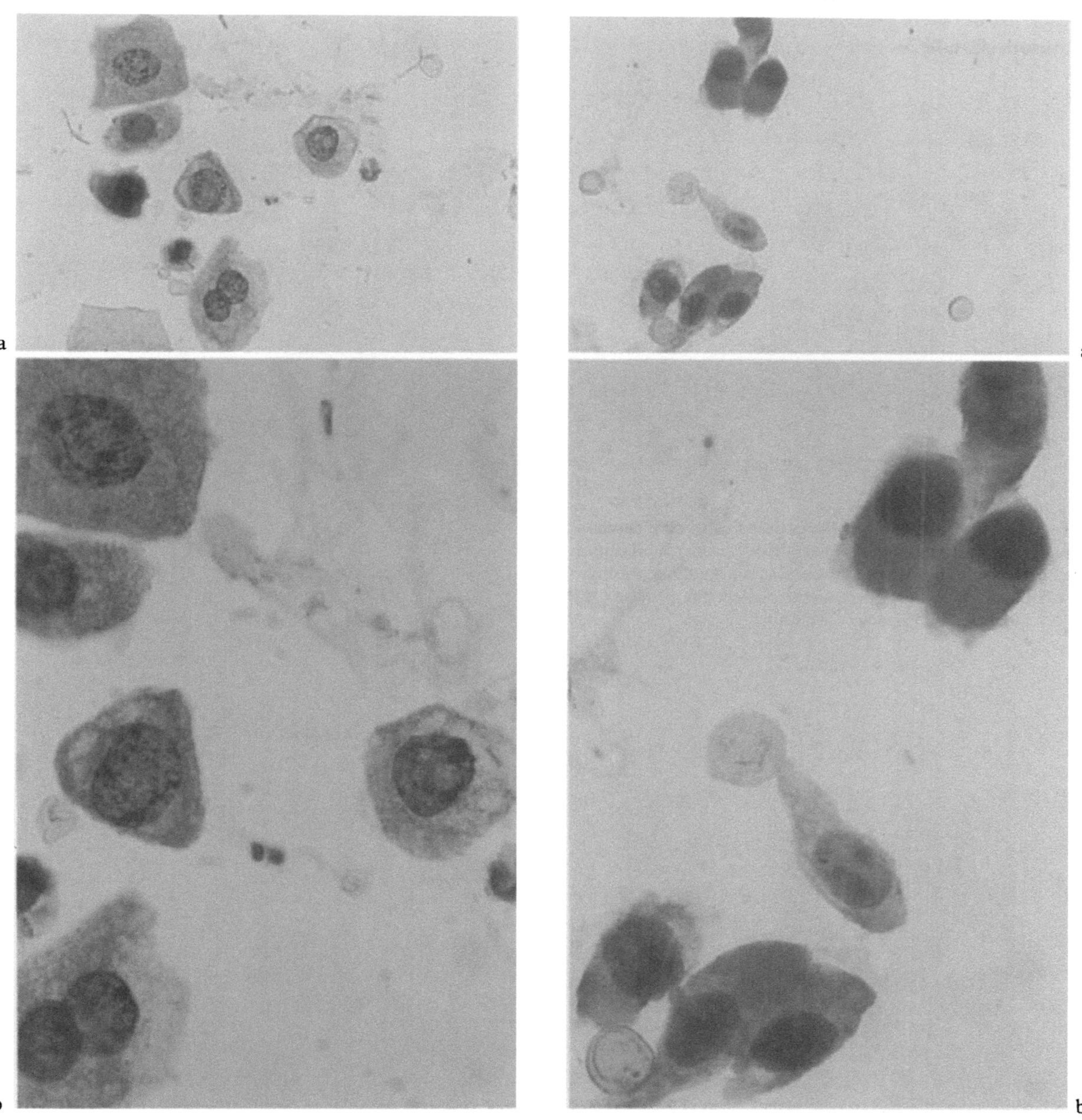

9.37

9.38

Figs. 9.37a,b and 9.38a,b. **Tumor-mimicking dysplasia in urolithiasis.** Although the primary cytologic impression was grade 1 carcinoma, endoscopy and uroradiology confirmed the absence of a tumor and the presence of urolithiasis (9.37 a and 9.38 a, × 340; 9.37 b and 9.38 b, × 850; Papanicolaou stain)

9.5.2 Moderately Differentiated Urothelial Tumors (Grade 2)

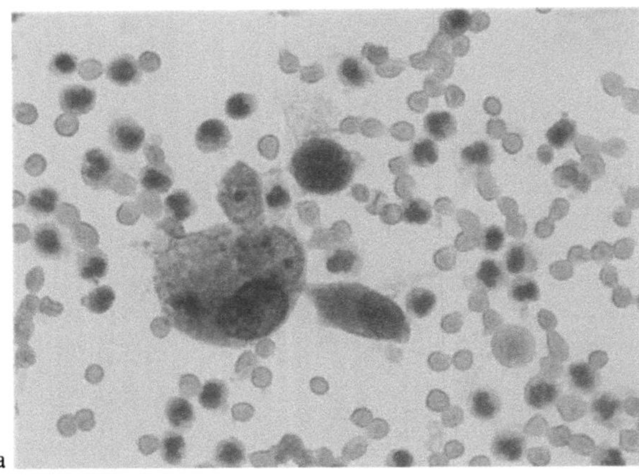

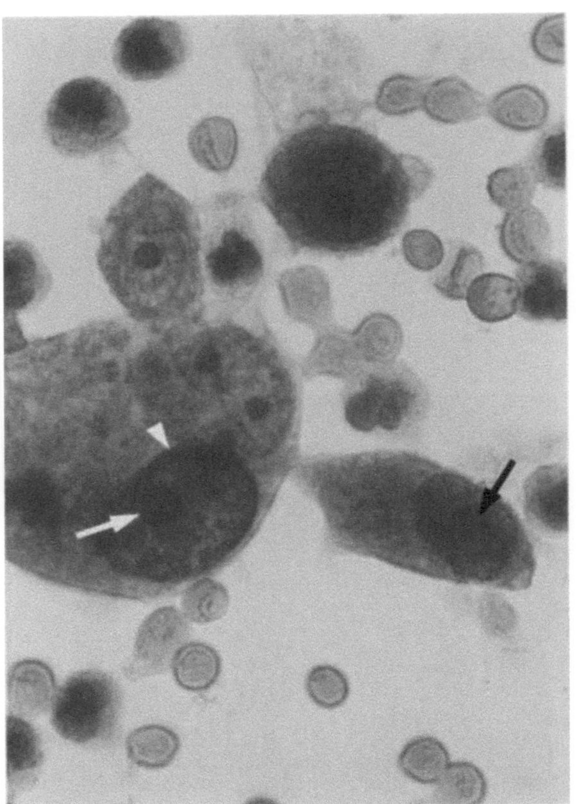

Fig. 9.39 a, b. **Moderately differentiated urothelial carcinoma (grade 2).** Leukocytes and erythrocytes surround tumor cells showing mild hyperchromasia, markedly pathologic nucleoli (← ←), and a prominent nuclear membrane (◄) (a, ×340; b, ×850; Papanicolaou stain)

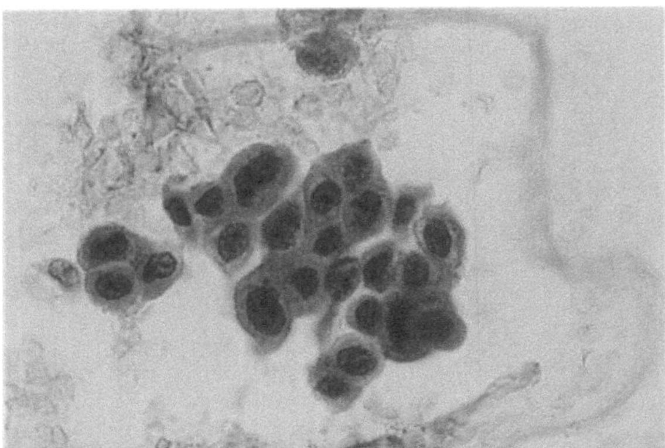

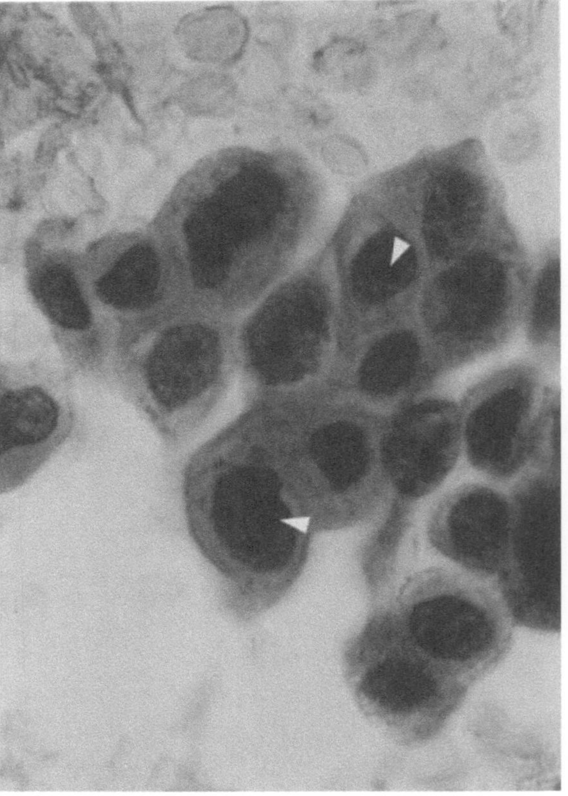

Fig. 9.40 a, b. **Moderately differentiated urothelial carcinoma (grade 2).** Besides increased chromatin (hyperchromasia with reduced nuclear transparency) and prominent nucleoli (◄ ◄), the malignancy criterion of nuclear pleomorphism is particularly well demonstrated in this specimen (a, × 340; b, ×850; Papanicolaou stain)

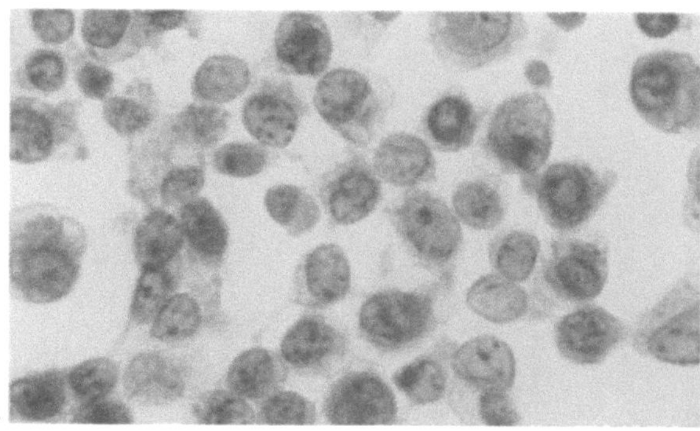

a

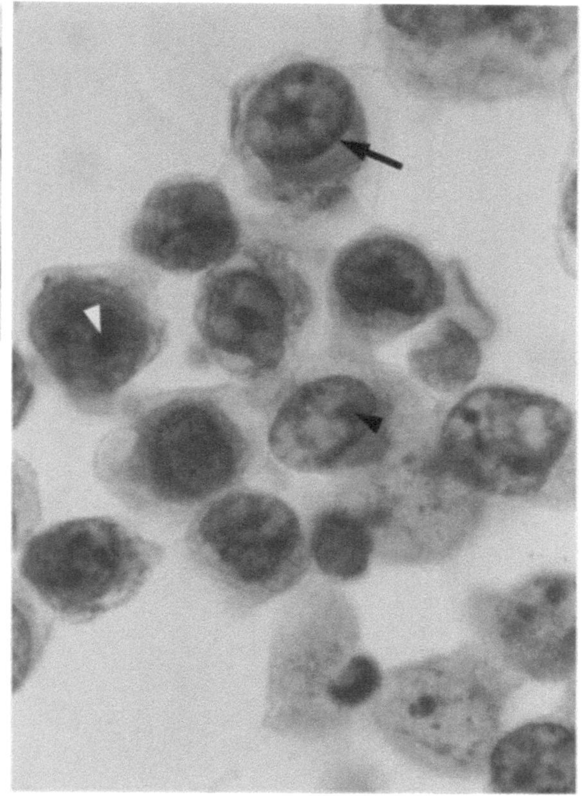

b

Fig. 9.41 a, b. **Moderately differentiated urothelial carcinoma (grade 2).** Very cellular specimen collected by irrigation cytology. There is conspicuous alteration of the nuclear–cytoplasmic ratio (*large nuclei*), a prominent nuclear envelope (←), and large nucleoli (◄ ◄) (a, × 340; b, × 850; Papanicolaou stain)

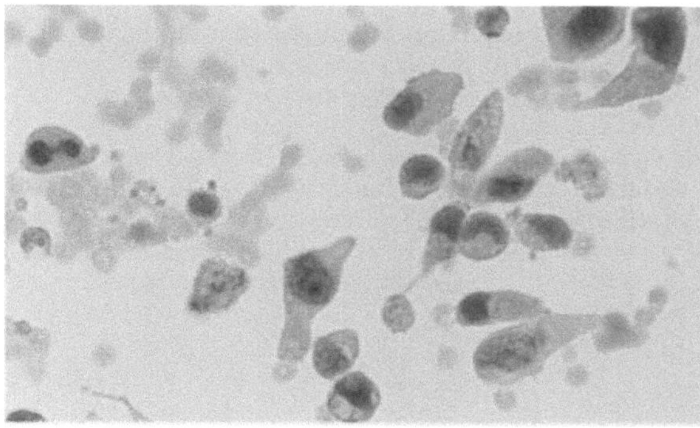

a

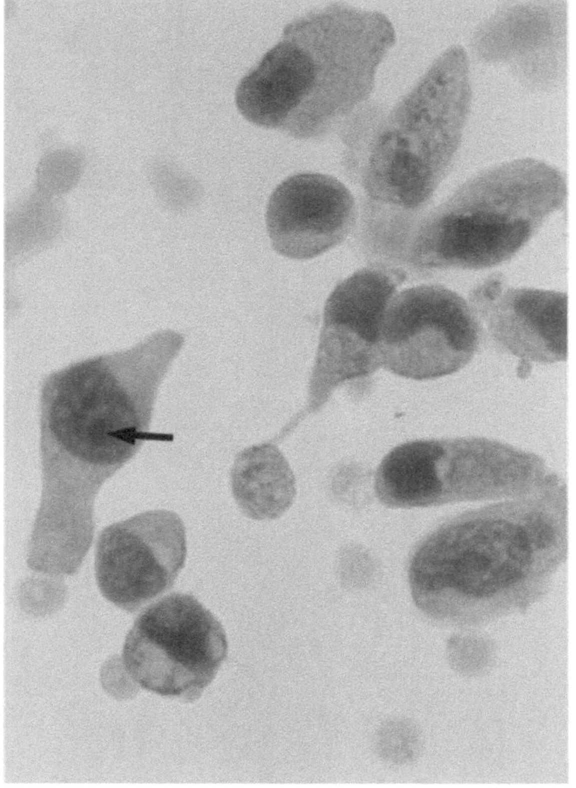

b

Fig. 9.42 a, b. **Moderately differentiated urothelial carcinoma (grade 2)** with enlarged, slightly pleomorphic nuclei and some prominent nucleoli (←) (a, × 340; b, × 850; Papanicolaou stain)

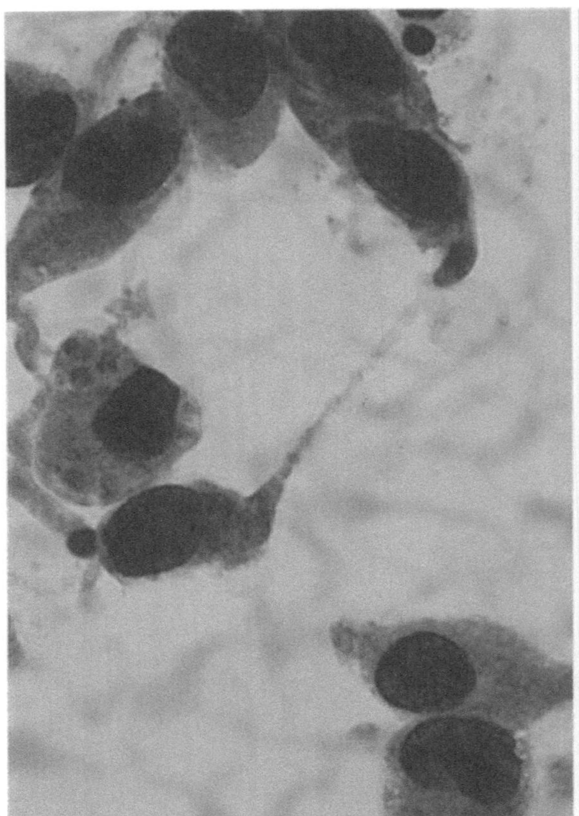

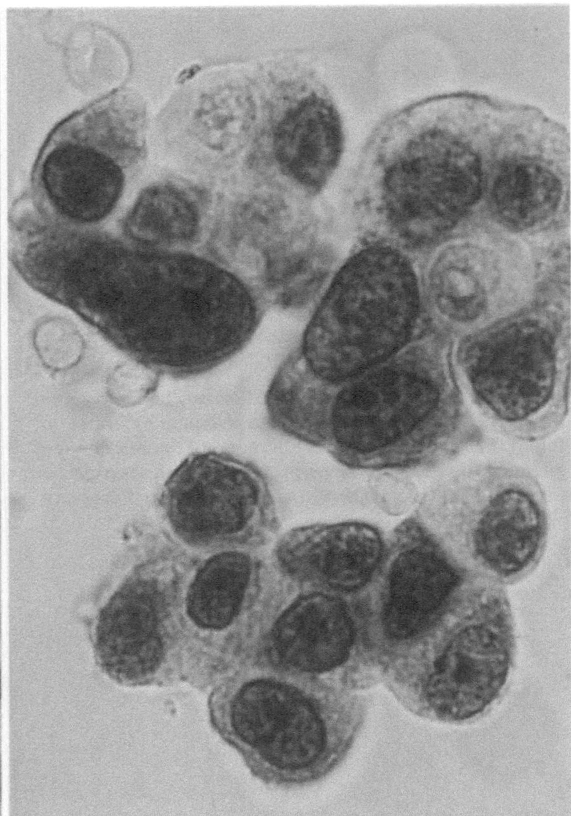

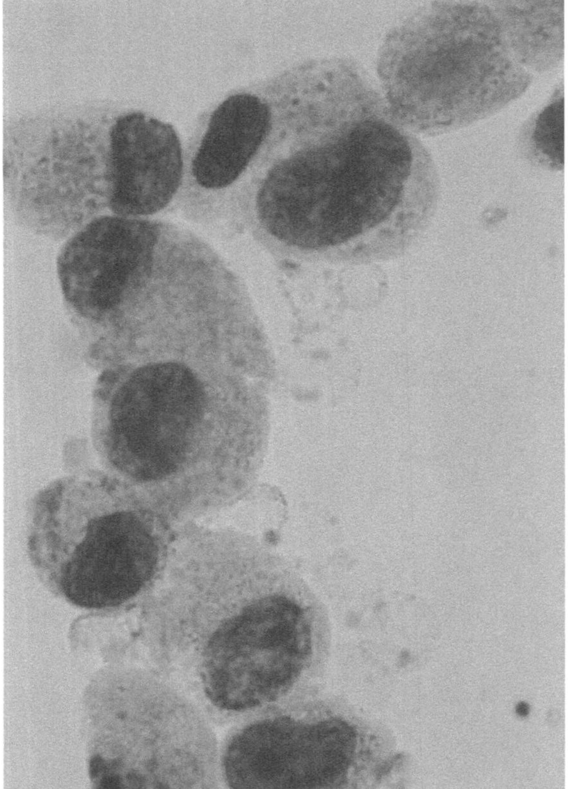

Fig. 9.43a–c. **Moderately differentiated urothelial carcinoma (grade 2)** with instant stain. Specimens seen in a and b were stained with Testsimplets (color differences due to different photographic filters) and c with methylene blue. Since dehydrating alcoholic stains were not used, there is no shrinkage of the diagnostically critical nuclei. Especially in b, there is excellent nuclear transparency with the malignancy criteria of a prominent nuclear membrane, a slightly altered nuclear–cytoplasmic ratio, and nuclear pleomorphism (×850)

Comparison with Other Findings

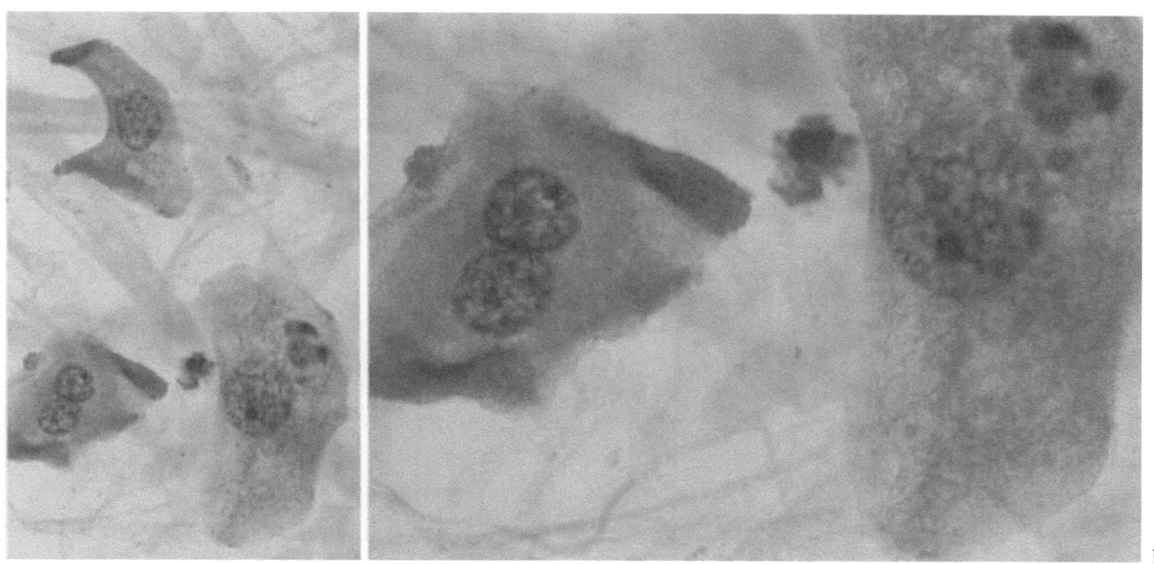

Fig. 9.44 a, b. **Normal urothelial cells** (a, × 340; b, × 850;
Papanicolaou stain)

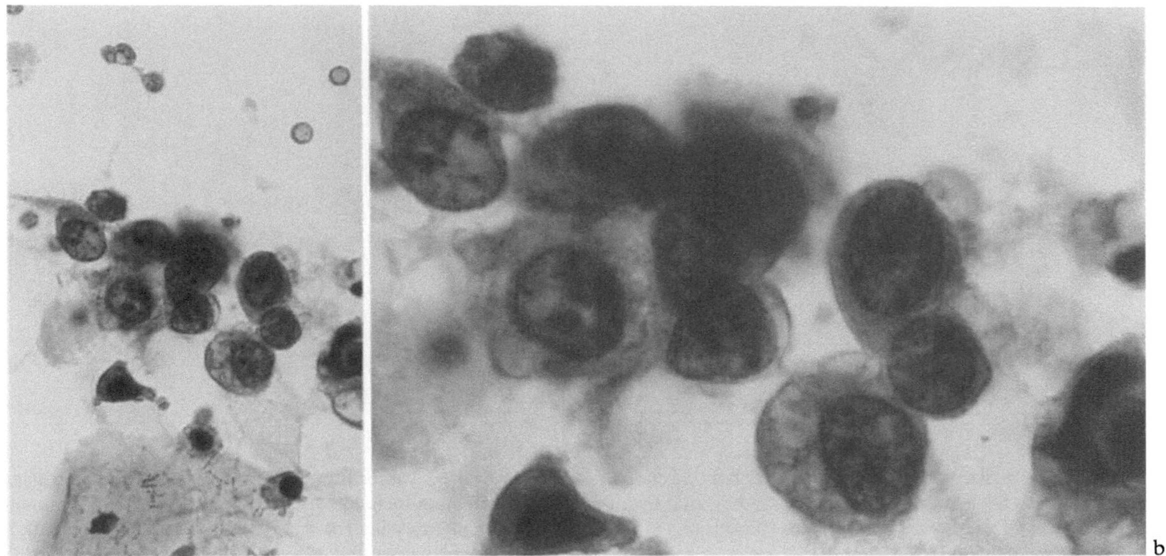

Fig. 9.45 a, b. **Poorly differentiated urothelial carcinoma
(grade 3)** with coarsely granular chromatin, hyperchromasia, a prominent nuclear membrane, and enlarged nucleoli
(a, × 340; b, × 850; Papanicolaou stain)

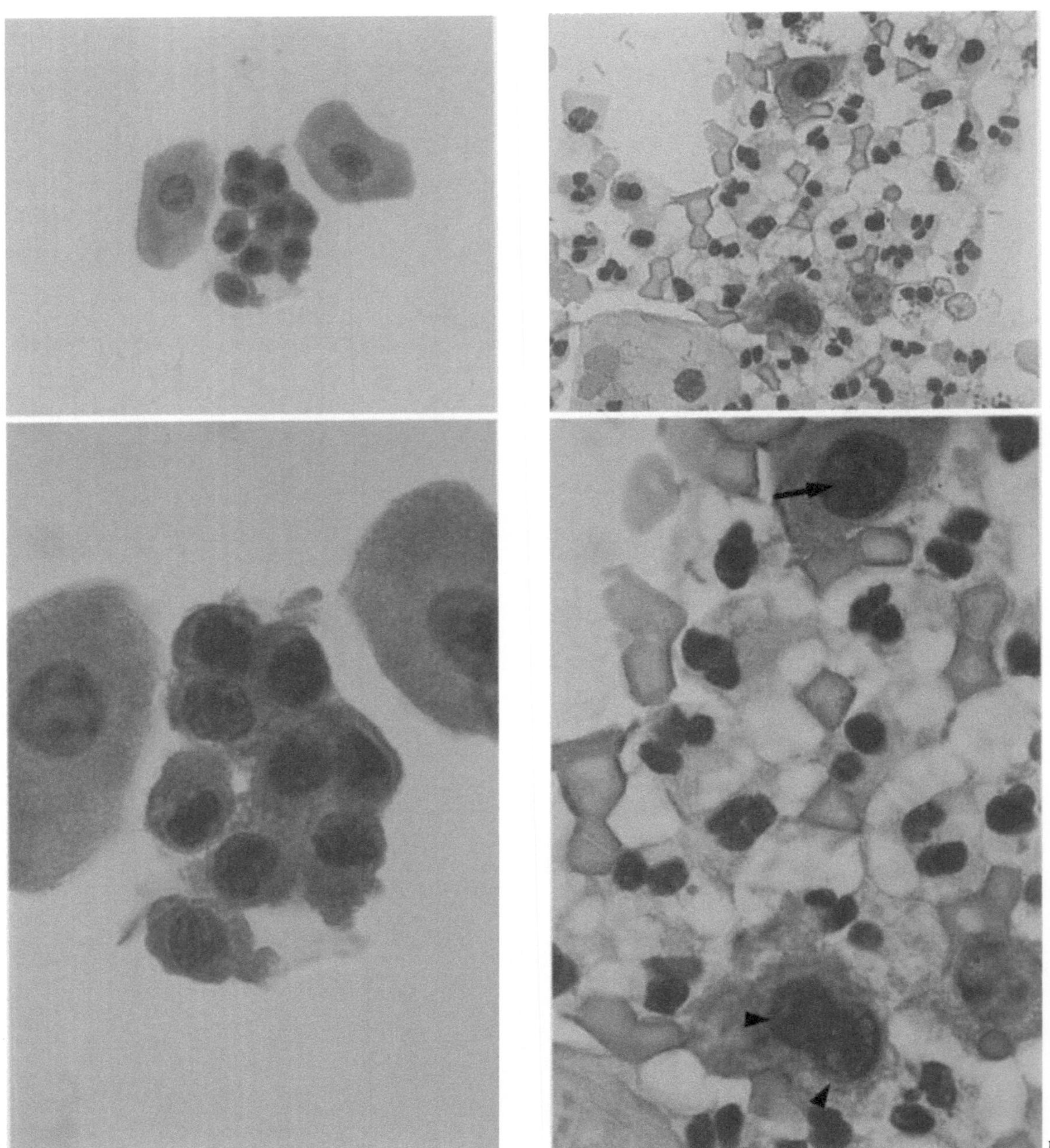

Fig. 9.46 a, b. **Well-differentiated urothelial carcinoma (grade 1)** with nuclear transparency reduced by mild hyperchromasia and slight nuclear pleomorphism (a, × 340; b, × 850; Papanicolaou stain)

Fig. 9.47 a, b. **Tumor-mimicking reactive changes in urinary tract infection** include hyperchromasia (←) and prominent nuclear membranes (◄◄). Although the massive leukocyturia and bacteriuria suggest infection, they could very well result from the necrotic degeneration of a tumor. These cases warrant at least a urinary cytologic follow-up after the infection has been cured (a, × 340; b, × 850; Papanicolaou stain)

9.5.3 Poorly Differentiated Urothelial Tumors (Grade 3)

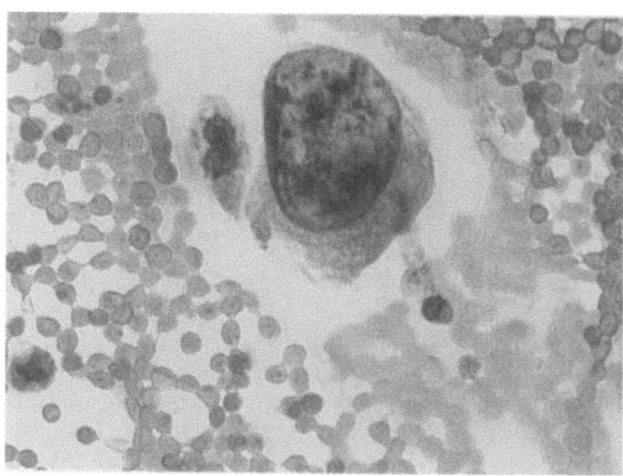

a

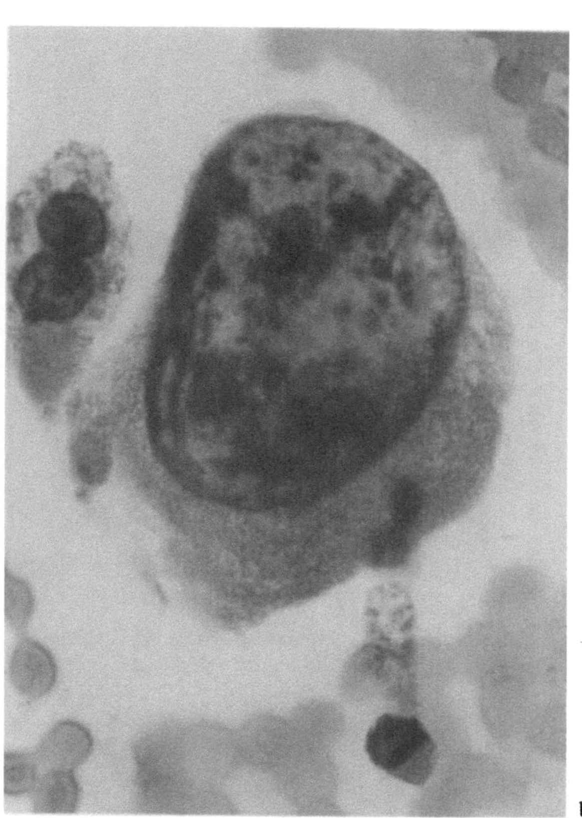

b

Fig. 9.48 a, b. **Poorly differentiated urothelial carcinoma (grade 3)** with a very pathologic giant nucleus. By "diluting" the hyperchromasia, the nuclear enlargement mitigates the apparent loss of nuclear transparency (a, × 340; b, × 850; Papanicolaou stain)

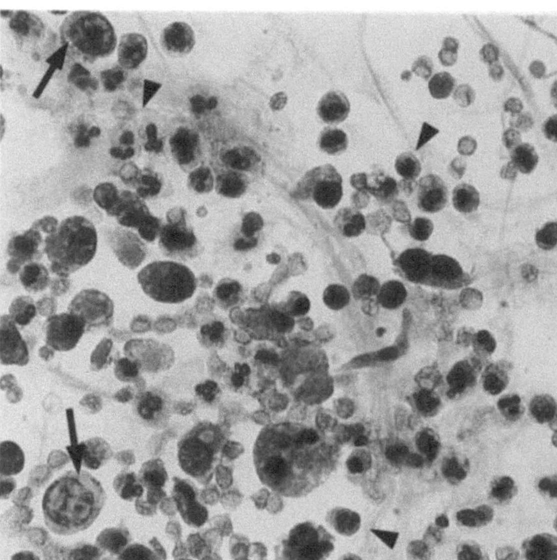

a

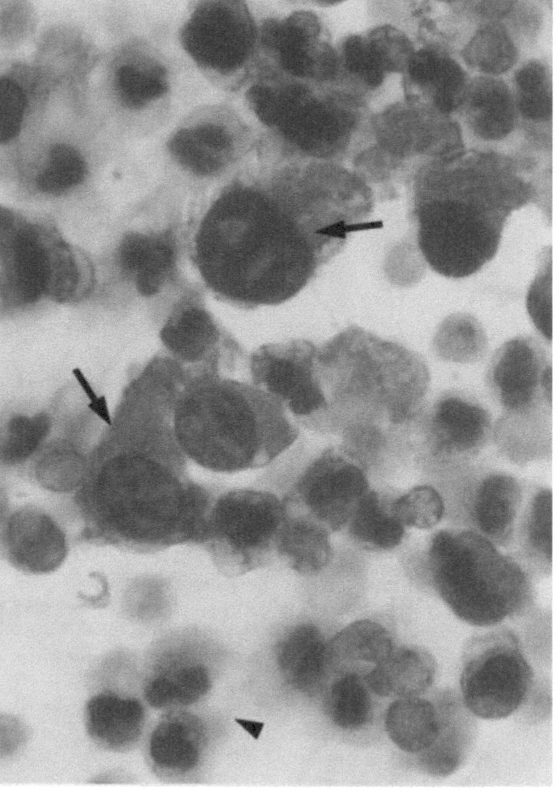

b

Fig. 9.49 a, b. **Poorly differentiated urothelial carcinoma (grade 3)** presenting as carcinoma in situ. Typically, endoscopy showed only a slight redness of the urothelium. Abundant segmented granulocytes (◀◀) and tumor cells (←). Malignant cell changes include an altered nuclear–cytoplasmic ratio and marked hyperchromasia with chromatin clumping (a, × 340; b, × 850; Papanicolaou stain)

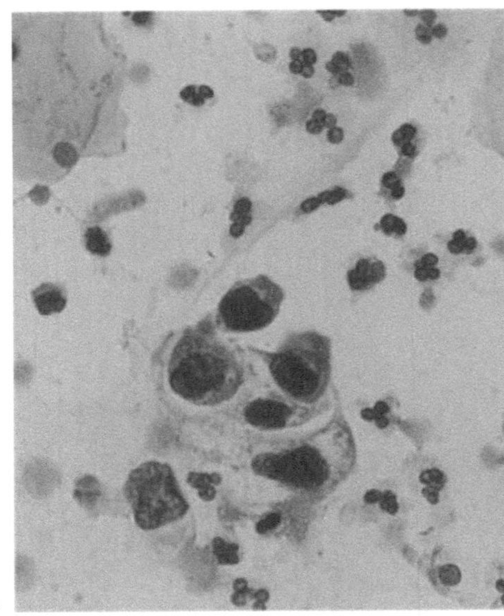
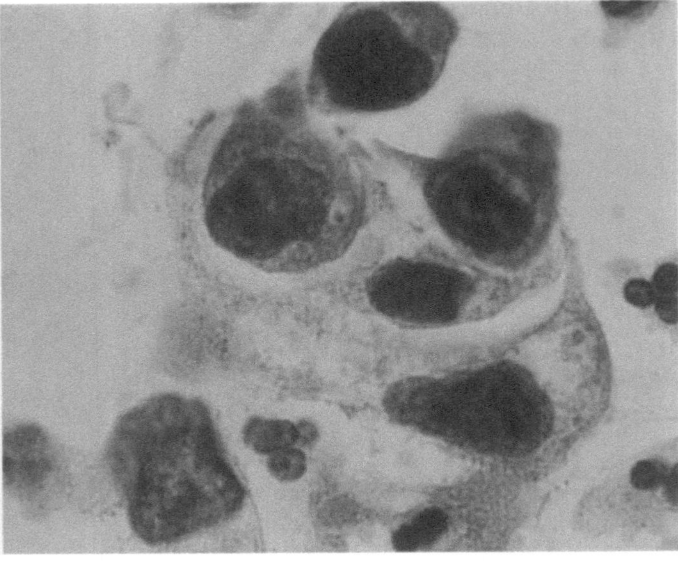

Fig. 9.50 a, b. **Poorly differentiated urothelial carcinoma (grade 3).** Besides hyperchromasia and nuclear enlarge-ment, nuclear pleomorphism is present as a significant criterion of malignancy (a, ×340; b, ×850; Papanicolaou stain)

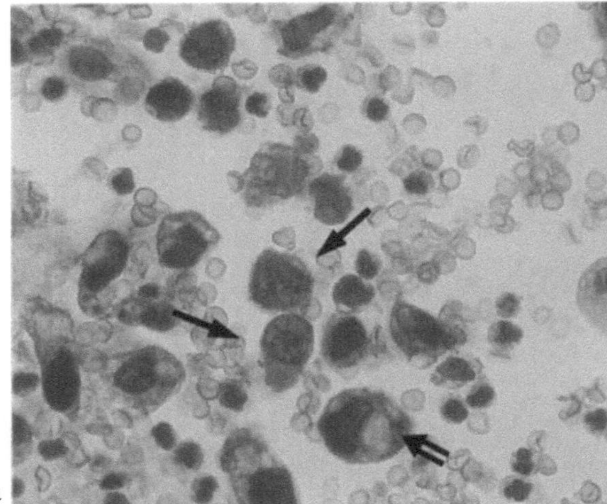
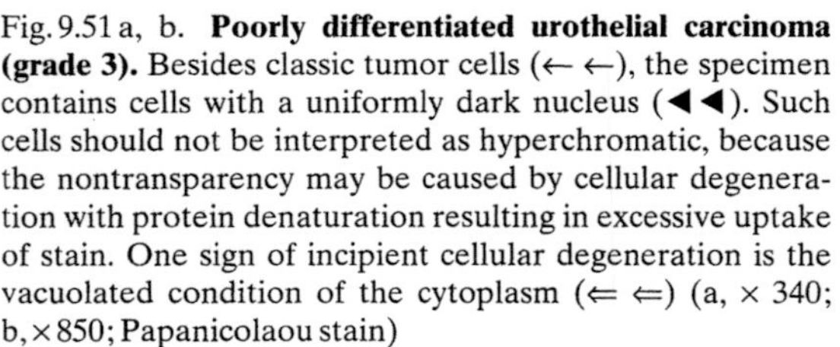
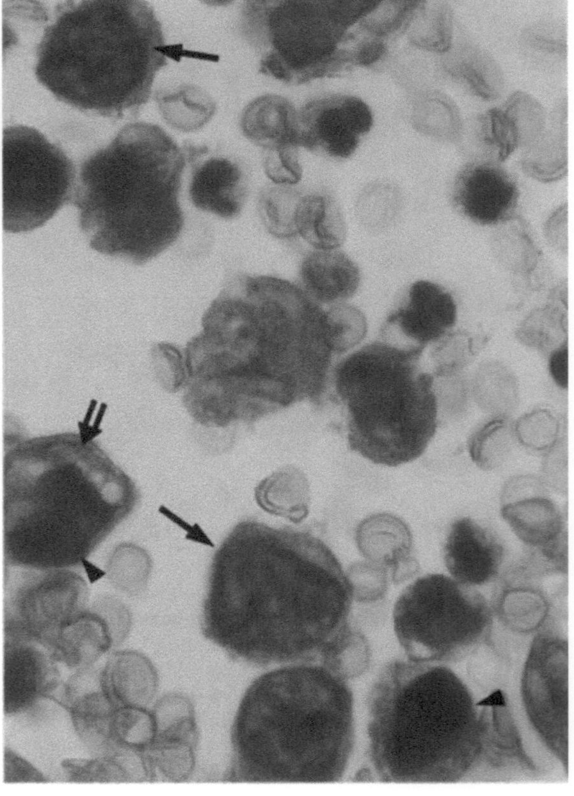

Fig. 9.51 a, b. **Poorly differentiated urothelial carcinoma (grade 3).** Besides classic tumor cells (← ←), the specimen contains cells with a uniformly dark nucleus (◄ ◄). Such cells should not be interpreted as hyperchromatic, because the nontransparency may be caused by cellular degeneration with protein denaturation resulting in excessive uptake of stain. One sign of incipient cellular degeneration is the vacuolated condition of the cytoplasm (⇐ ⇐) (a, × 340; b, × 850; Papanicolaou stain)

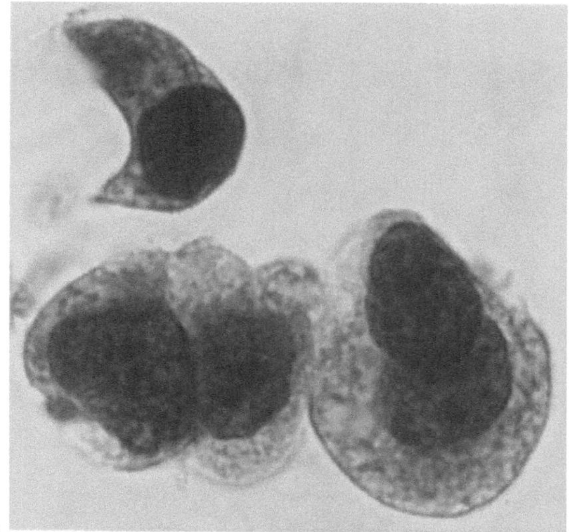

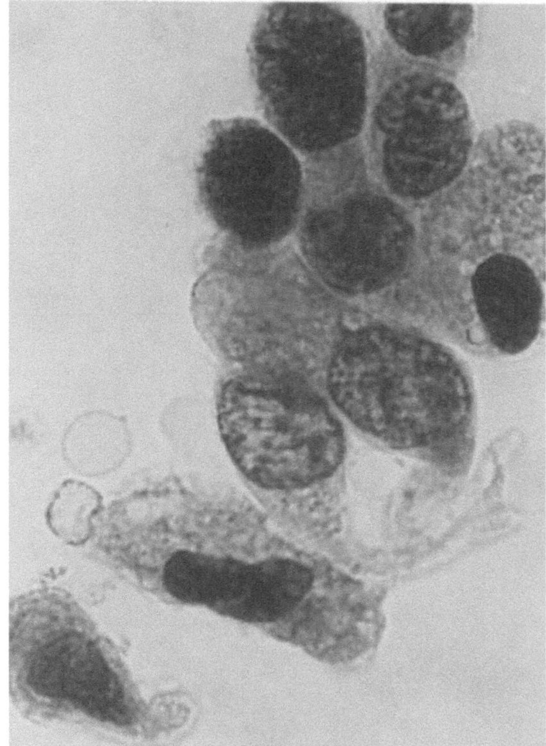

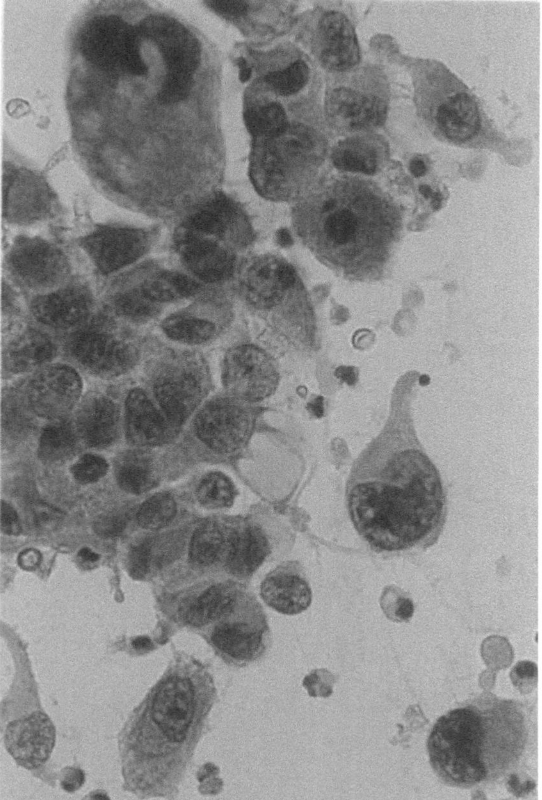

Fig. 9.52 a, b. **Poorly differentiated urothelial carcinoma (grade 3)** stained with Testsimplets (×850)

Fig. 9.53 a, b. **Poorly differentiated urothelial carcinoma** ▷ **(grade 3)** with enlarged and pleomorphic nuclei. Due to the nuclear size, the hyperchromasia causes little reduction in nuclear transparency (a, ×340; b, ×850; Papanicolaou stain)

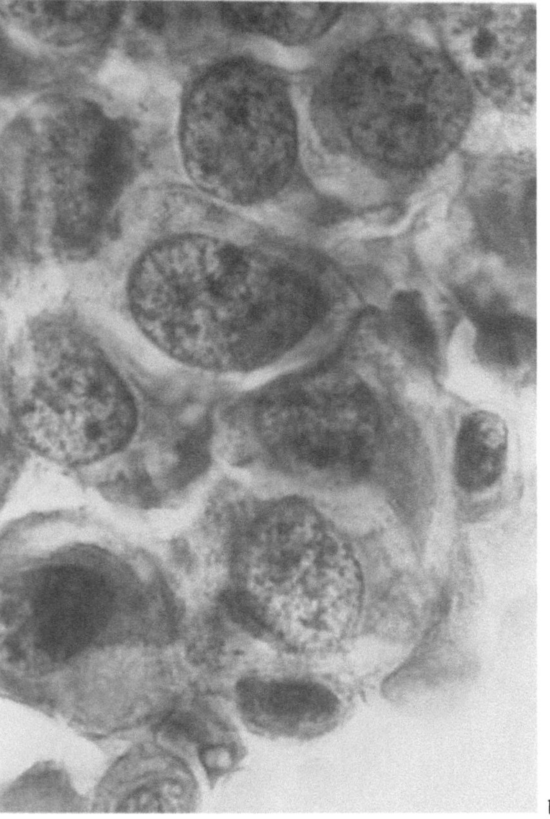

Comparison with Other Findings

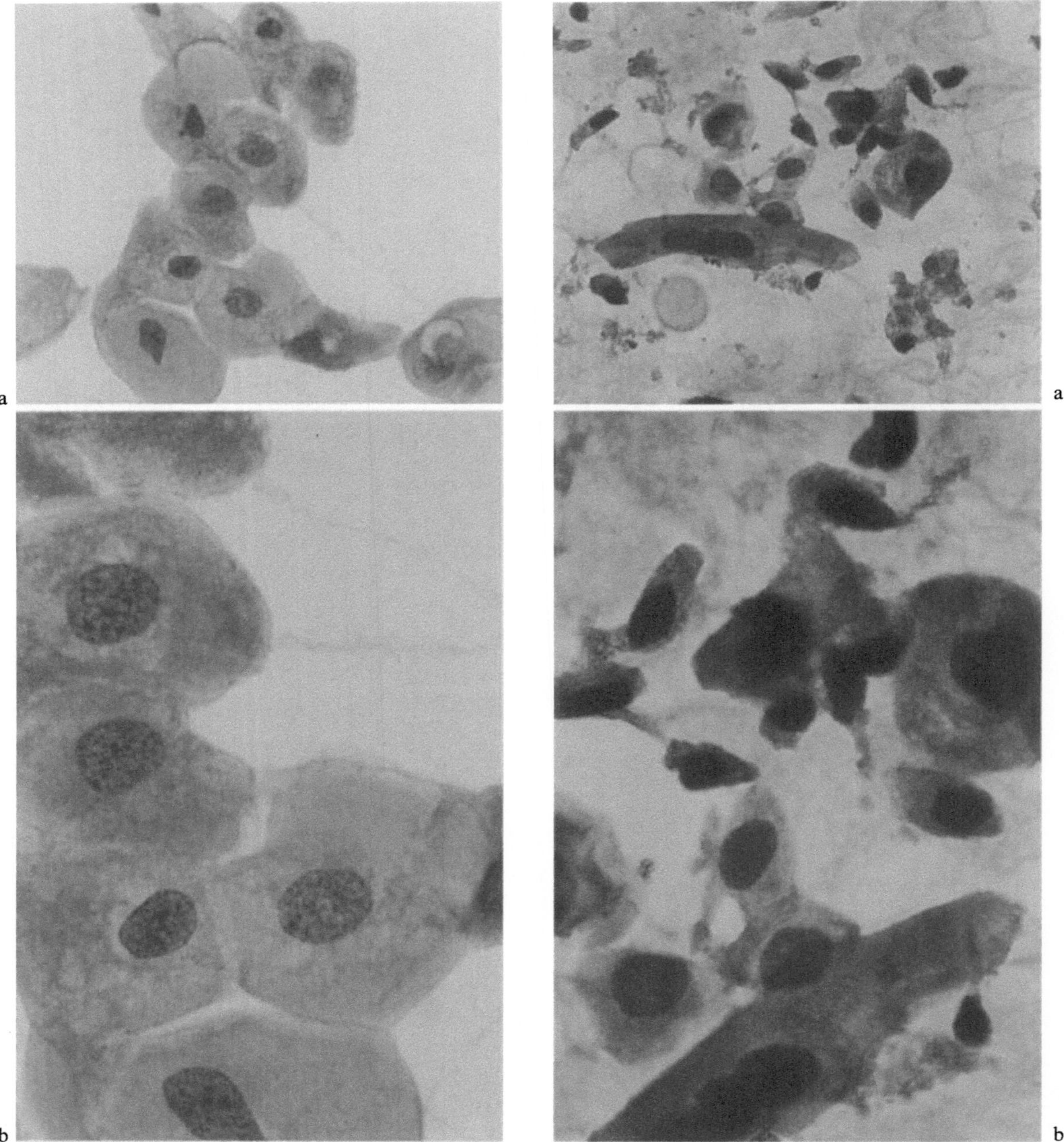

Fig. 9.54 a, b. **Normal urothelial cells** (a, × 340; b, × 850; Papanicolaou stain)

Fig. 9.55 a, b. **Moderate urothelial dysplasia** with no endoscopic or biopsy evidence of carcinoma. Although the nuclei do not show classic hyperchromasia with coarse granularity and clumping, they appear abnormally dense. This presentation cannot be reliably differentiated from carcinoma (a, × 340; b, × 850; Papanicolaou stain)

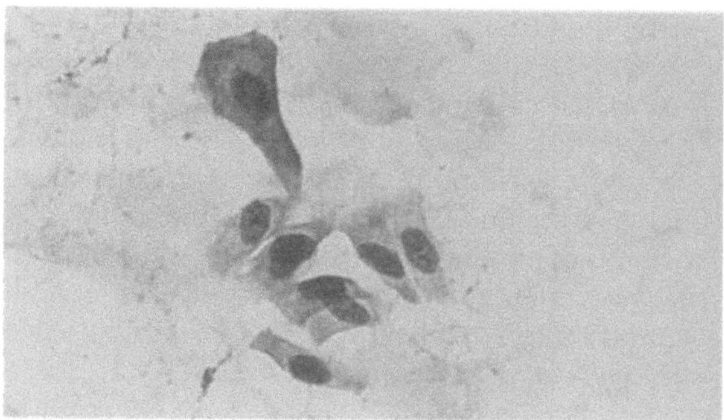

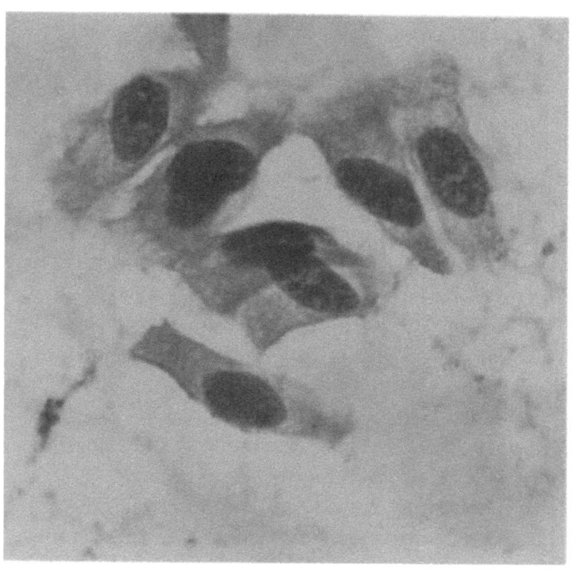

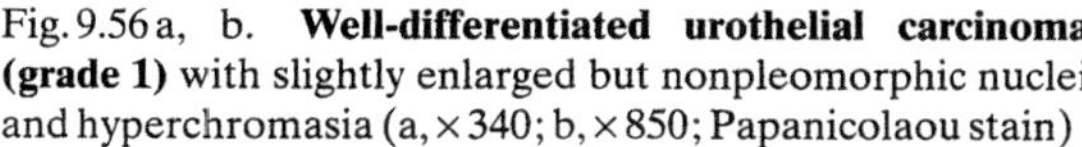

Fig. 9.56 a, b. **Well-differentiated urothelial carcinoma (grade 1)** with slightly enlarged but nonpleomorphic nuclei and hyperchromasia (a, ×340; b, ×850; Papanicolaou stain)

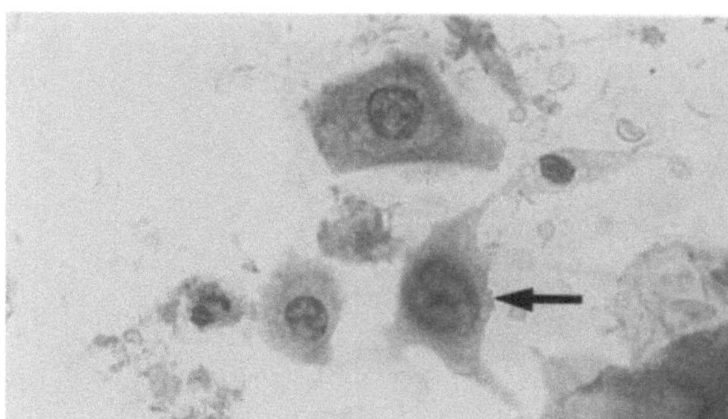

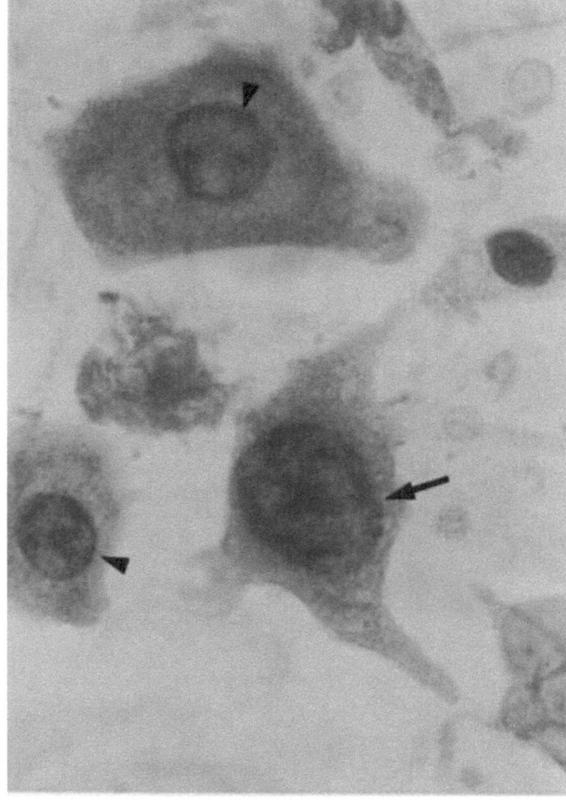

Fig. 9.57 a, b. **Reactive urothelial changes in lithiasis** (mild dysplasia). The cells exhibit nuclear enlargement (← ←) with slight prominence of the nuclear membrane (◀◀). This finding is cytologically indistinguishable from well-differentiated carcinoma (a, × 340; b, × 850; Papanicolaou stain)

9.6 Irrigation Cytology of the Upper Urinary Tract

The evaluation of urothelial cells from the upper urinary tract, usually collected by irrigation cytology, is made *difficult* by the reactive cell changes induced by the manipulation. Frequently the cells present the characteristics of highly differentiated urothelial tumor cells, creating a potential for overgrading or a false-positive interpretation. If the *diagnosis of well-differentiated bladder tumors* by urinary cytology is already uncertain, diagnosis by *retrograde irrigation cytology is virtually impossible* based on the lack of reliable differentiating criteria for reactive changes (Koss 1981).

A *multistep technique* of specimen collection is recommended to improve diagnostic certainty (see Sect. 8.3.2, and Fig. 8.9). In this technique the examiner collects the washings themselves (using an isotonic solution to avoid additional hypo- or hyperosmolar cell changes) after first obtaining a sample of bladder urine and a sample from the retrograde-catheterized urothelial segment before starting the irrigation. Especially with questionable well- to moderately differentiated neoplastic changes, comparison of the actual washings with the two preliminary urine fractions can be helpful in establishing a possible reactive, manipulative etiology for the cellular changes.

Normal Cytologic Findings

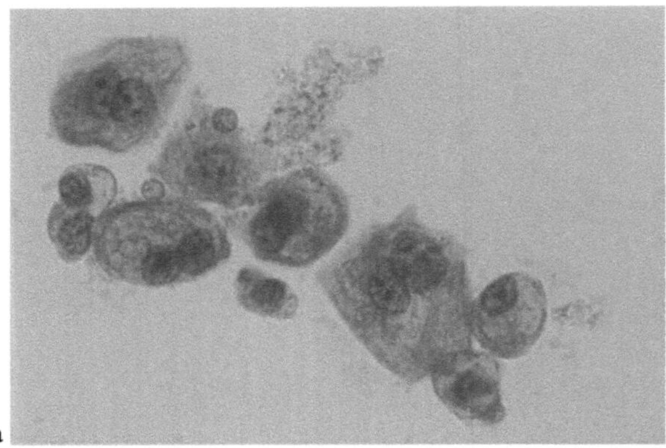

a

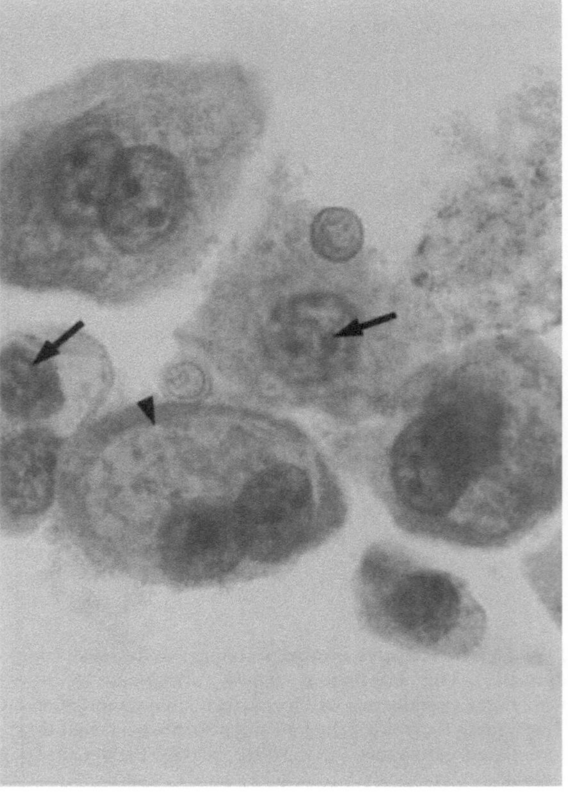

b

Fig. 9.58 a, b. **Normal urothelial cells in upper urinary tract washings.** The slight prominence of the nuclear membrane and especially of the nucleoli (← ←) is a normal response to the irritation of the collection procedure. The slightly vacuolated cytoplasm (◄) is also typical (a, × 340; b, × 850; Papanicolaou stain)

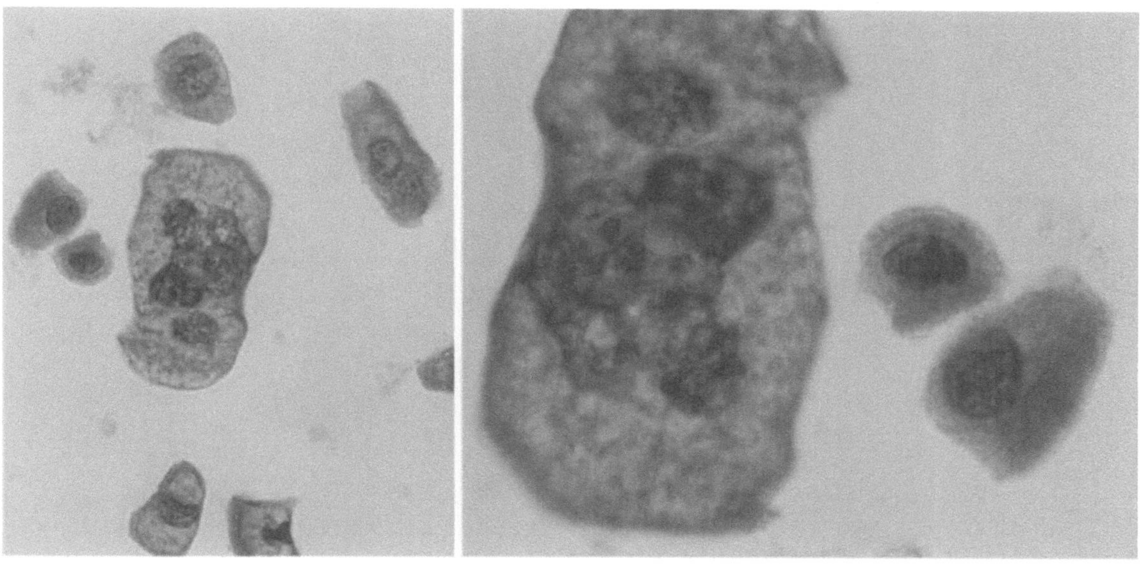

a

b

Fig. 9.59 a, b. **Multinucleated giant cell following retrograde catheterization.** These giant forms of umbrella cells, which may contain up to several dozen benign nuclei, are a relatively common finding. They have no pathologic significance (a, × 340; b, × 850; Papanicolaou stain)

Suspicious Cytologic Findings

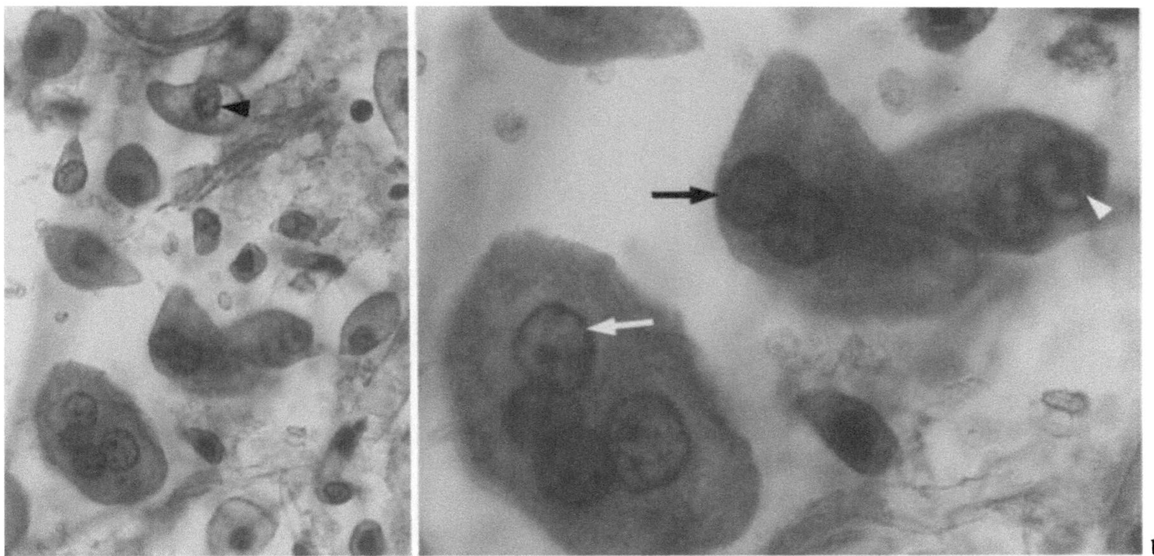

a

b

Fig. 9.60 a, b. **Changes simulating a well-differentiated tumor after retrograde irrigation.** Besides slight coarse granularity of the chromatin (◀ ◀), there is conspicuous thickening of the nuclear membranes (← ←). Further diagnostic procedures did not confirm the suspicion of neoplasia (a, × 340; b, × 850; Papanicolaou stain)

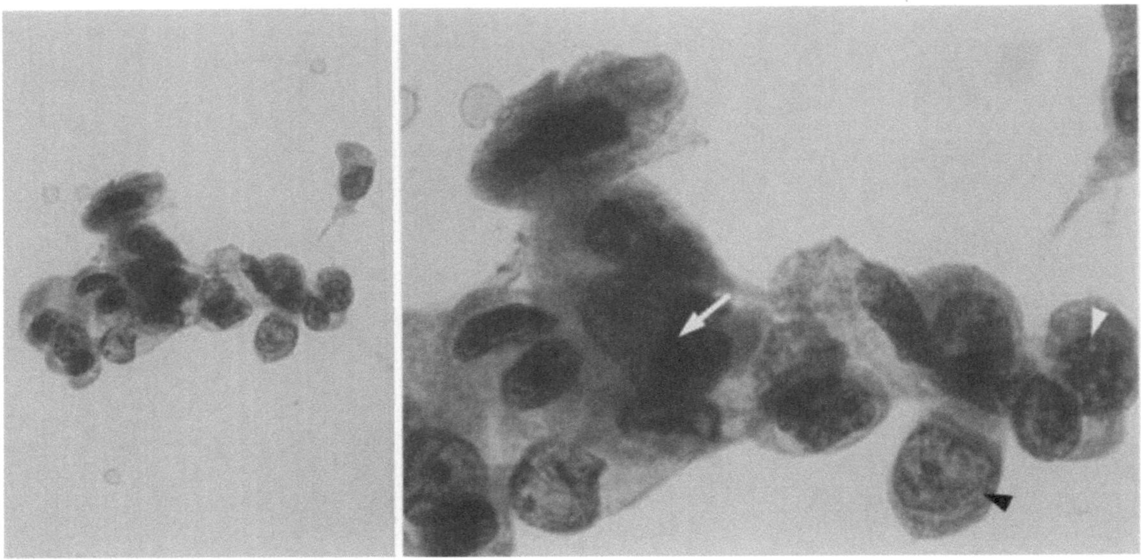

Fig. 9.61 a, b. **Tumor-mimicking changes after retrograde irrigation.** The marked nuclear enlargement (unfavorable nuclear–cytoplasmic ratio; ◄ ◄) with mild hyperchromasia raised suspicion of a well to moderately differentiated carcinoma, but further diagnostic procedures did not confirm this. The "apparent hyperchromasia" at the center of the cell cluster (←) is an overlap effect (a, × 340; b, × 850; Papanicolaou stain)

Positive Cytologic Findings

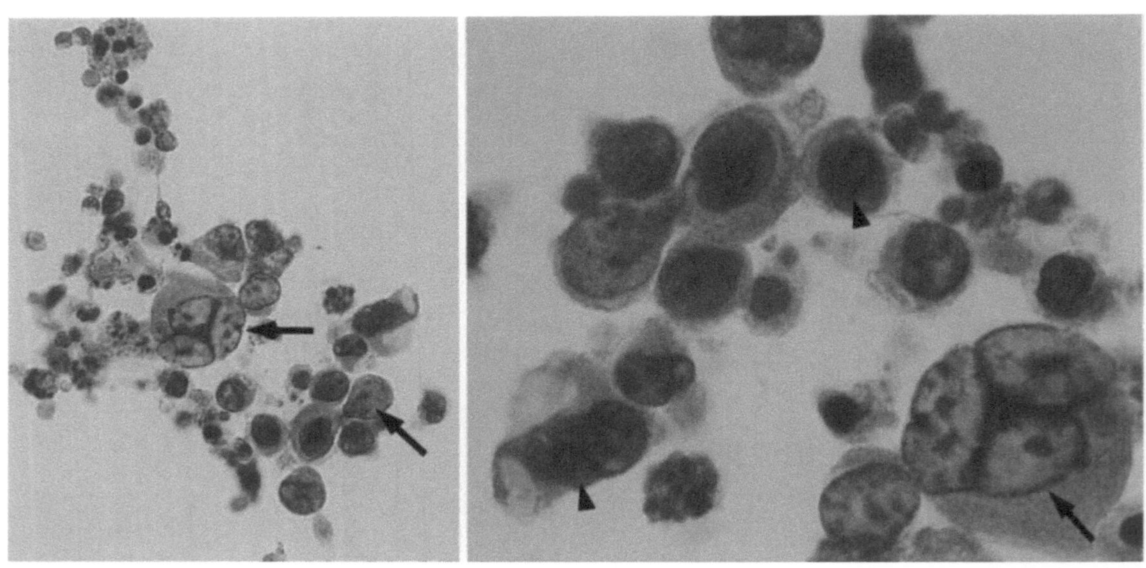

Fig. 9.62 a, b. **Poorly differentiated carcinoma of the renal pelvis (grade 3).** Besides greatly enlarged nuclei (← ←) with prominent nucleoli, there are nuclei with markedly pathologic hyperchomasia (◄ ◄). Changes of this magnitude cannot be explained in terms of an irrigation response or irritation by other external agents (e.g., a stone) and must have a malignant etiology (a, × 340; b, × 850; Papanicolaou stain)

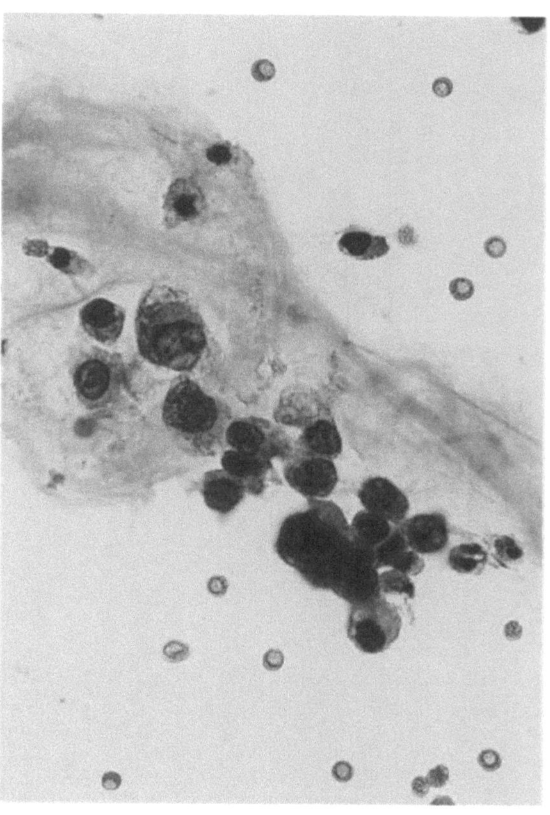

Fig. 9.63. **Poorly differentiated carcinoma of the ureteral remnant (grade 3)** (×340; Papanicolaou stain)

9.7 Cytologic Evaluation of Treatment Response in Urothelial Tumor Patients

9.7.1 Urine Cytology After TUR and Laser Vaporization

Urinary cytology is widely recognized as an *essential component* in the postoperative surveillance of urothelial carcinomas (Rübben et al. 1989). *Postoperative urinary cytology is more sensitive than multiple intraoperative biopsies* for diagnosing *carcinoma in situ* coexistent with an exophytic bladder tumor (Harving et al. 1988). Although useful results can be achieved as early as 3 days postoperatively, despite the reactive degenerative cell changes induced by the surgery (Müller et al. 1985), it is prudent to *defer urinary cytology for at least 7 days* in order to minimize false-positive findings.

Normal Postoperative Findings

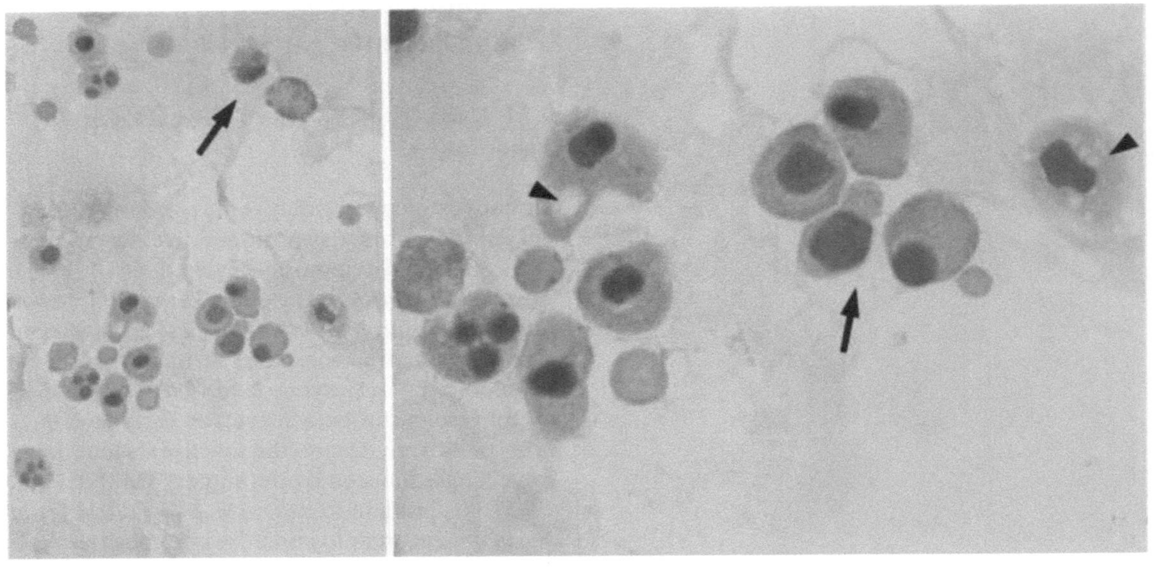

Fig. 9.64 a, b. **Reactive cellular changes 8 days after TUR.** Besides segmented leukocytes and intact erythrocytes, the specimen contains urothelial cells with dense, slightly enlarged nuclei (← ←). These nuclei do not show typical coarsely granular hyperchromasia, however, but appear homogeneous. The vacuolated cytoplasm (◄ ◄) is attributed to residual degeneration (a, × 340, b, × 850; Papanicolaou stain)

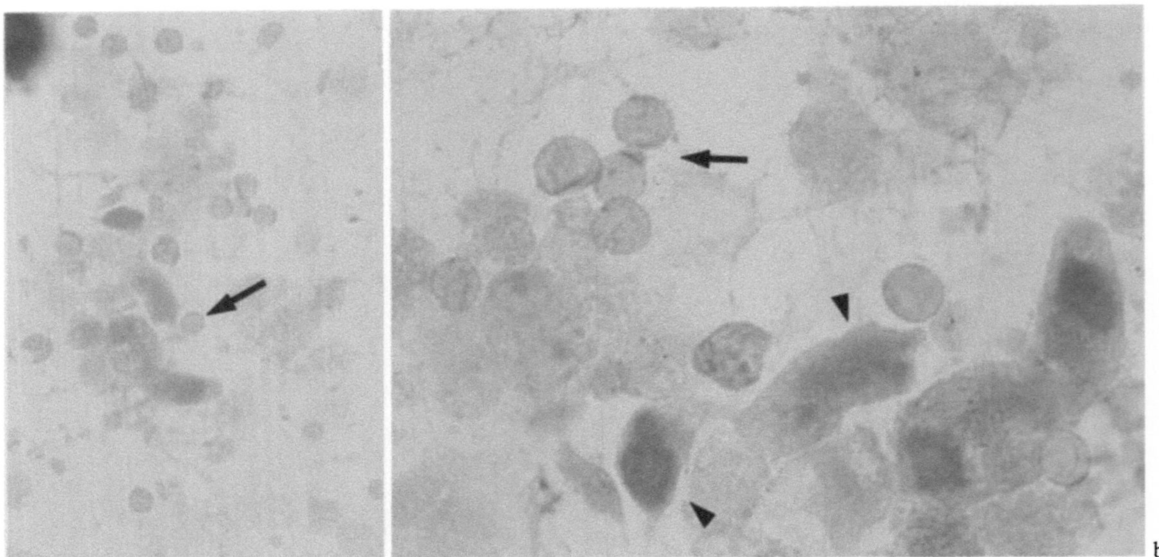

Fig. 9.65 a, b. **Reactive cellular changes 8 days after laser vaporization.** The intact erythrocytes (← ←) prove that the specimen has been satisfactorily preserved, despite its generally lytic appearance. The nuclear enlargement and pleomorphism (◄ ◄) are reactive alterations (a, × 340; b, × 850; Papanicolaou stain)

Suspicious Postoperative Findings

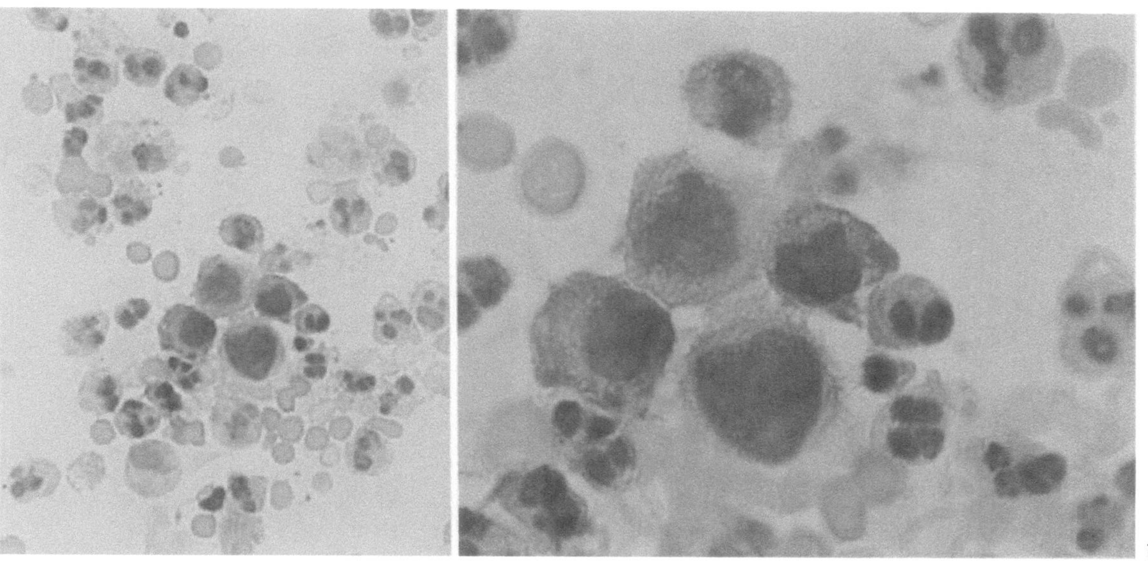

a   b

Fig. 9.66 a, b. **Suspicious urothelial cells 8 days after bladder tumor resection (grade 2).** With the markedly altered nuclear–cytoplasmic ratio and nuclear pleomorphism, it is unclear cytologically whether we are dealing with resection-induced degeneration or persistent neoplasia. At least a short-term follow-up should be maintained in these cases (a, × 340; b, × 850; Papanicolaou stain)

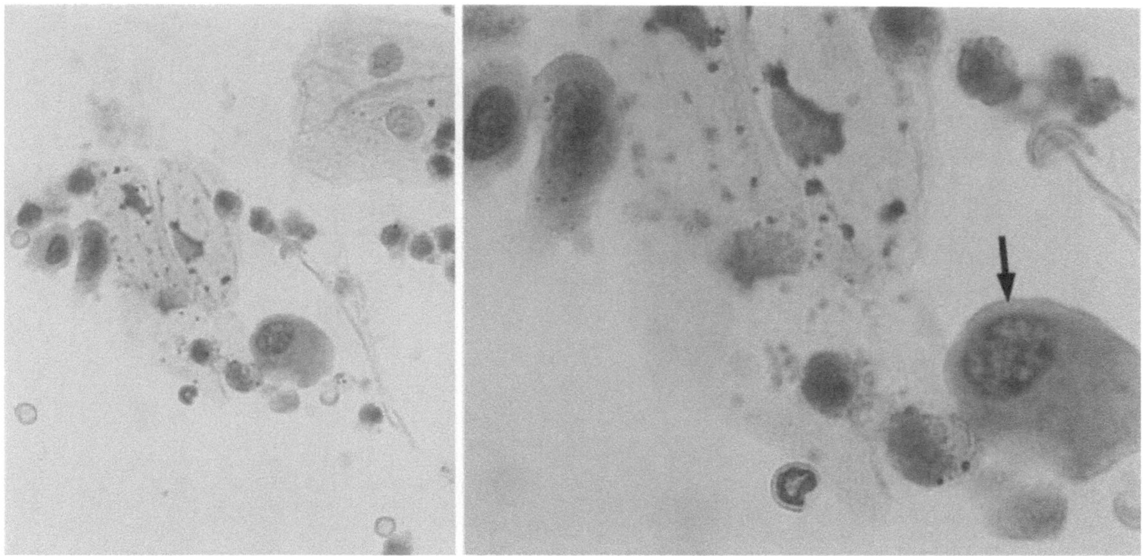

a   b

Fig. 9.67 a, b. **Suspicious urothelial cells 14 days after bladder tumor resection (grade 2).** Besides the prominent nuclear membrane and the increased, coarsely granular chromatin (←), the absence of other degenerative residua (e.g., cytoplasmic vacuoles) warrants classifying the cells as suspicious (a, × 340; b, × 850; Papanicolaou stain)

Positive Postoperative Findings

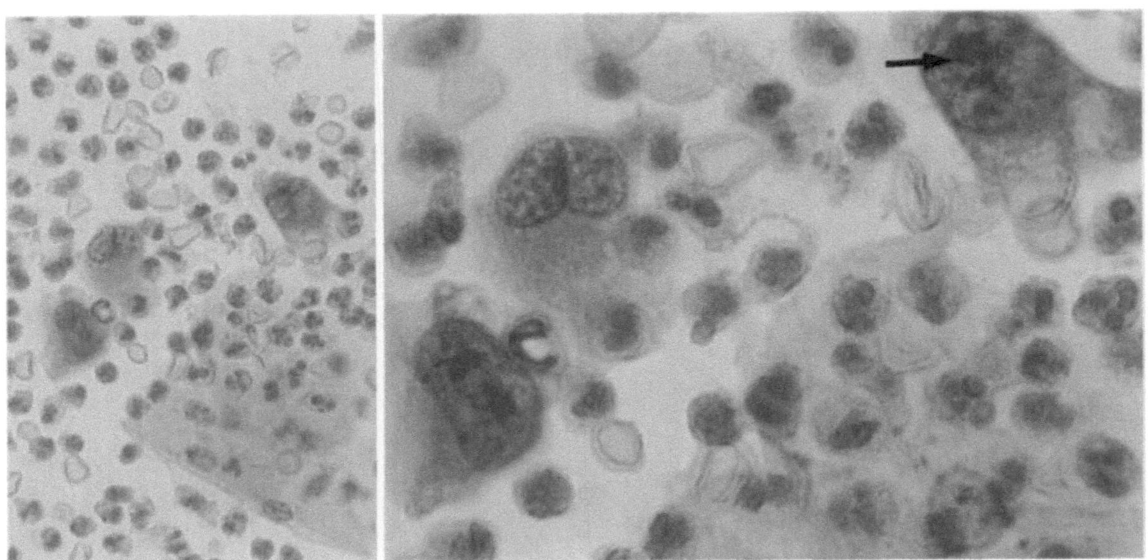

Fig. 9.68 a, b. **Incompletely resected urothelial carcinoma (grade 2).** Eight days after TUR, the typical leukocytic response is accompanied by pathologic urothelial cells with extreme nuclear enlargement and large nucleoli (←). The absence of bacteriuria rules out the possibility of an infectious process inciting reactive cell changes (a, × 340; b, × 850; Papanicolaou stain)

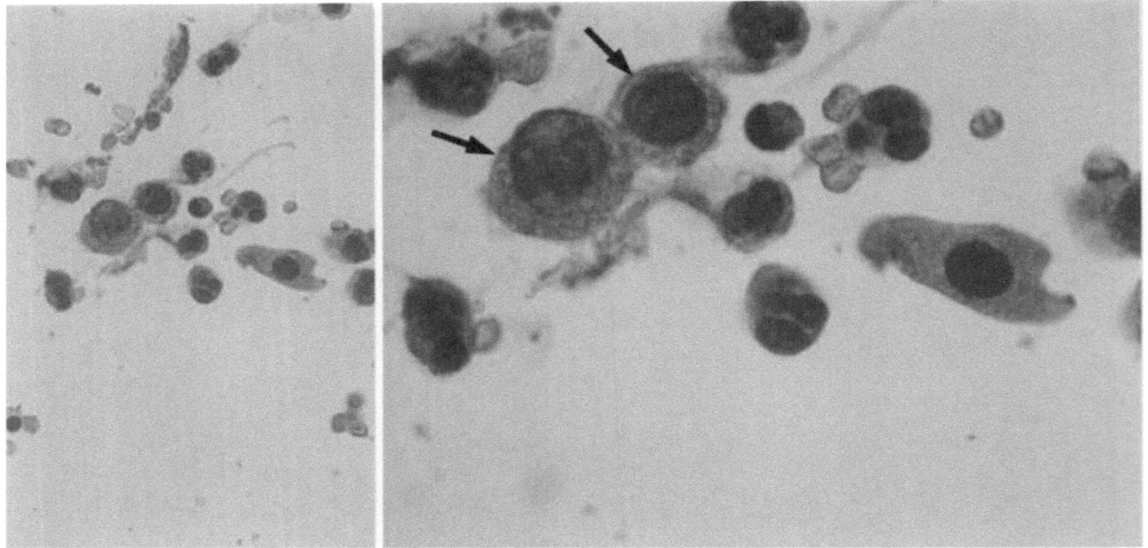

Fig. 9.69 a, b. **Incompletely resected urothelial carcinoma (grade 2).** This specimen shows markedly enlarged nuclei with coarsely granular chromatin (← ←) (a, ×340; b, ×850; Papanicolaou stain)

9.7.2 Urethral Lavage After Cystectomy

Based on the *4%–18% incidence of carcinoma of the urethral remnant* following radical cystectomy and the inadequate sensitivity of endoscopic diagnosis (see references in Chap. 2), urethral lavage has become an essential part of the postoperative follow-up of these cases. This surveillance must be maintained *indefinitely*, since urethral stump recurrences can develop more than 10 years after the cystectomy (Stöckle et al. 1990).

Technically, it is better to catheterize the urethra with a thin-gauge catheter and then irrigate with an isotonic solution rather than attempt to irrigate through an olive placed against the urethral meatus (see Chap. 8).

The interpretation of urinary cytologic washings also *requires some experience*. This is due both to the reactive cell changes incited by the irrigation and to the fact that cellular atrophy or degeneration appears to result from the disruption in the passage of urine.

Normal Urethral Washings

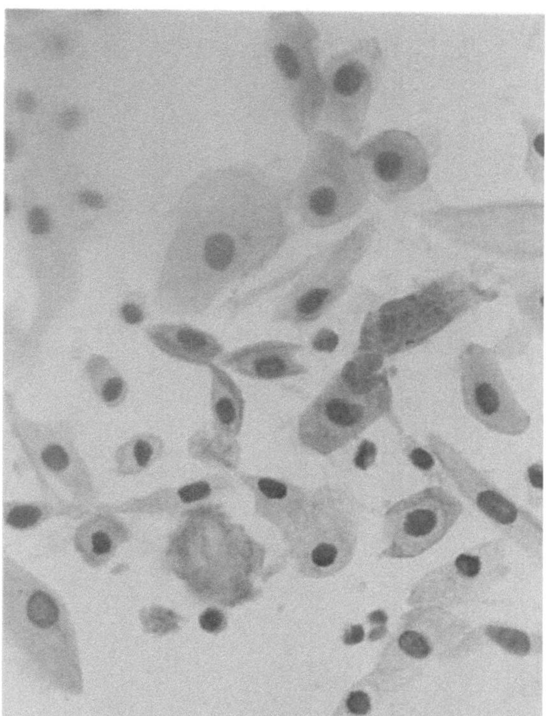

a

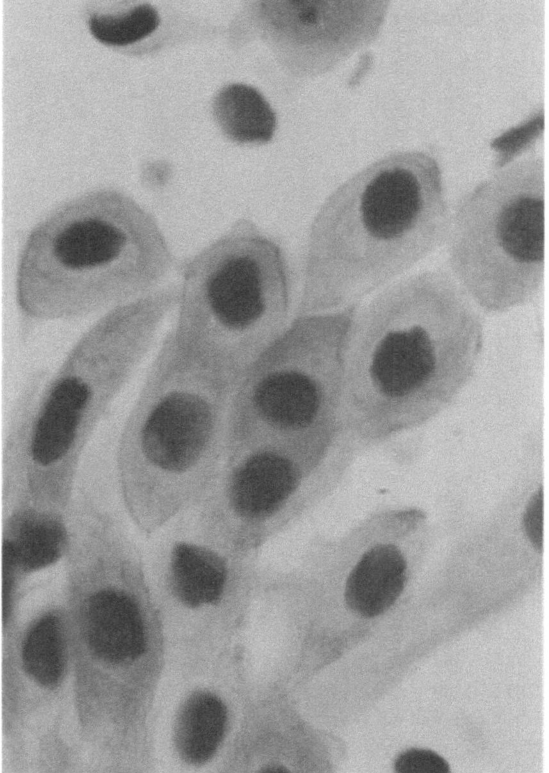

b

Figs. 9.70 a,b and 9.71a,b. **Normal urethral washings.** A typical feature besides the cellularity of the specimen is the chromatin-dense appearance of the nuclei without coarse granularity (Figs. 9.70 a and 9.71a, × 340; Figs. 9.70 b and 9.71 b, × 850; Papanicolaou stain)

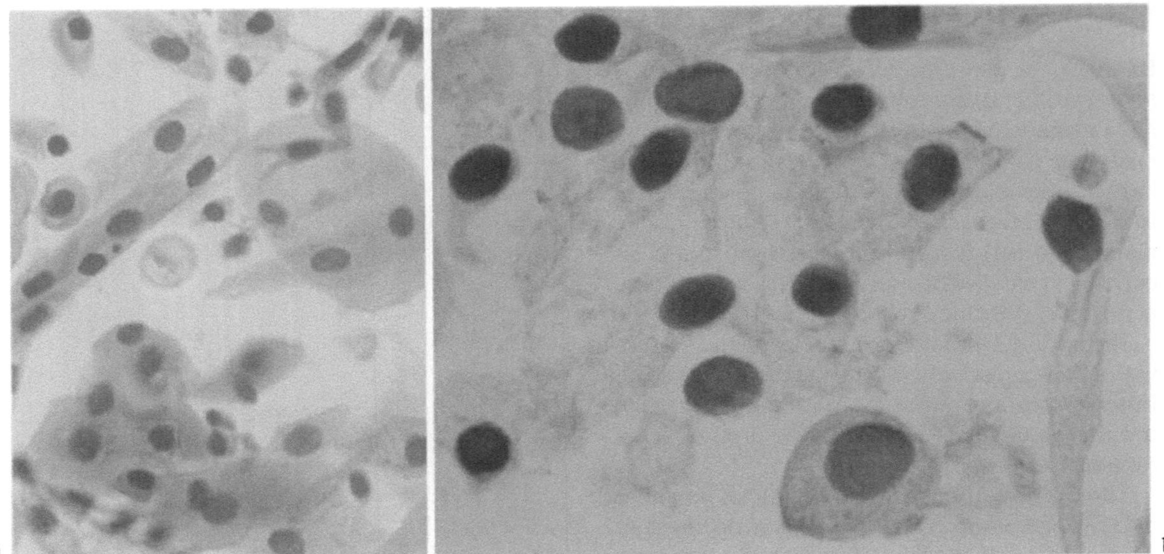

9.71

Suspicious Urethral Washings

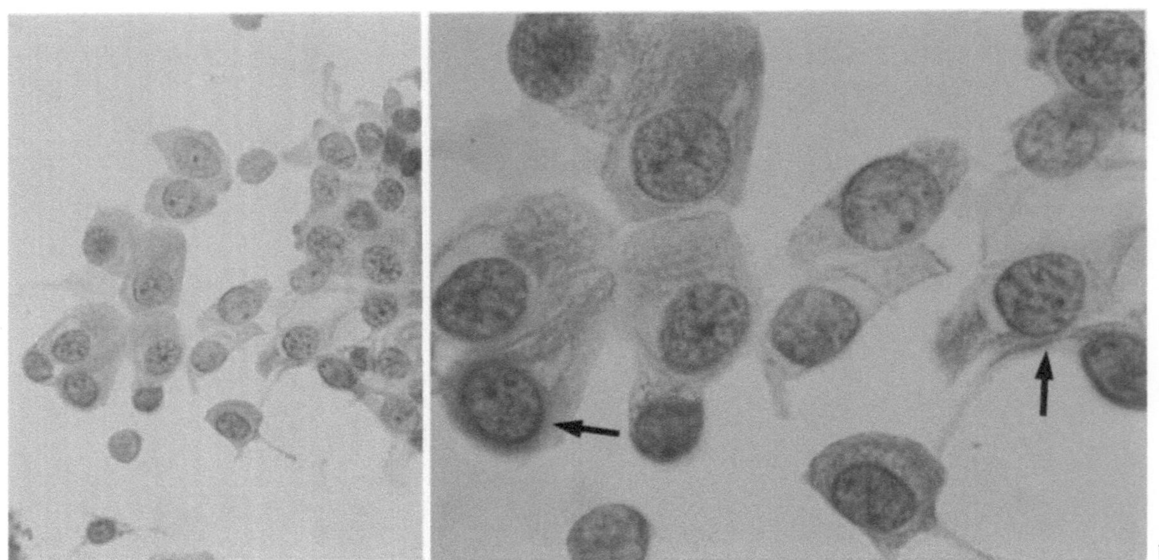

Fig. 9.72 a, b. **Cytologically suspicious urethral washings.** Very cellular specimen with enlarged nuclei and slightly prominent chromatin (← ←). Endoscopy revealed papillary polyps/tumors, but the patient refused operative treatment (a, × 340; b, × 850; Papanicolaou stain).

Positive Urethral Washings

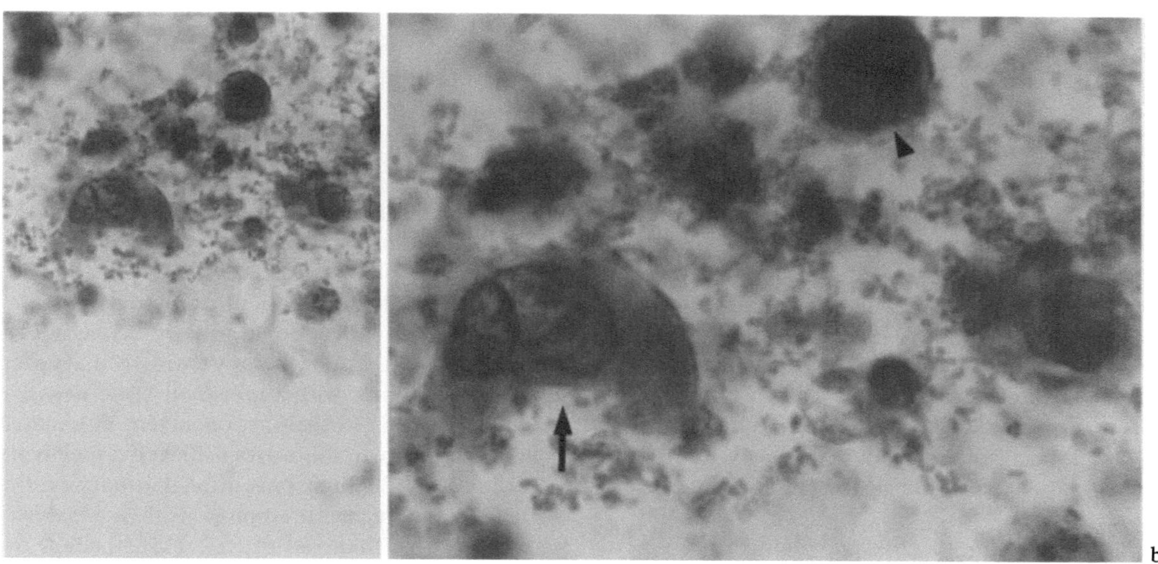

a

b

Fig. 9.73 a, b. **Poorly differentiated recurrent carcinoma of the urethral remnant (grade 3).** Besides massive bacterial infection, there are tumor cells (◀) showing extreme nuclear enlargement (←) and hyperchromasia (a, × 340; b, × 850; Papanicolaou stain)

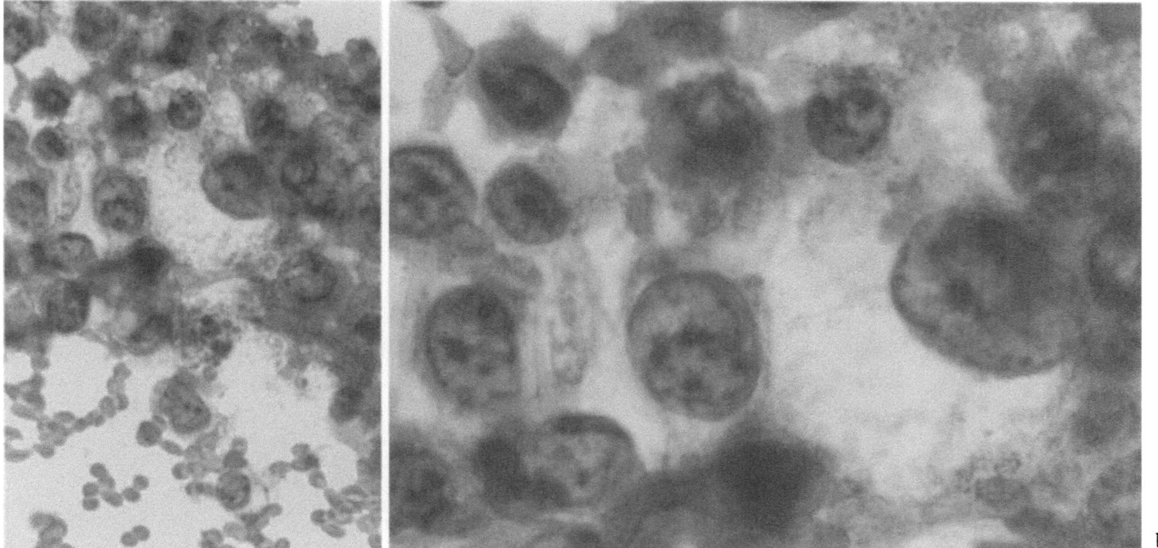

a

b

Fig. 9.74 a, b. **Moderately differentiated recurrent carcinoma of the urethral remnant (grade 2)** shows marked nuclear enlargement with very prominent nucleoli (a, × 340; b, ×850; Papanicolaou stain)

9.7.3 Urine Cytology During Intravesical Chemo- and Immunotherapy

The intravesical instillation of cytostatic drugs and/or immunotherapeutic agents (BCG) can evoke marked *cytotoxic reactions* which alter cellular morphology and can significantly increase the difficulty of cytodiagnosis. The reactive changes provoked by these therapies affect not only the cytoplasm but also the nucleus, which of course is critical for the diagnosis of malignancy (Roth and Rathert 1989). This fact *qualifies the significance* of the *classic malignancy criteria* of urinary cytology such as an altered nuclear–cytoplasmic ratio and nuclear pleomorphism. An excessive false-positive rate can be avoided by becoming familiar with relevant aspects of the patient's history and taking them into account during the cytologic interpretation.

Although the reactive cell changes are reversible, the time factor is not precisely known. For practical purposes, we recommend performing the *cytologic examination about 4 weeks after the instillation*. In this way the specimen is collected immediately before the next instillation when the usual schedule of instillation cycles is followed.

Suspicious or positive urinary cytologic findings are an indication for:

▶ Short-term endoscopic and biopsy follow-ups
▶ Uroradiologic evaluation of the upper urinary tract
▶ A change of chemotherapeutic or immunotherapeutic agent

Despite problems of interpretation, the *efficacy* of cytologic follow-up in this setting is *well documented*. For example, Bretton et al. (1989) showed in a follow-up study of 65 patients treated by BCG instillation that conventional urinary cytology performed 3 months after therapy correctly diagnosed absence of tumor with a *specificity* (true-negative findings) of 81% (29/36). Concurrent automated flow cytometry in this series achieved a specificity of only 56% (20/36). Conventional urinary cytology detected recurrent tumors with a *sensitivity* (true-positive findings) of 55% (19/29). Flow cytometry was superior to conventional cytology in this regard, achieving a sensitivity of 69% (20/29), but this is a highly specialized procedure that is not used for primary diagnosis. Moreover, the correct identification of tumor absence (specificity) is as important as the detection of recurrence in terms of avoiding unnecessary invasive diagnostic procedures.

Normal Reactive Urothelial Changes

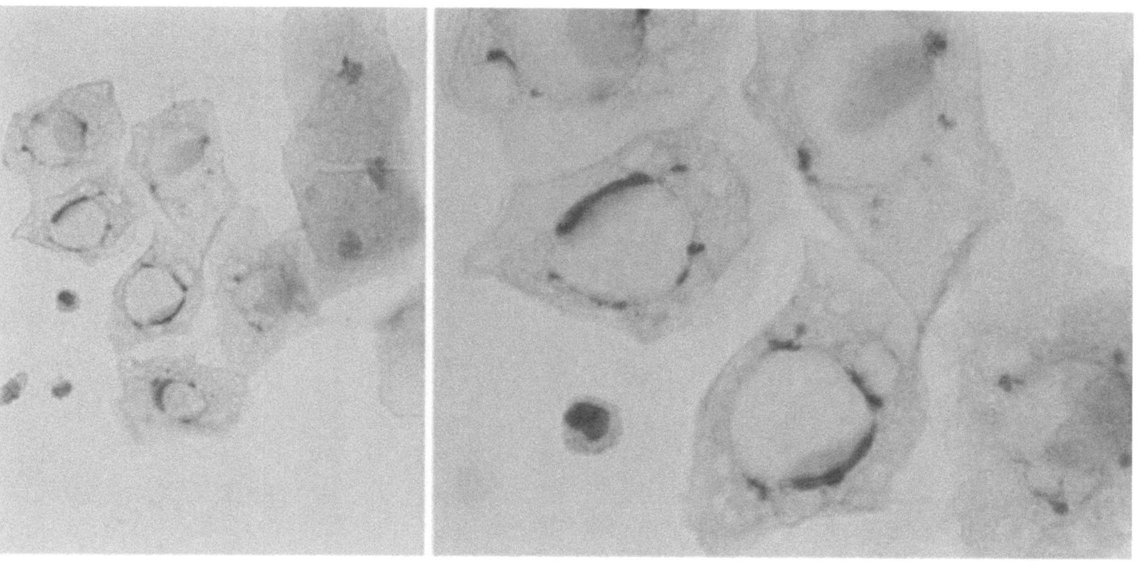

a   b

Fig. 9.75 a, b. **Reactive vacuolation** of the cytoplasm and nuclei secondary to the instillation of mitomycin (a, × 340;
b, × 850; Papanicolaou stain)

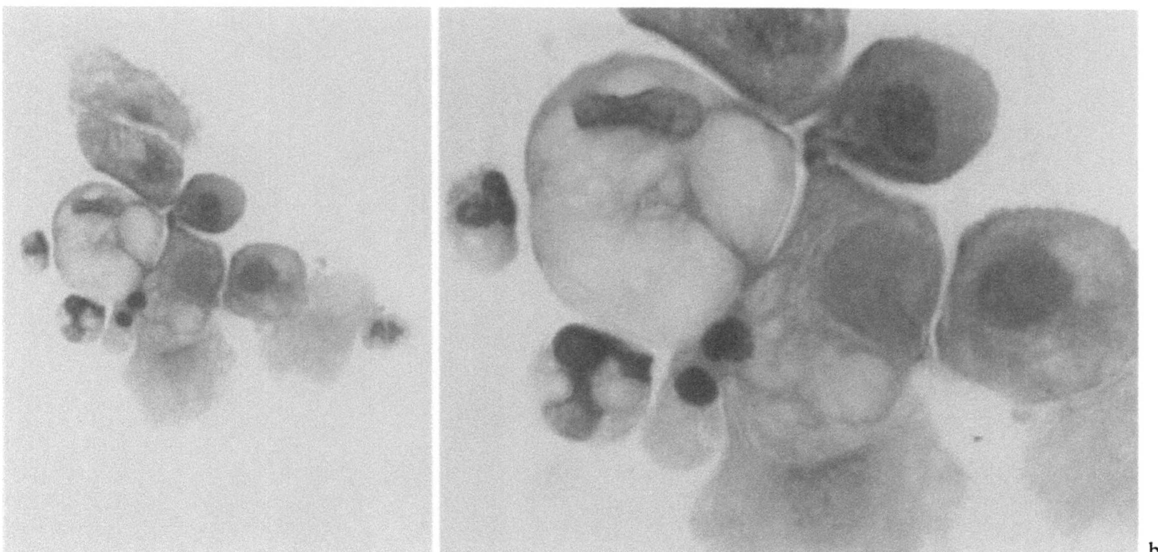

a   b

Fig. 9.76 a, b. **Reactive cytotoxic cell degeneration** during
Adriablastin instillation. There is vacuolation in addition to
nuclear pleomorphism, which should not be interpreted as a
sign of malignancy (a, × 340; b, × 850; Papanicolaou stain)

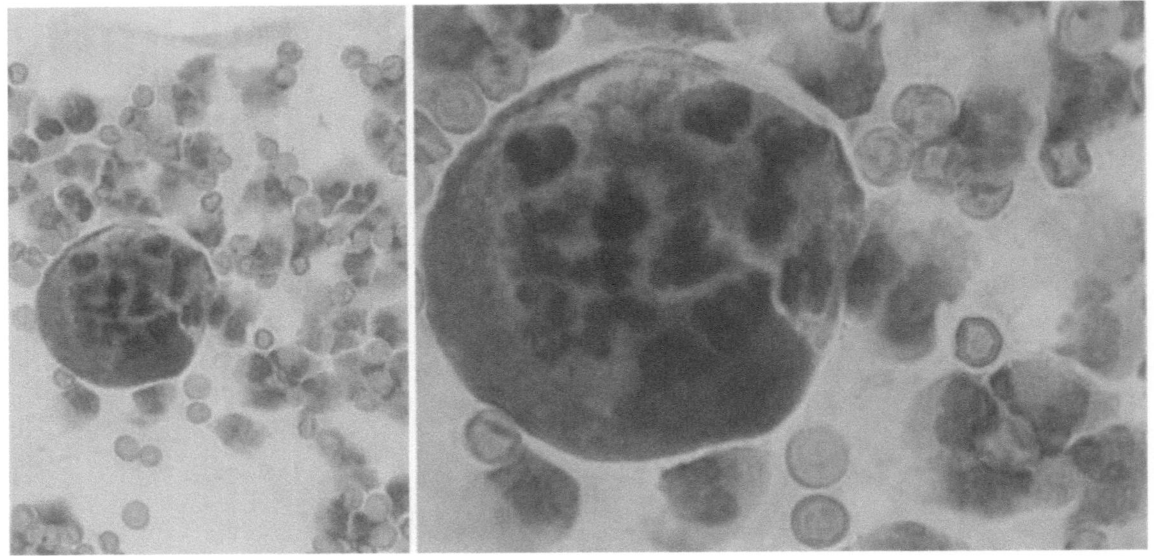

a   b

Fig. 9.77 a, b. **Cytotoxic nuclear degeneration** secondary to BCG instillation ("effective cytostasis"). The nucleus is not hyperchromatic but shows marked lytic disruption. An intact nuclear membrane is no longer present. There is also the typical chemocystitic accompaniment of "sterile leukocyturia" and hematuria (a, × 340; b, × 850; Papanicolaou stain)

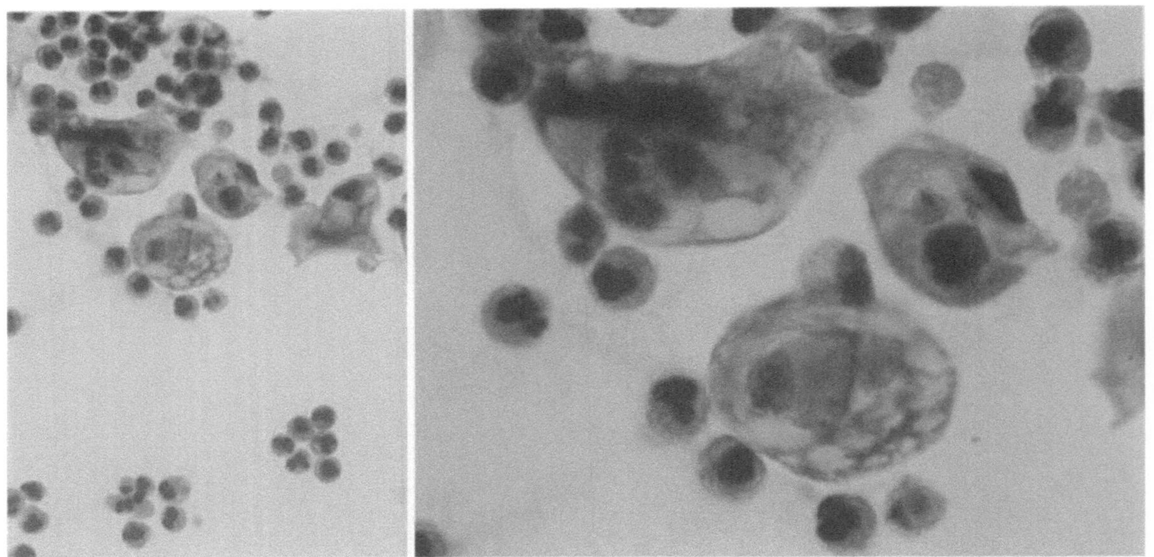

a   b

Fig. 9.78 a, b. **Cytotoxic degeneration of tumor cells** following BCG therapy for carcinoma in situ ("effective cytostasis") (a, × 340; b, × 850; Papanicolaou stain)

Figs. 9.80 and 9.81. **Cytologically suspicious findings during instillation therapy.** Besides the marked destruction of urothelial cells consistent with a morphologically effective cytostasis (← ←), there are tumor-suspicious cells that are largely intact (◄ ◄) (× 850; Papanicolaou stain) ▷

Suspicious Urothelial Changes

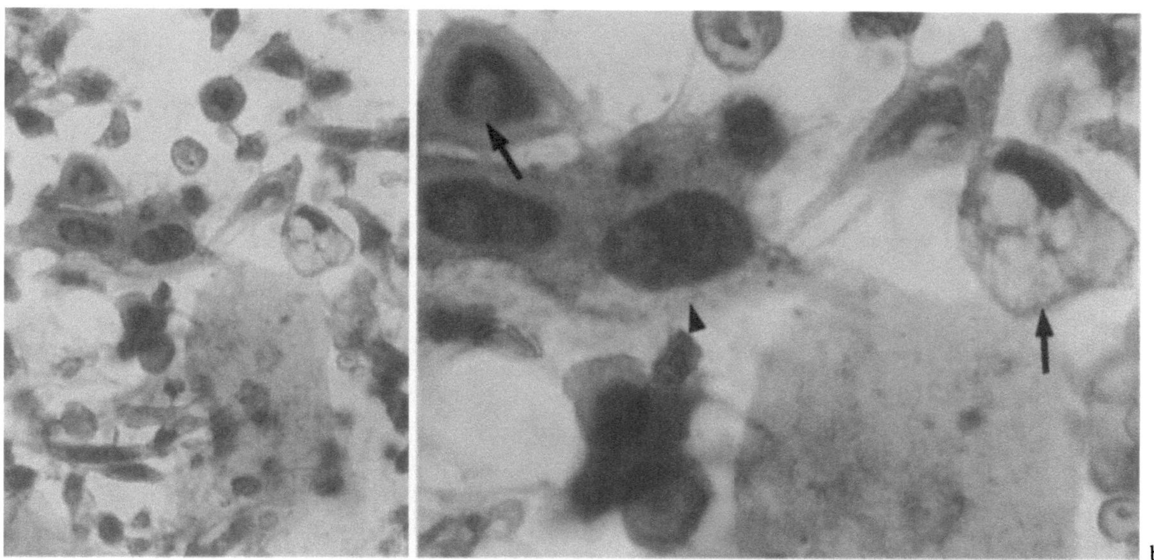

a   b

Fig. 9.79 a, b. **Cytologically suspicious finding** in a patient treated with mitomycin. The very cellular irrigation specimen contains degenerative urothelial cells (← ←) along with cells showing suspicious hyperchromasia (◀) with no cytotoxic alterations (a,× 340; b, x 850; Papanicolaou stain)

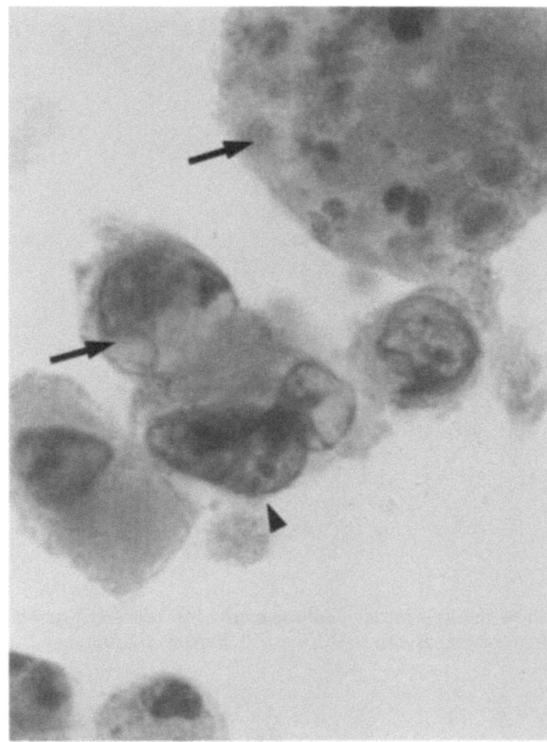

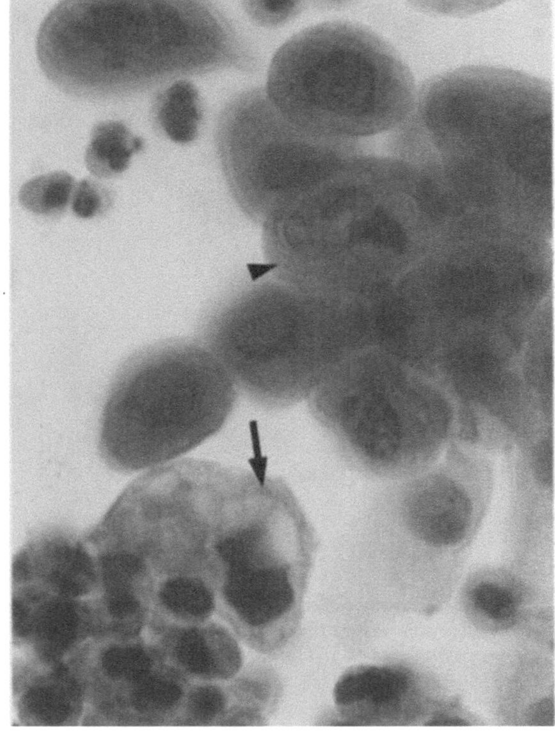

9.80   9.81

Recurrence of Urothelial Carcinoma

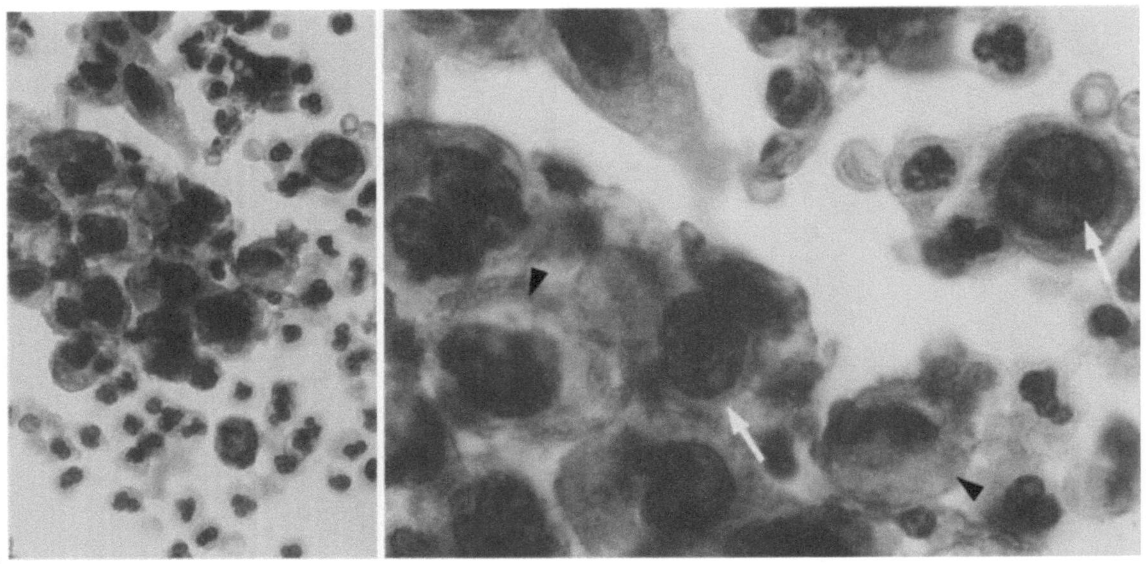

Fig. 9.82 a, b. **Persistent carcinoma in situ** during BCG therapy. Despite the slight cytotoxic vacuolation (◀ ◀), the majority of nuclei are intact and display chromatin clumping and condensation (← ←) (a, × 340; b, × 850; Papanicolaou stain)

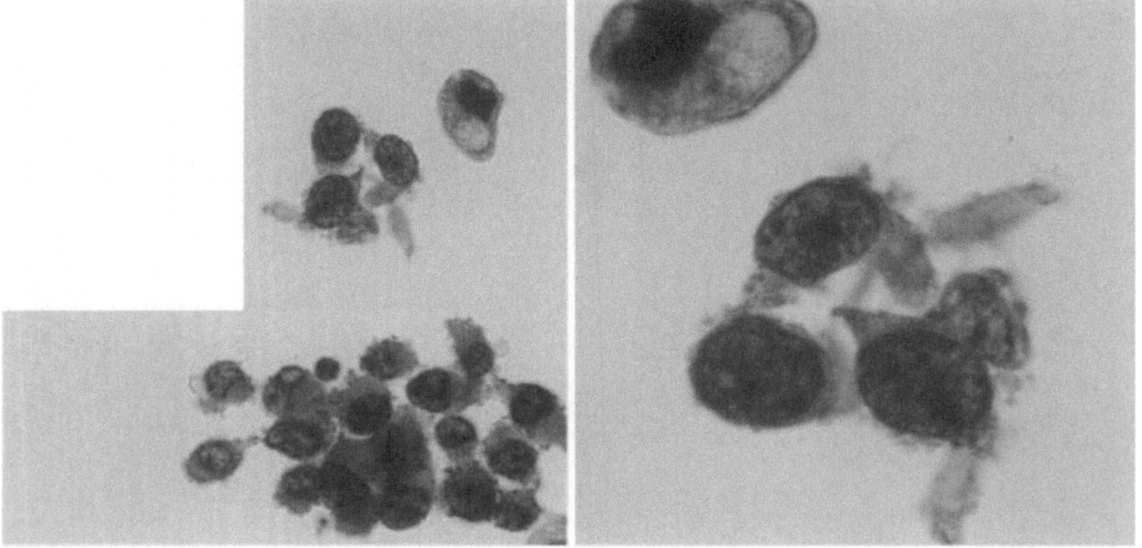

Fig. 9.83 a, b. **Poorly differentiated carcinoma (recurrent)** 8 weeks after mitomycin therapy. There is no chemocystitic accompanying reaction. The largely intact urothelial cells show marked nuclear enlargement with coarsely granular hyperchromasia (a, × 340; b, × 850; Papanicolaou stain)

9.7.4 Urine Cytology After Radiation and Systemic Chemotherapy

Normal Reactive Findings

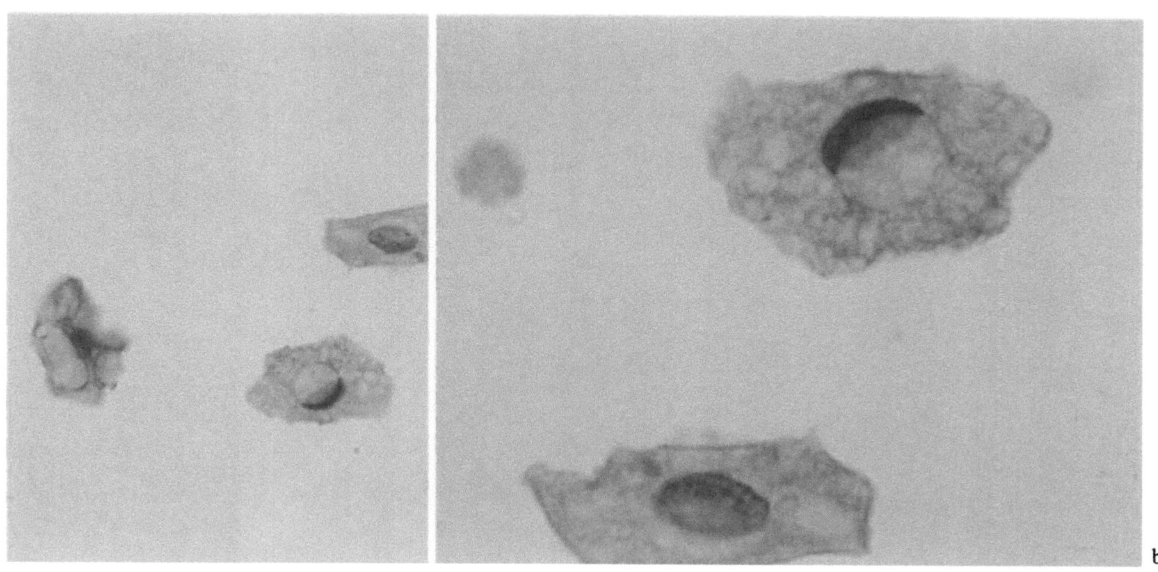

a   b

Fig. 9.84 a, b. **Degenerative urothelial cells following radiotherapy**, no evidence of malignancy. The radiation-induced changes present chiefly as cytoplasmic and nuclear vacuolation. These changes may persist for years (a, × 340; b × 850; Papanicolaou stain)

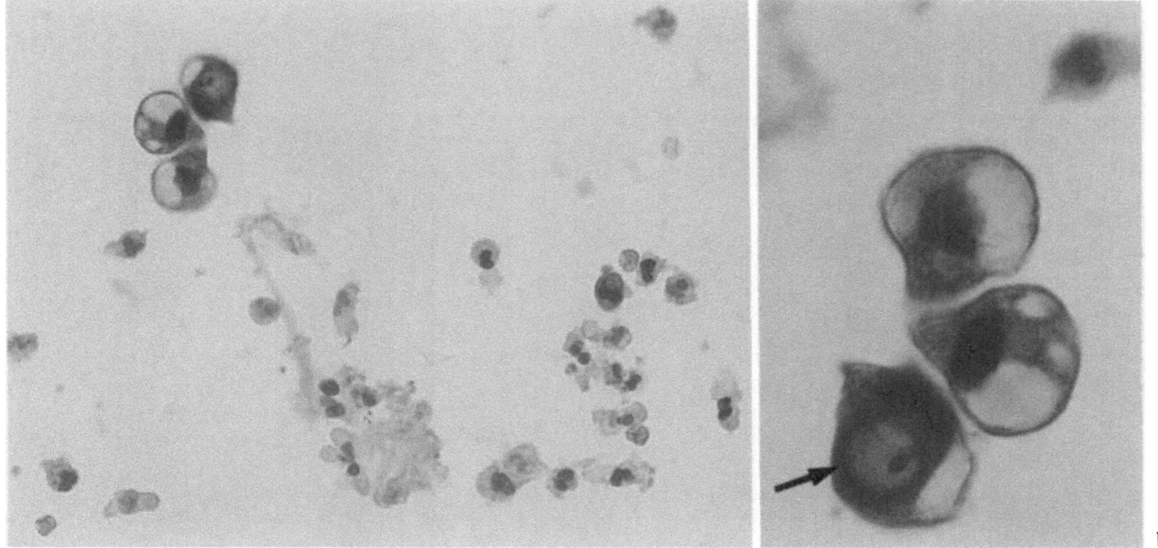

a   b

Fig. 9.85 a, b. **Radiogenic degeneration of urothelial cells** with cytoplasmic and nuclear (←) vacuolation (a, × 340; b, × 850; Papanicolaou stain)

Suspicious Findings

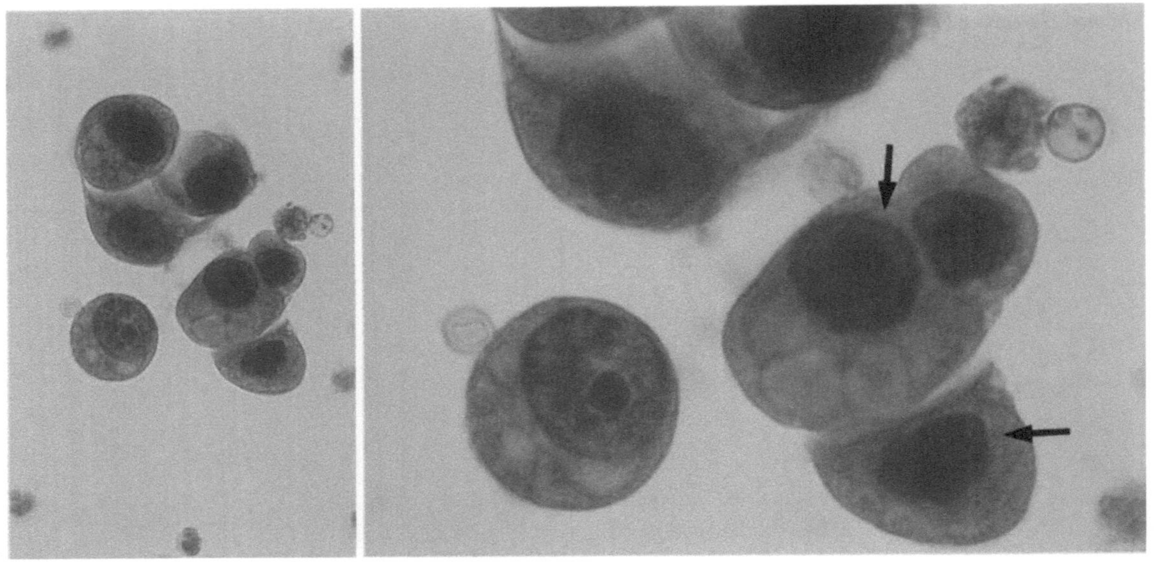

Fig. 9.86 a, b. **Tumor-suspicious urothelial changes** following radiochemotherapy of a recurrent muscle-infiltrating carcinoma. Despite the vacuolated cytoplasm, the nuclei do not have the uniformly dark appearance of devitalized nuclei overstained due to protein denaturation. Rather, the residual transparency of the nuclei (←) is suggestive of pathologic hyperchromasia (a, × 340; b, × 850; Papanicolaou stain)

Recurrence of Urothelial Carcinoma

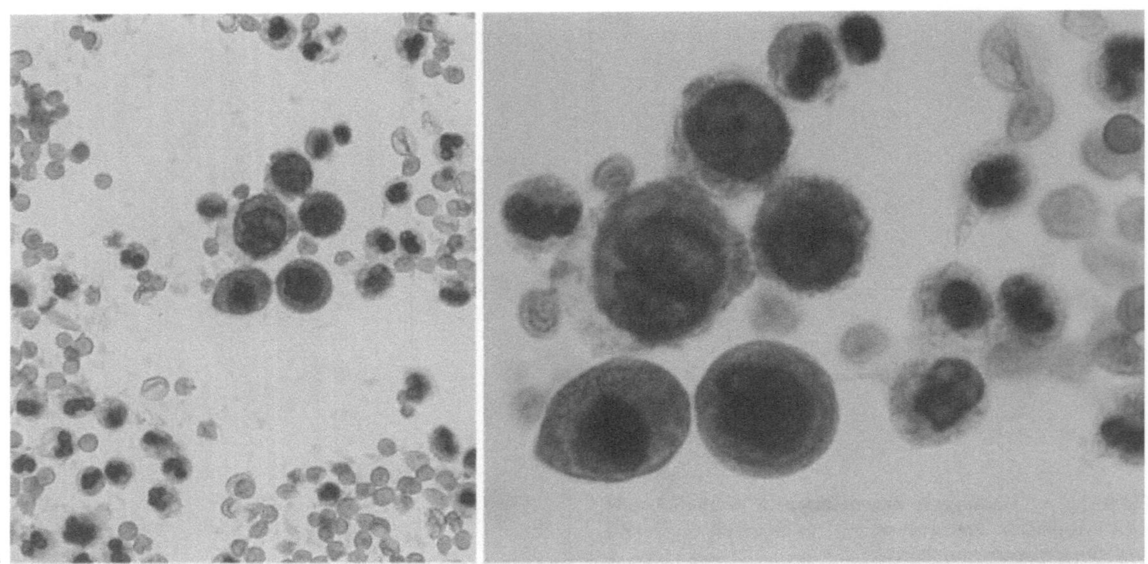

Fig. 9.87 a, b. **Persistent poorly differentiated urothelial carcinoma** during systemic M-VAC chemotherapy. The cells show no evidence of cytotoxic degeneration. The nuclei are markedly enlarged, pleomorphic, and hyperchromatic (a, × 340; b, × 850; Papanicolaou stain)

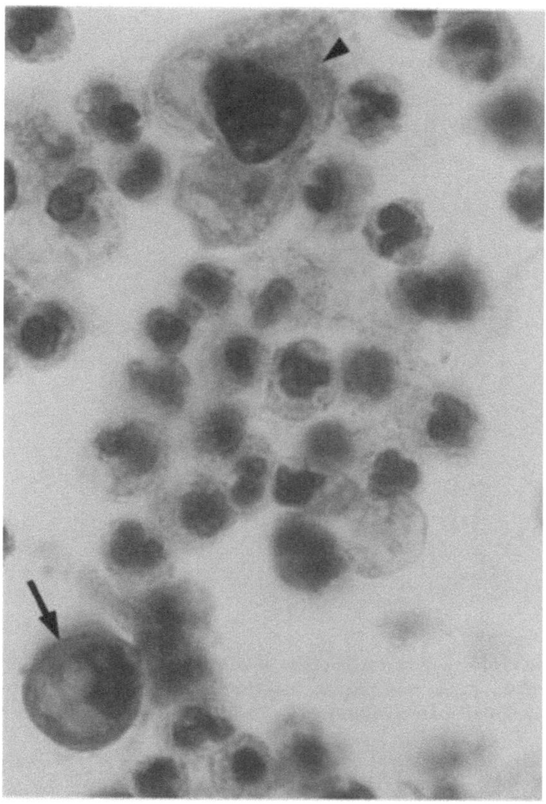

Fig. 9.88. **Recurrence of invasive bladder cancer (grade 2)** following a course of systemic chemotherapy (CISCA). One cell shows cytotoxic degeneration (← ←), and there is a largely intact tumor cell with a pathologic nucleus (◄) (×, 850; Papanicolaou stain)

9.7.5 Urine Cytology in Patients with an Ileal Conduit and Reconstructed Bladder

Because *urothelial carcinoma* is characterized as a *panurothelial tumor*, the upper urinary tract should be monitored for recurrent neoplasia following radical cystectomy with supravesical diversion and also after the construction of a substitute bladder. Although the presence of degeneratively altered intestinal cells makes the evaluation much more difficult, urinary cytology is still an essential component of aftercare, especially in patients who are no longer acceptable candidates for retrograde contrast radiography or endoscopy.

Normal Reactive Findings

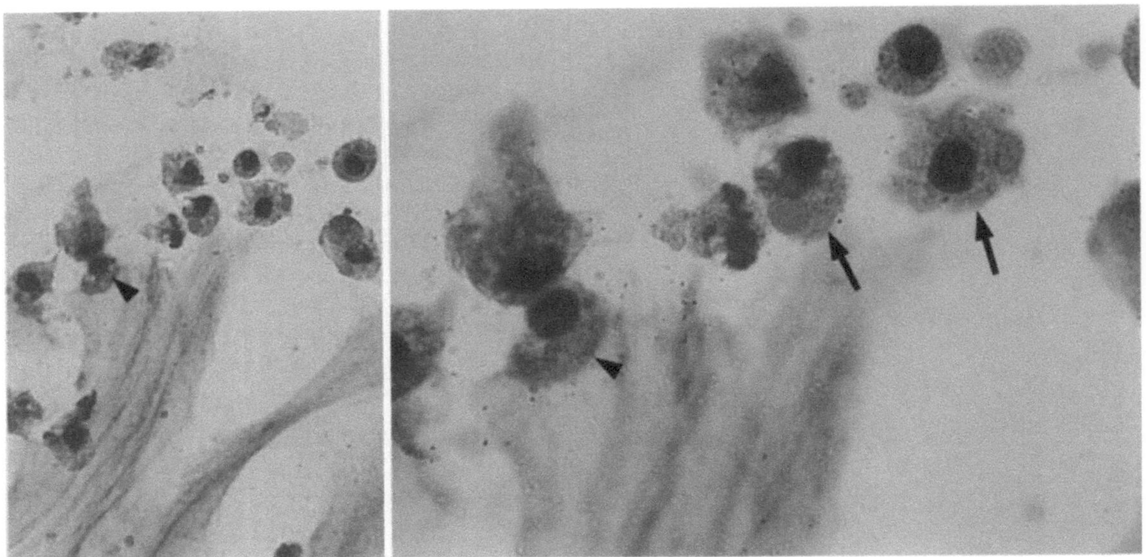

Fig. 9.89 a, b. **Typical normal degenerative changes** in ileal conduit urine. Besides transparent mucosal elements, the specimen contains degeneratively altered intestinal cells (← ←) presenting a characteristic rounded shape (Koss 1981). The degenerative cells contain eosinophilic inclusions (◄◄) as a nonspecific associated finding (a, × 340; b, × 850; Papanicolaou stain)

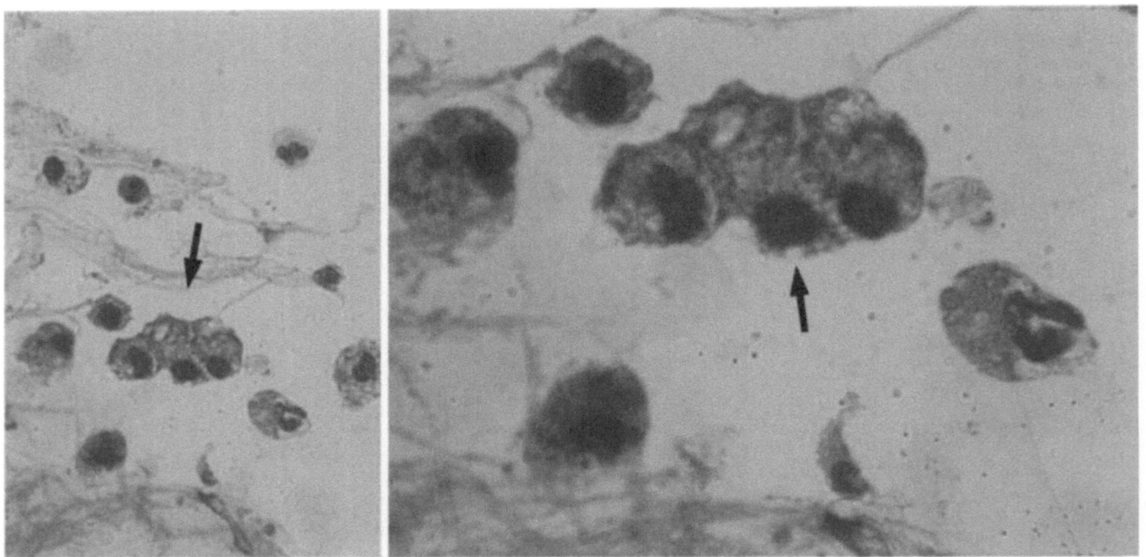

Fig. 9.90 a, b. **Normal findings in an ileal neobladder.** Degeneratively altered urothelial cells (← ←) with cytoplasmic vacuolation and increased nuclear density but no true hyperchromasia (a, × 340; b, × 850; Papanicolaou stain)

Recurrence of Urothelial Carcinoma

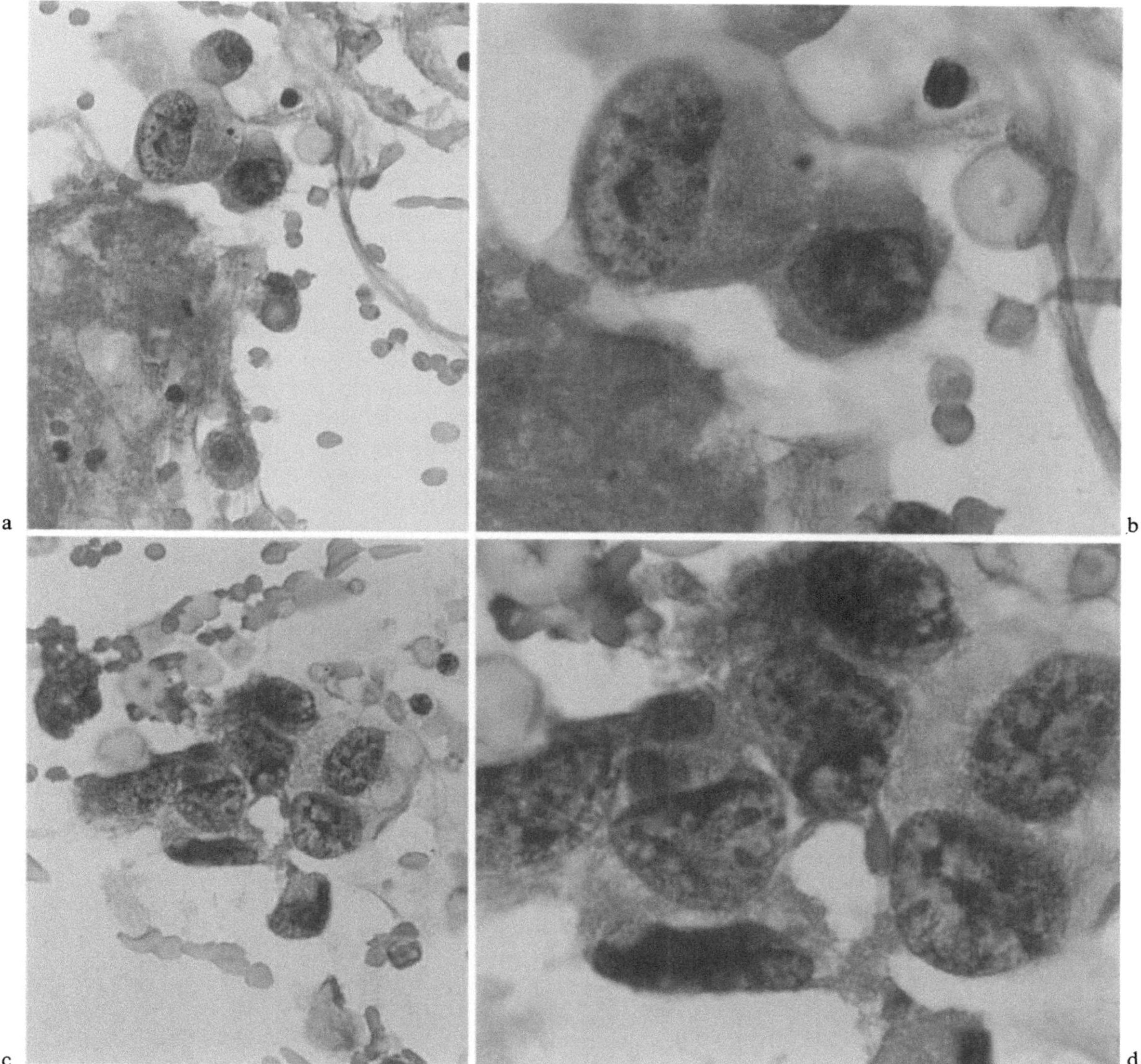

Fig. 9.91a–d. **Poorly differentiated recurrent carcinoma** of
the upper urinary tract following ileal conduit diversion. The
giant nuclei with coarse chromatin permit a definitive cytologic diagnosis (a and c, × 340; b and d, × 850; Papanicolaou
stain)

S. Roth and P. Rathert

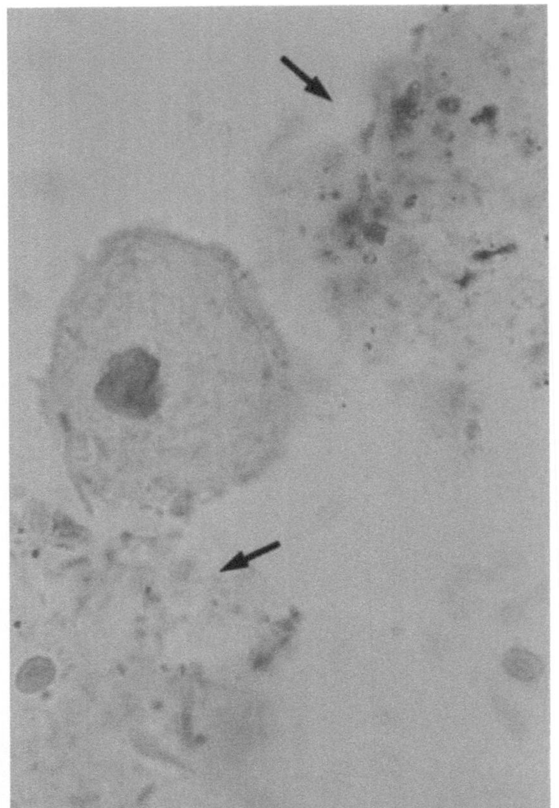

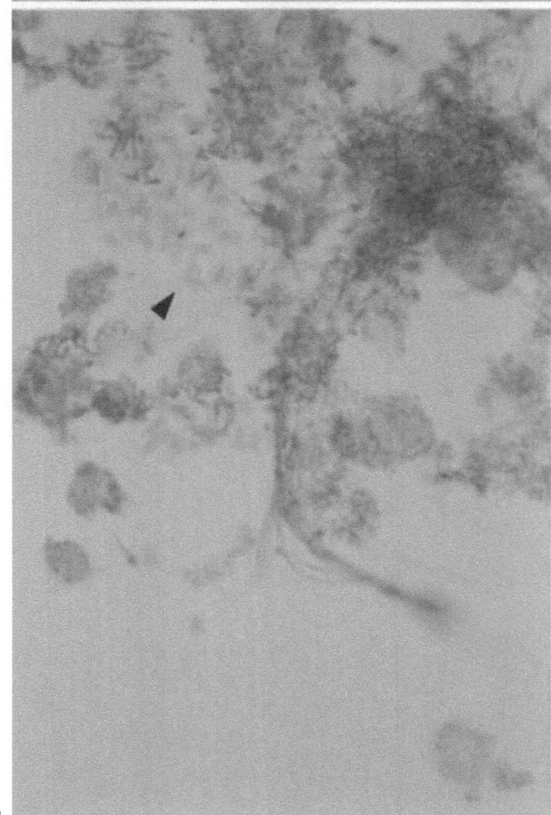

9.8 Rare Urinary Cytologic Findings

9.8.1 Vesicoenteric Fistulas

The diagnosis of a vesicoenteric fistula as a prelude to surgical treatment is often difficult even when the full spectrum of urologic diagnostic procedures are applied. This is due to the *frequently poor diagnostic yield of endoscopic and uroradiologic examinations*. Despite success with computed tomography and the "Bourne test," involving oral barium administration and subsequent radiographic detection of barium in the sediment of a 24- to 48-h urine sample (Roth and Rathert 1988), urinary cytology also should be considered as a diagnostic tool – especially when we note that vesicoenteric fistulas typically run an intermittent course due to spontaneous opening and closure of the fistulous communication and often necessitate *multiple examinations. Urinary cytology* is *ideal* in this regard owing to its convenience and noninvasiveness.

Fig. 9.92 a, b. **A typical finding with vesicoenteric fistulas** is the presence of undigested fiber remnants (← ←) and coliform bacteria (◄). Individual "bulky" fiber remnants in the specimen probably represent contamination during cytopreparation; the fiber remnants of vesicoenteric fistulas tend to appear "frayed" and "digested" (×, 850; Papanicolaou stain)

9.8.2 Parasitic Infections

9.8.2.1 Schistosomiasis of the Urogenital Tract

Of all the tropical parasitic diseases, schistosomiasis (bilharziosis) most commonly affects the urogenital system. It is chiefly endemic to the subtropical and tropical regions of Africa, Asia, and South America, but the scope of international tourism makes it advisable to consider schistosomiasis generally as a potential cause of chronic bladder inflammations that are refractory to treatment.

The causative parasites of schistosomiasis are trematodes. The most common causative organism of urinary schistosomiasis is *Schistosoma haematobium*, known also as the "human bladder fluke."

Endoscopic inspection of the bladder submucosa reveals "bilharziomas," which are also described as "sandy deposits." The diagnosis is ultimately established by biopsy, although direct detection by urinary cytology is also possible. If schistosomiasis is suspected, the *terminal urine fraction* should be examined as it will contain the largest number of schistosome ova expelled by bladder contraction (Fig. 9.93).

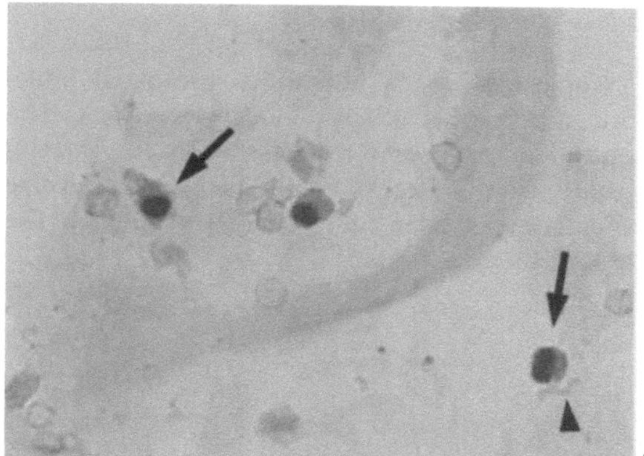

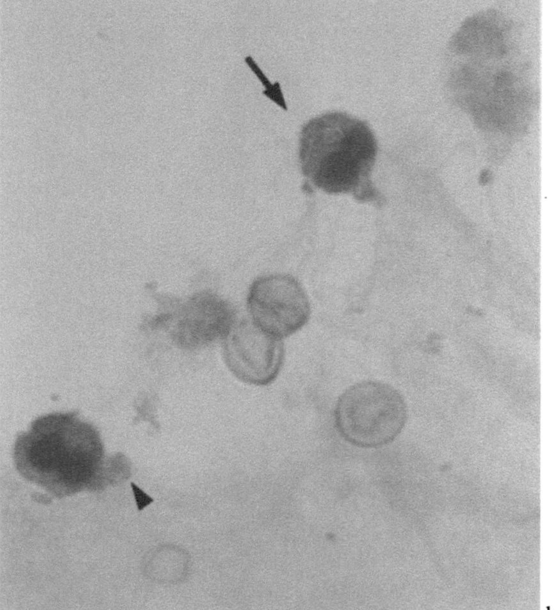

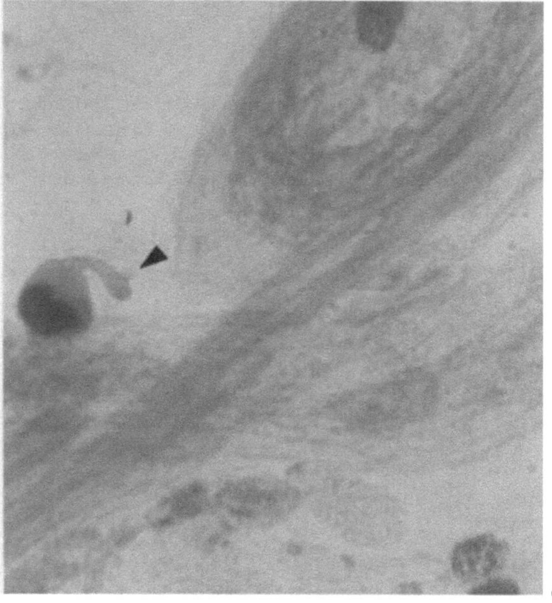

Fig. 9.93a–c. **Schistosomiasis of the bladder.** The schistosome ova (← ←) are markedly smaller than urothelial cells and are easily mistaken for segmented granulocytes. The key differentiating feature is the "terminal spine," appearing as a thin cytoplasmic projection from the ovum (◀◀). A terminal spine is not consistently observed on all schistosome ova, however (a, ×340; b, ×850; Papanicolaou stain)

9.8.2.2 Toxoplasmosis

Toxoplasmosis is a ubiquitous protozoal infection that produces few clinical symptoms. The causative organism of toxoplasmosis, *Toxoplasma gondii*, chiefly attacks the brain and the cardiac and skeletal muscle but rarely may be found in the urogenital tract.

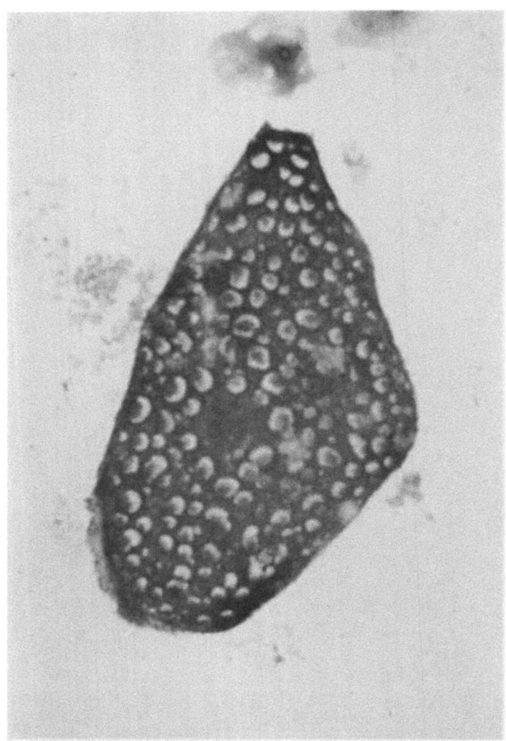

Fig. 9.94. **Toxoplasmosis** in a 2-day-old infant. The infected cell shows pseudocystic alterations and contains reddish structures (parasites) surrounded by a clear zone (× 850; May-Grünwald/Giemsa stain). (From de Vooget et al. 1979)

9.8.3 Viral Infections of the Urinary Tract

In contrast to viral infections of the external genitalia, little research has been done on the significance of viral infections involving the excretory portion of the urinary tract. One reason for this is the difficulty of making a confident diagnosis of viral diseases in that region.

Viruria is known to be a feature of numerous viral diseases (Nafe 1989). In the past, urologic symptoms in generalized infections were interpreted mainly as a result of immunobiologic processes rather than a direct result of infection by viruses. But in recent years there has been growing speculation about a viral etiology of various urologic and nephrologic disorders. The problem of urinary tract viral infections will likely become more pressing in connection with the possible activation of latent forms in patients who are immune-suppressed as a result of AIDS infection or organ transplantation.

Urinary cytology can be particularly valuable when there is question of urothelial involvement by *papilloma virus* and for the differential diagnosis of *cytomegalovirus infection* and *acute graft rejection after renal transplantation*.

9.8.3.1 Cytomegalovirus Infection and Renal Graft Rejection

Winkelmann et al. (1985), by examining Papanicolaou-stained specimens, were able to differentiate between acute allograft rejection and a cytomegalovirus infection provoked by the use of immunosuppressive agents in 33 patients who underwent renal transplantation.

The typical cytomorphologic criteria of *graft rejection* were large numbers of renal tubular cells, lymphocytes, mitoses, and increasing erythrocyturia. By contrast, specimens from six of seven patients with serologically diagnosed *CMV infection* contained typical "owl's-eye cells" with a ground-glass nucleus (Fig. 9.95). There were no cytologic criteria of graft rejection, so the findings were not misinterpreted. Urinary cytology has thus proven to be a valuable tool in the follow-up of renal transplant patients. It should be added, however, that multiple cytologic examinations are often required when a CMV infection is suspected (Koss 1981).

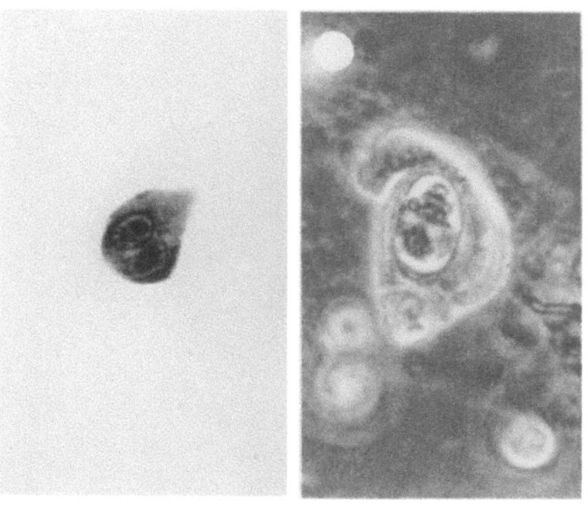

a
b

Fig. 9.95 a, b. In **cytomegalovirus infection,** the intranuclear inclusions impart a characteristic "owl's eye" appearance to the nucleus (a, × 400, May-Grünwald/Giemsa stain; b, × 400, phase contrast microscopy). (From de Vooget et al. 1979)

9.8.3.2 Urothelial Involvement by Human Papillomavirus

The condylomatous infections caused by human papillomaviruses (HPV) are among the most common sexually transmitted diseases. Besides the familiar exophytic acuminate form and the flat condylomata affecting the external genitalia, the *urethral meatus* should also be integrated into the diagnostic workup as a potential *reservoir for autoinfection* (Roth et al. 1990).

Urethroscopy is problematic due to the risk of iatrogenic spread of virus to higher levels, so an ENT speculum has been recommended as an aid to inspection. Although few studies have been done on the efficacy of urine cytology in the diagnosis of vesical and urethral HPV infections, it is reasonable to undertake a urinary cytologic examination given the lack of practical alternatives and the fact that we do not yet have an accurate molecular virologic procedure that can routinely detect an HPV infection independent of visible morphologic changes.

Cellular damage by HPV is characterized chiefly by a conspicuous cytoplasmic alteration in fully mature squamous epithelial cells. This includes the formation of a clear perinuclear "halo" with more or less pronounced nuclear atypia, frequently com-

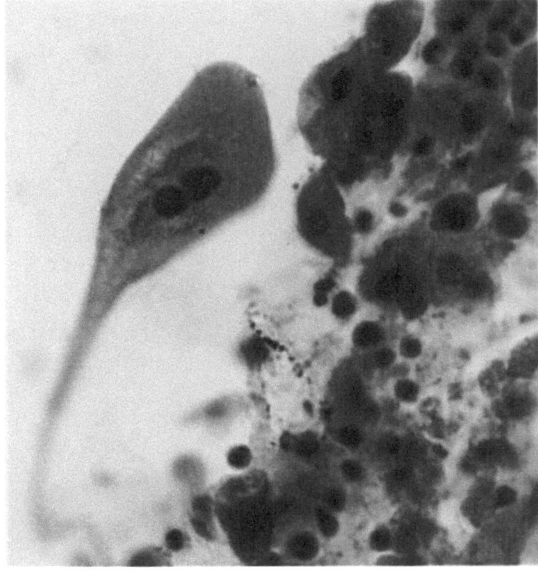

a

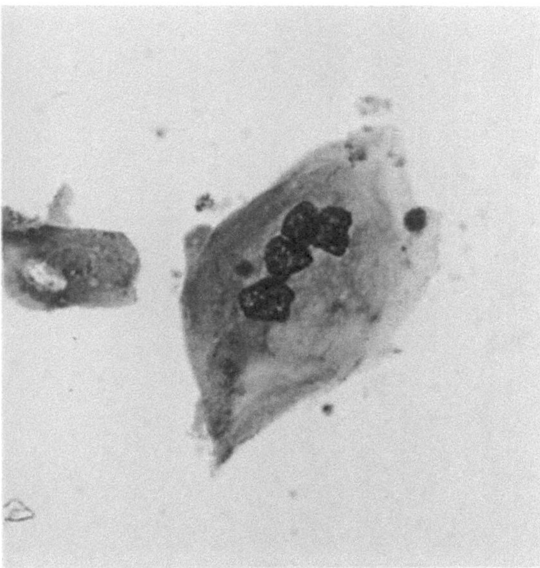

b

Fig. 9.96 a, b. **Urethral condylomata acuminata** in a male. Besides nuclear atypias with mild hyperchromasia and chromatin condensation, there is conspicuous multinucleation of the squamous epithelial cells. A typical perinuclear halo is faintly visible in b (× 475; Papanicolaou stain). (From de Vooget et al. 1979) bined with nuclear degenerative signs. Koss coined the term "koilocytes" for cells that manifest these HPV-specific changes (Koss and Durfee 1956).

9.8.3.3 Herpesvirus Infection

In patients with persistent dysuric complaints and no significant detectable bacterial infection, the possibility of herpes cystitis should be considered in the differential diagnosis once carcinoma and other organic diseases have been excluded. This type of infection is rare (Masukawa et al. 1972; Person et al. 1973), so correlation with immuno-suppressive therapy can provide an important clue.

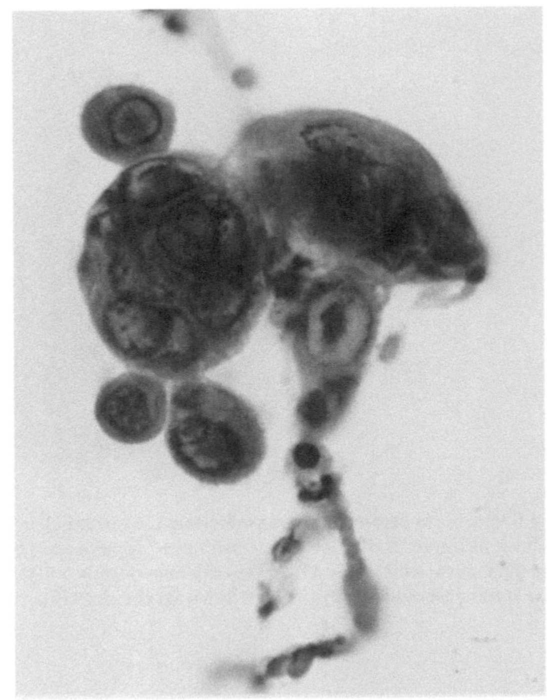

Fig. 9.97. **Herpesvirus infection** with "virocytes." Striking features are the multinuclearity of the cells and the ground-glass appearance of the nuclei. Koss describes the cells typifying this pattern as "multinucleated with nuclear crowding and reciprocal nuclear indentations" (Koss 1981); (× 475; Papanicolaou stain). (From de Vooget et al. 1979)

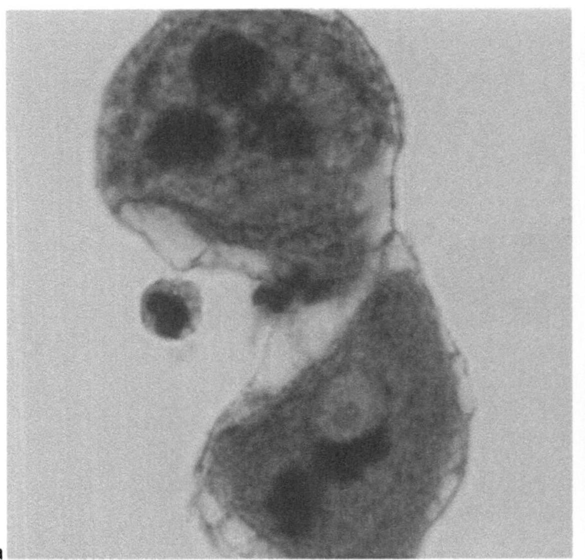

a

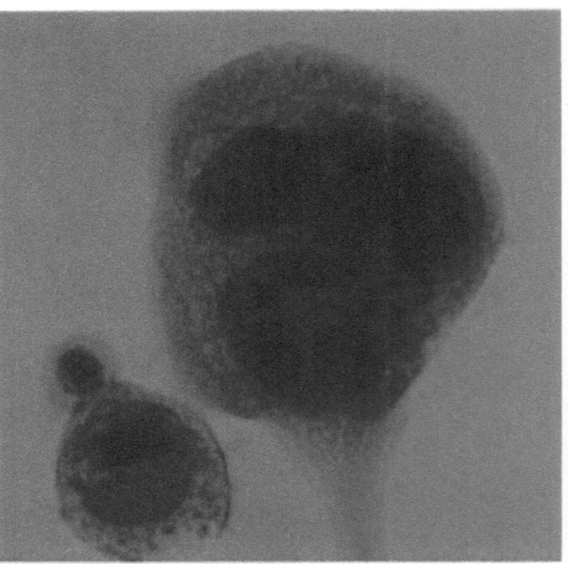

b

Fig. 9.98 a, b. **Virocytes of unknown origin resembling tumor cells.** This patient had gross hematuria with bilateral renal pain and fever lasting 3 days. The cytologic impression of a urothelial tumor (b) was refuted by uroradiology, en-doscopy, and close follow-up. While some cells showed atypical plasma vacuoles (a), others were morphologically indistinguishable from tumor cells (b); (× 850; Papanicolaou stain)

References

Beyer-Boon ME (1979) Urinzytologie und ihre Beziehung zur Histologie des Harntraktes. In: Voogt HJ de, Rathert P, Beyer-Boon ME (eds) Praxis der Urinzytologie. Springer, Berlin Heidelberg New York, p 34

Bretton PR, Herr HW, Kimmel M, Fair WR, Whitemore WF Jr, Melamed MR (1989) Flow cytometry as a predictor of response and progression in patients with superficial bladder cancer treated with bacillus calmetteguerin. J Urol 141: 1332

Esposti PL, Edsmyr B, Tribukait B (1978) The role of exfoliative cytology in the management of bladder carcinoma. Urol Res 6: 197

Harving N, Wolf H, Melsen F (1988) Positive urinary cytology after tumor resection: a indicator for concomitant carcinoma in situ. J Urol 140: 495

Jakse G, Hofstädter F, Marberger H (1980) Wert der Harnzytologie und Quadrantenbiopsie bei oberflächlichen Blasenkarzinomen. Akt Urol 4: 309

Jakse G, Hufnagel B, Hofstädter F, Rübben H (1986) Sequentielle Blasenschleimhautbiopsie beim Urothelkarzinom der Harnblase. Verh Dtsch Ges Urol 37: 200

Koss LG (1981) Cytology of the urinary tract and its histologic bases. Revised 2nd ed Tutorials of cytology, Chicago

Koss LG, Durfee GR (1956) Unusual patterns of squamous epithelium of uterine cervix: cytologic and pathologic study of koilocytotic atypia. Ann NY Acad Sci 63: 1245

Koss LG, Deitch D, Ramanathan AB, Sherman AB (1985) Diagnostic value of cytology of voided urine. Acta Cytol 29: 810

Lopez-Cardozo (1976) Zit. nach Voogt HJ de, Rathert P, Beyer-Boon ME (1979) Praxis der Urinzytologie. Springer, Berlin Heidelberg New York

Masukawa T, Garancis JC, Rytel MW (1972) Herpes genitalis virus isolation from human bladder urine. Acta Cytol 16: 416

Müller F, Kraft R, Zingg E (1985) Exfoliative cytology after transurethral resection of superficial bladder tumors. Br J Urol 57: 530

Murphy WM, Emerson LD, Chandler RW, Moinuddin SM, Soloway MS (1986) Flow cytometry versus urinary cytology in the evaluation of patients with bladder cancer. J Urol 136: 815

Nafe R (1989) Virusinfektionen der ableitenden Harnwege. Akt Urol 20: 187

Person DA, Kaufman RH, Gardner HL (1973) Herpes virus type 2 in genitourinary tract infections. Am J Obstet Gynecol 116: 993

Roth St, Rathert P (1988) Vesiko-enterale Fisteln: Lösungswege eines diagnostischen Dilemmas. Urologe [A] 27: 142

Roth St, Rathert P (1989) Cytological surveillance of carcinoma in situ during and after intravesical chemotherapy. In: Murphy GP, Khoury S (eds) Therapeutic progress in urological cancers. Liss, New York, p 523

Roth St, Brandt H, Rathert P (1990) Subklinische Condylomatainfektionen des männlichen Genitales – Erweiterte Lokalisationsdiagnostik und therapeutische Konsequenzen. Urologe [B] 30: 9

Rübben H, Rathert P, Roth St, Hofstädter F, Giani G, Terhorst B, Friedrichs R (1989) Exfoliative Urinzytologie, 4th edn. Harnwegstumorregister, Fort- und Weiterbildungskommission der Deutschen Urologen, Arbeitskreis Onkologie, Sektion Urinzytologie

Stöckle M, Gökcebay E, Riedmiller H, Hohenfellner R (1990) Urethral tumor recurrence after radical cystoprostatectomy: the case for primary cystoprostatourethrectomy? J Urol 143: 41

Voogt HJ de, Rathert P, Beyer-Boon ME (1979) Praxis der Urinzytologie. Springer, Berlin Heidelberg New York

Winkelmann M, Grabensee B, Pfitzer P (1985) Differential diagnosis of acute allograft rejection and CMV-infection in renal transplantation by urinary cytology. Pathol Res Pract 180: 161

10 Diagnostic DNA Cytometry of the Urothelium

A. Böcking

CONTENTS

10.1 Introduction

The subjective cytologic examination of urinary sediment by no means offers an ideal solution to the diagnosis of bladder carcinoma:

– The *sensitivity* is only 70%, depending on the grade of tumor malignancy.
– The *specificity* is only 80%–95% (Murphy et al. 1986; Rübben et al. 1989).
– The *reproducibility* of malignancy grading, at approximately 60%–70%, is poor (Ooms et al. 1989).
– The *prognostic relevance* of the cytologic tumor grade is not adequate to have a definite impact on treatment planning (Böcking et al. 1990).

– The *prospective biologic behavior* of dysplastic cells cannot be accurately predicted.
– The *labor cost* is high for the screening of large sample volumes in subjects at risk.

Since the 1960s, the latest *cytometric techniques* have been applied in an effort to furnish objective, reproducible data on urothelial cells, and thus help solve the problems listed above (Tavares et al. 1966; Koss et al. 1975; Foss and Kaalhus 1976a,b; Tribukait and Esposti 1978). The earliest results, and still the most promising to date, have been supplied by *DNA cytometry*, which was formerly done using microphotometers and today employs TV image analysis systems or flow cytometers.

Tumor cytogenetics forms the biological basis for *diagnostic DNA cytometry*, which involves the detection of aneuploid chromosome sets in neoplasia and relates the variability of this aneuploidy to the grade of tumor malignancy. Attempts have also been made to use TV image analysis (morphometry) of the urothelial cells to find solutions to the foregoing problems. Here we shall discuss the scientific results of these investigations in terms of their *relevance to routine cytologic diagnosis* and its clinical application.

10.2 Morphometry of the Urothelium

Koss et al. (1975), unlike Fossa and Kaalhus (1976a), were unable to find a significant difference in nuclear sizes between normal and malignant urothelial cells. In a sophisticated image-analysis study of 117 bladder carcinomas, these authors found no significant correlation between various nuclear parameters (size and its variability, chromatin density and distribution) and clinical tumor stage or length of patient survival. Aikens and Liedtke (1982) used a high-resolution microscopic cytophotometric system in an attempt to discriminate between normal and malignant urothelial cells. The results were ultimately disappointing. Even more recent studies with TV image analysis systems (Montironi et al. 1985) can at best demonstrate significant differences between the large groups of all benign and malignant urothelial changes but cannot furnish results relevant to the diagnosis of individual cases. Consequently, morphometric image analysis is no longer being seriously considered as a tool for the evaluation of problem cases or the grading of malignancy in urinary cytology. At the same time, image-analysis parameters can be successfully employed in automated systems when the goal is to replace the screening function of a human assistant by a machine, rather than to enhance the confidence and prognostic relevance of a cytologic diagnosis already made by an examiner (see Sect. 10.7).

10.3 Biologic Principles of DNA Cytometry

10.3.1 Discrimination

Except for gametes, the nuclei of human cells each contain two sets of 23 chromosomes (=2c). Four sets of chromosomes (=4c) are present in the G2 phase of the cell cycle prior to cell division. A multiple of the chromosome set corresponding to whole-number powers of the 2c value (i.e., 4c, 8c, 16c, 32c) regularly occurs as a physiologic phenomenon in some tissues. This process is called *euploid polyploidization*. Polyploid chromosome sets are consistently found, for example, in thyroid epithelium, seminal vesicle epithelium, hepatic epithelium, mesothelial cells, cardiac muscle fibers, and also in urothelial cells. As early as 1959, Walker demonstrated this in cytogenetic studies of the mouse urothelium. Several authors have used DNA cytometry to detect polyploidization of up to 8c in human urothelial cells (Levi et al. 1969; Fossa 1975; Farsund and Hostmark 1983). We detected 4c and 8c ploidies in 45 of 50 normal urothelial populations (Biesterfeld et al. 1992). Figure 10.1 shows a DNA summation histogram of these normal urothelial cell populations from spontaneously voided urine samples. Some cytologically detectable viral infections, such as HPV, can likewise induce a polyploid nuclear DNA content (Chatelain et al. 1989a), although this has not yet been demonstrated for the urothelium. In the synthesis phase of the cell cycle, a limited percentage of the cells (<10%) may show DNA contents that are between the integral powers of the 2c value (Sandritter 1981).

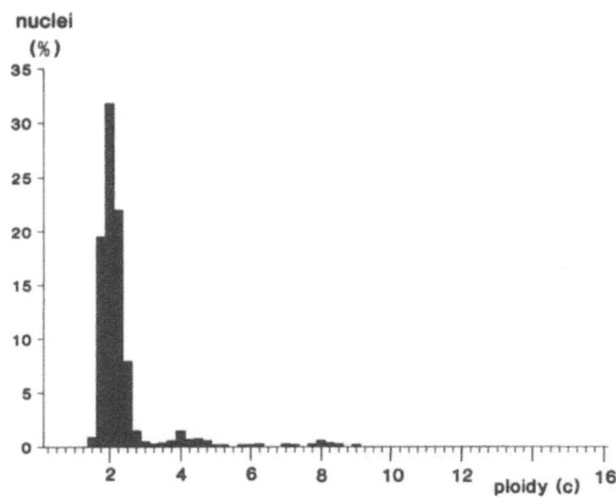

Fig. 10.1. **Summation histogram of DNA distribution in 5000 normal urothelial cells from urine samples of 50 different subjects.** For each subject 100 Feulgen-stained cells were measured using the MIAMED system (Wild-Leitz, Wetzlar); granulocytes served as reference cells ($\times 1.20$)

Aneuploidy refers to numerical or structural deviations from these normal chromosome sets and corresponding DNA contents (Fig. 10.2). Heim and Mittelman (1987) describe a 5p isochromosome, monosomy X, and trisomy 7 as primary chromosomal abnormalities associated with carcinoma of the urinary bladder. Pauwels et al. (1987), on the other hand, describe a loss of chromosome 5 and the deletion of a fragment from chromosome 19 as characteristic anomalies in urothelial neoplasms. A tumor that exhibits this aneuploidy can be identified as originating from the urothelium. This characteristic chromosomal aneuploidy is not readily detectable by DNA cytometry, however, because it corresponds to a loss of only about 2.5% of the nuclear DNA, or 0.05c. This is why *low-grade urothelial tumors are usually not identified as "aneuploid" by DNA cytometry, even though they are cytogenetically aneuploid.* This applies to almost all tumors, and there are virtually no "diploid" urothelial tumors. But cytometry is sometimes incapable of demonstrating chromosomal aneuploidies, leading many cytometrists to speak incorrectly of "diploid" tumors. Structural and numerical aberrations of chromosomes 1, 7, 11, 17, and others as well as marker chromosomes may appear as secondary abnormalities during the course of tumor progression, but they are not characteristic of urothelial tumors. Once these secondary chromosomal anomalies have appeared,

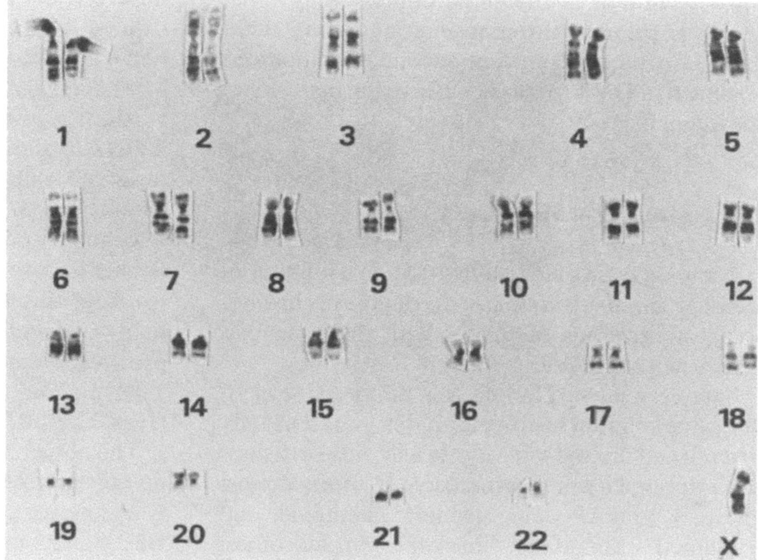

Fig. 10.2. **Karyotype of a representative tumor cell from a human grade 1 urothelial carcinoma.** Monosomy X, translocation of a fragment from chromosome 12 to 1, trisomy 20. (Specimen courtesy of Dr. Ott, Dr. Offner, Institute for Pathology, RWTH Aachen, FRG)

the nuclear DNA content can be successfully identified as abnormal or "aneuploid" even by cytophotometric analysis.

Chromosomal aneuploidy is viewed in cytogenetics as a *hard marker* for the neoplastic transformation of cells (Heim and Mittelman 1987). This means that the detection of cells with an aneuploid chromosome set is equivalent to the detection of neoplastically transformed cells (Atkin 1971; Huber 1986; Pauwels et al. 1987; Smeets 1987). For example, the cytogenetic detection of aneuploid cells is used as a marker for malignancy in the cytologic diagnosis of pleural effusions (Heim and Mittelman 1987). Some benign cells also may exhibit aneuploidy as a result of exposure to ionizing radiation or cytostatic agents. If the single-cell interpretation of DNA cytometric values is to be useful for detecting aneuploidy, these effects must be excluded.

Due to the *high cost of chromosomal preparations*, attention has shifted in the past 40 years from cytogenetic analysis to DNA cytometry as a means of evaluating cells for aneuploidy to detect neoplastic transformation (Atkin 1971; Sandritter and Böhm 1975; Ploem-Zaaijer et al. 1979; Barlogie et al. 1980; Wied et al. 1983; Hedley et al. 1984; Barlogie 1981). After classic Feulgen staining (Feulgen and Rossenbeck 1924), in which DNA is combined with a violet dye (parafuchsin) in a stoichiometrically precise fashion, the DNA content can be determined with reasonable accuracy by microdensitometric measurement of the dye content per nucleus and calibration with nuclei having a known, normal 2c content (reference cells).

Although aneuploidy is a chromosomal phenomenon of the individual cell, classic DNA cytophotometry relates this term to the most frequently occurring (modal) value in a cell population (the highest peak in the *DNA histogram*). In the case of a tumor, this value corresponds to the *"stemline"* (Seidel and Sandritter 1963). A single-cell interpretation of aneuploidy can be accomplished by DNA cytometry only in non-polyploidizing tissues (Böcking et al. 1984; Böcking 1990). If the modal value or tumor stemline is located in an "aneuploid region" of the DNA histogram (outside 2c, 4c, 8c, 16c), aneuploidy of the entire population is assumed, and the tumor is said to be "aneuploid" even if most and not all of the nuclei have an aneuploid DNA content. In polyploidizing tissues such as the urothelium, we classify a modal value outside of 2c, 4c, 8c, 16c ± 2c, 4c, 8c or 16c × CV as "aneuploid." Because CV is the coefficient of variation in the reference cell population (e.g., lymphocytes), less variation in the reference cell population means a better chance for the successful DNA cytometric detection of aneuploidy. This conservative DNA cytometric interpretation of aneuploidy is *very specific* but *relatively insensitive*, because aneuploidy is assumed to be present only if the majority of tumor cells display a quantitatively significant alteration in their chromosome sets. Generally this is the case only if the *quantity of DNA is increased by more than 10%*, corresponding to a modal value greater than 2.2c. In practical terms, this means that an urothelial carcinoma can be diagnosed by the detection of an aneuploid stemline only if, beyond the primary chromosomal mutations, there have also been significant secondary mutations leading to a DNA increase >2.2c in the majority of the tumor cells.

10.3.2 Grading of Malignancy

It is known from cytogenetics that, in a number of solid carcinomas in humans, the degree of chromosomal aberrations correlates with the prognosis (Heim and Mittelman 1987).

Sandberg showed in 1986 that noninvasive bladder carcinomas of histologic grades 1 and 2 usually had normal diploid karyotypes and only occasionally exhibited marker chromosomes, whereas most grade 3 tumors were strongly aneuploid and contained abundant marker chromosomes. Superficially invasive grade 2 tumors displayed more aberrations than noninvasive tumors. The presence of *marker chromosomes* appears to be an absolute prerequisite for *invasive tumor growth* in bladder cancer patients (Summers et al. 1989). Massive chromosomal abnormalities were found mainly in aggressive tumors. In patients with noninvasive papillary tumors, recurrence developed in 90% of cases with marker chromosomes but in only 5% of cases with no chromosomal anomalies.

The *malignant potential of urothelial carcinomas* depends less on the most frequently occurring chromosomal anomaly (modal anomaly), as is commonly believed, than on the *variability ("range") of the secondary anomalies*. This results from the genetic instability of the tumor. The more the secondary chromosome anomalies vary from cell to cell in a urothelial carcinoma, the more malignant and aggressive the tumor (Pauwels et al. 1988). Modal chromosome counts <49 occurred in 97% of all grade 1 tumors but in only 49% of grade 2 and 8% of grade 3 tumors. While all infiltrating bladder carcinomas showed cells with more than 49 chromosomes, this was noted in only 22% of the noninfiltrating tumors (Pauwels et al. 1988). The chromosomal range of aneuploidies in DNA cytometry corresponds to the variance about the normal 2c value (=2c deviation index, 2cDI; Böcking et al. 1984). The higher the measured values and the greater their variability, the greater will be the variance about the 2c value. The loss of a stemline is considered a particularly unfavorable prognostic sign, as it also leads to a high variance of DNA values and to very high individual values. Variance is too abstract a prognostic index for the clinician,

however, and it is more useful to work with a logarithmic conversion of the 2c deviation index (2cDI) in which the *DNA malignancy grade* (DNA MG) is assigned a value on a 0–3 scale. In this conversion the lowest possible variance of 0 represents a DNA malignancy grade of 0, while the highest observed value of 51 (for osteosarcoma) corresponds to a DNA malignancy grade of 3.0 (Böcking and Auffermann 1986). Follow-up studies to date have confirmed the *prognostic relevance* of this DNA malignancy grade for malignant lymphomas and for carcinomas of the prostate, breast, and bladder (Böcking et al. 1985, 1986a,b, 1988, 1989, 1990; Auffermann et al. 1986) (Figs. 10.3–10.5).

The above observations from tumor cytogenetics are consistent with observations of a statistically significant prognostic improvement in patients with bladder carcinomas that display a DNA stemline in the "diploid" or "tetraploid" range, as opposed to those with aneuploid stemlines in the 3c–6c range (Tavares et al. 1973, 1986).

At the same time, a review of the literature shows little consistency in the prognostically relevant interpretation of the DNA distribution in statistical and flow-cytometric DNA analyses. Frequently the authors confine their efforts to the subjective description of DNA histograms. Many nonstandard terms are employed in an effort to describe the characteristics of the DNA distribution, although most authors focus their attention on the position of the stemline. This refers to the most frequent (modal) value that is accompanied by a doubling peak corresponding to cells in the G2/M phase (Sandritter and Carl 1966). Authors have attempted to characterize the modal value with terms such as diploid, triploid, tetraploid, hyperpentaploid, near-hexaploid, hyperoctoploid, etc., without defining precisely what these terms mean. Other authors have described various prognostically relevant types of histogram in which the DNA distribution must be subjectively evaluated (Auer et al. 1980). Pfitzer et al. (1976) were among the first to emphasize the poor reproducibility of the subjective interpretation of histograms. Besides the 2c deviation index mentioned above, a variety of other indices have been devised for the more precise and objective interpretation of DNA histograms:

– The DNA index of the modal value (Barlogie et al. 1980)
– The mean nuclear DNA content (Sprenger et al. 1974)

- The diploid deviation quotient: DDQ (Fossa and Kaalhus 1977), which also is an expression of mean ploidy
- The percentage of nuclei with a DNA content >5c (Ploem-Zaaijer et al. 1979)
- The Z value (Sprenger et al. 1974), which expresses the ratio of "euploid" to "aneuploid" DNA values

The most widely used, prognostically relevant *differentiation into "diploid" and "aneuploid" tumors has no biological, cytogenetic basis*, because virtually all tumors are aneuploid at the chromosomal level. Since DNA cytometry reveals any number of transitional states between slight and profound deviations of the stemline modal value from the normal 2c value, an arbitrary division into "diploid" and "aneuploid" tumors makes little sense. It would be more useful to establish empirically derived, prognostically relevant thresholds of the modal value for specific tumor entities.

10.4 Specimen Processing

10.4.1 DNA Single-Cell Cytometry

DNA single-cell cytometry can be performed on routine cytologic material that has been processed by any standard cytopreparatory technique. However, the cells in the voided urine or washings should be fixed by the addition of 50% alcohol immediately after collection. Also, the cellular material should be concentrated in a monolayer to facilitate individual cell detection. Cytocentrifuge specimens are particularly suitable, but sediment smears are also acceptable.

Feulgen stain, necessary for *DNA cytometry* in visible light, can be applied to *any specimen regardless of its prior fixation or staining*. It is necessary only to (post)fix the specimen for at least 2 h in 4% formaldehyde in phosphate buffer at pH 7.0 prior to hydrolysis in hydrochloric acid. If cytologic specimens can be immediately fixed by this technique, postfixation is not required. Pure alcohol fixation adversely affects the proportionality of hydrolysis between the different cell types (Böhm et al. 1968). The former practice of adding acetic acid is not advised, as it leads to an increased scatter of measured values (Roels 1990). Besides forming free aldehyde groups on purine bases, the hydrochloric acid hydrolysis serves to destain previously stained cells. Coverslipped specimens

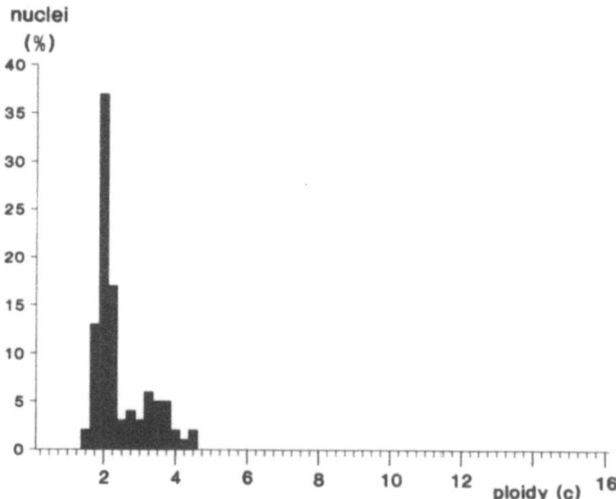

Fig. 10.3. **DNA histogram of a grade 1 urothelial carcinoma,** DNA malignancy grade 0.35

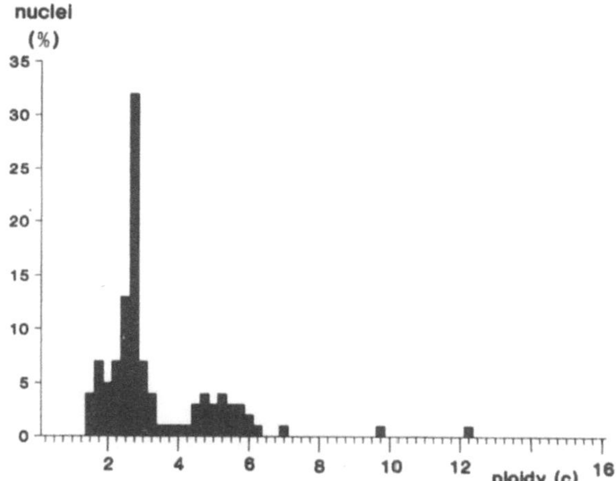

Fig. 10.4. **DNA histogram of a grade 2 urothelial carcinoma,** DNA malignancy grade 1.28

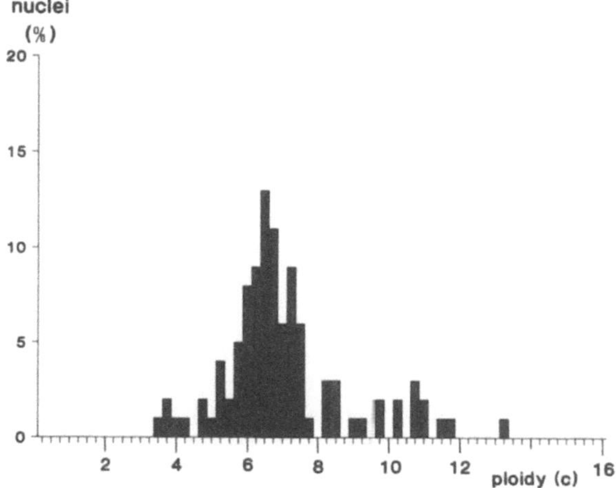

Fig. 10.5. **DNA histogram of a grade 3 urothelial carcinoma,** DNA malignancy grade 2.60

also can be used, as several hours' soaking in xylene will remove the coverslip while leaving the cells adherent to the slide. DNA cytometry can even be performed on tissue material fixed with formalin and embedded in paraffin, since there is a relatively simple technique for recovering individual cells from the embedded tissue (Delgado et al. 1984). Precise, stoichiometric, automated Feulgen staining can be performed overnight with a temperature-controlled staining machine (Chatelain et al. 1989b).

The use of fluorescent stains for interactive DNA single-cell cytometry is not recommended, because the fading effect makes it impossible to repeat the measurements on the same nuclei. This is occasionally necessary, though, for reasons of quality control. Cytometry in UV light also must be performed in a darkened room, requiring prolonged adaptation of the eyes.

Thus, a *special preparatory technique is not required* for DNA single-cell cytometry since any specimen, regardless of its previous fixation or staining, can be processed by applying appropriate postfixation and destaining procedures. The standard procedure for Feulgen staining is outlined in Table 10.1.

10.4.2 DNA Flow Cytometry

Cell samples for flow cytometric analysis are processed differently than samples for conventional urinary cytology.

Following immediate centrifugation (1500 rpm for 5 min), the sediment can be resuspended in 3 ml of phosphate buffer and either stained at once or fixed prior to staining. Several hours' fixation of the sediment in 95% alcohol is recommended to prevent autolysis and standardize the results of the measurement. Fluorescent staining of the resuspended sediment is performed overnight by adding 5 ml of propidium iodide or DAPI. The addition of a detergent (e.g., Triton X-100) destroys the cells while leaving their nuclei largely intact (Ratliff et al. 1985). Before cytometry, the cell suspension is sieved through a 40–70 µm nylon mesh filter. Deparaffinated tissue samples originally fixed in formalin can be analyzed by DNA flow cytometry following combined enzymatic-mechanical cell retrieval (Coon et al. 1986).

Table 10.1. Steps in the Feulgen staining procedure

No.	Time (min)	Contents of the cuvet	Function
1	60	4 % Formaldehyde in Sörensen buffer, pH 7.0	Fixation
2	5	Tap water, flowing	Rinse
3	2	Distilled water	Rinse
4	55	4 N HCl, 27.5 °C	Hydrolysis
5	5	Tap water, flowing	Stop hydrolysis
6	2	Distilled water	Rinse
7	60	Schiff's reagent[a]	Stain
8	10	SO$_2$ water[a]	Wash out excess dye
9	10	SO$_2$ water	Wash out excess dye
10	10	SO$_2$ water	Wash out excess dye
11	5	Tap water, flowing	Rinse
12	2	Distilled water	Rinse
13	5	70 % ethanol	Dehydration
14	5	96 % ethanol	Dehydration
15	5	100 % ethanol	Dehydration
16	5	Xylene	Clearing

[a] Prepared by method of Graumann (1953).

10.5 DNA Single-Cell Cytometry

10.5.1 Measurement Systems

Until a few years ago, the DNA content of individual cells was usually measured with *scanning microphotometers* (Deeley 1955) using photomultiplier tubes. Either the object is moved beneath the measuring beam (e.g., UMSP, Zeiss, Oberkochen, FRG), or a moving mirror is used to scan the beam across the object (integrated microdensitometer, e.g. Vickers M 86, Vickers, York, U.K.). The main advantage of these systems is their high gray-scale resolution; their main disadvantage is the relatively slow and complicated measurement process. Each individual nucleus must be manually positioned within the aperture of a mask. The subjective and thus imprecise determination of the individual background (blank value) of the nuclei represents a further limitation of this method (Auffermann et al. 1984). About 3–4 h is needed for the measurement of approximately 20 reference cells and 150 "diagnostic" cells. These instruments are poorly suited for routine diagnostic use, therefore.

Since the 1980s, scanning cytophotometers have increasingly been replaced by *TV image analysis systems* that employ a video camera connected to a microscope (Bachmann and Hinrichsen 1979;

Auffermann et al. 1984; Böcking et al. 1987) (Fig. 10.6). The image from the TV camera is processed by an image analysis computer, whose functions include the densitometric measurement of individual cells (Fig. 10.7). With this type of system, even nuclei that are closely adjacent or in contact with one another can be concurrently measured and their morphometric parameters determined. An individual background (blank) measurement can be performed for each nucleus. This method is so rapid and convenient that 20 reference cells and approximately 250 diagnostic cells can be measured in about 20 min. Potential disadvantages of nuclear DNA measurement with TV image analysis systems are shading (nonuniform image illumination and camera sensitivity) and glaring (scattered light in the microscope beam path). Both effects, however, can be corrected to a degree by appropriate hardware and software selection.

Unlike flow cytometric systems, *TV image analysis systems can perform measurements on routine cytologic and histologic specimens*. This eliminates the need for separate or repeat specimen collection with further processing. Feulgen staining is essential for DNA measurements, however. A suitable individual- and tissue-specific reference cell population can be selected for each specimen (e.g., squamous epithelial cells in the urine). Since each individual cell can be identified and classified prior to measurement, the relevant cells can be selectively evaluated without the concomitant measurement of artifacts. This also allows for "rare event detection," meaning the isolation and measurement of rare cells that are of special diagnostic interest. Thus, dysplastic cells in a bladder irrigation specimen containing various cell populations can be measured as selectively for benign/malignant discrimination as tumor cells can for the grading of malignancy. In this way various cell populations can be measured concurrently within the same sample. The immediate assignment of quantitative results to specific cells on the monitor provides for an immediate feedback of measured values (see Fig. 10.7). The ability to check results by relocating individual cells and remeasuring them offers a decisive advantage in terms of quality control. These systems also permit the determination of morphometric parameters such as nuclear shape, size (area), chromatin pattern and derivative parameters as well as the quantification of immunohistochemical reactions.

Although the number of measurable cells is somewhat decreased compared with flow cytometry, rapid advances in image analysis hardware have made it possible to measure 300 cells, including references cells, within a 20-min period

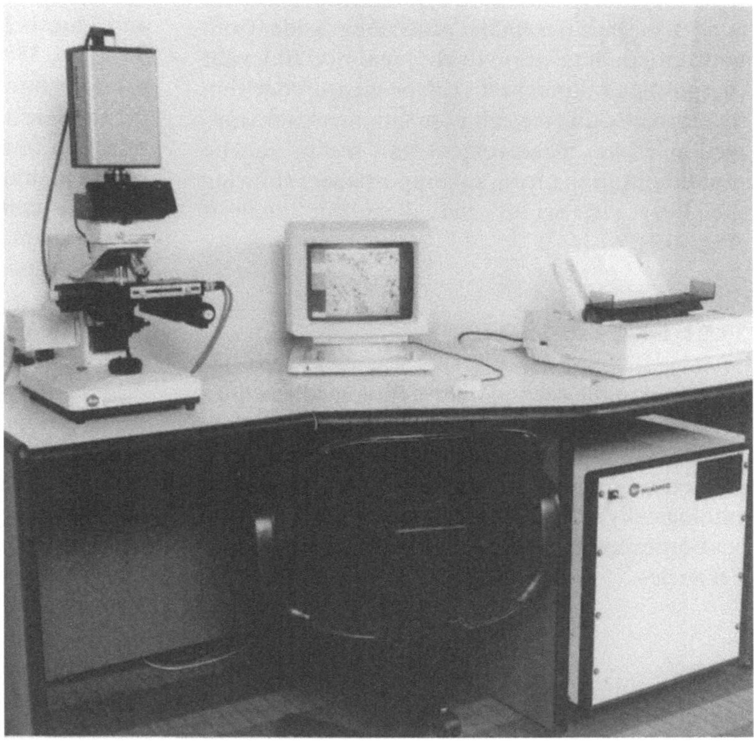

Fig. 10.6. **Interactive TV image analysis system** includes an automatic microscope for semiautomated diagnostic DNA measurements and the determination of nuclear morphometric parameters (MIAMED, Wild-Leitz, Wetzlar, FRG)

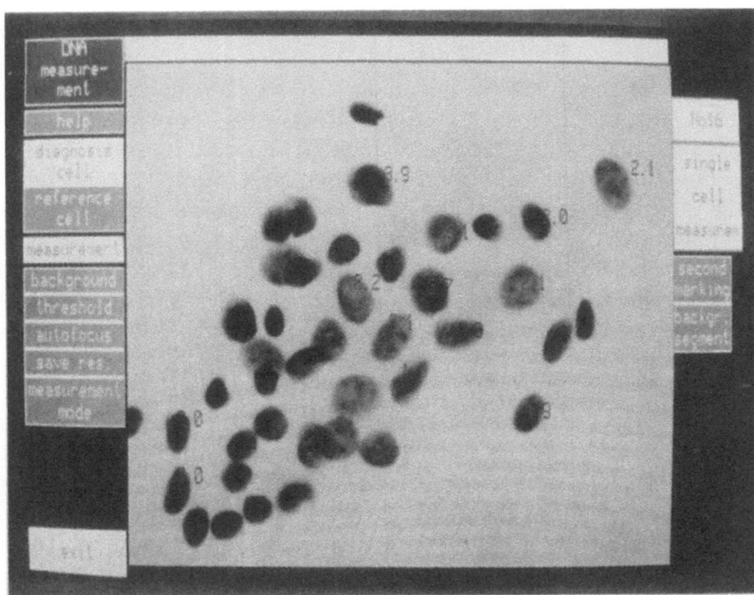

Fig. 10.7. **Interactive monitor of the MIAMED TV image analysis system** with screen-edge sensor fields and display of measured tumor-cell nuclei and their DNA contents in c (Feulgen stain)

(Böcking et al. 1992). Measurements of larger cell numbers do not significantly alter the observed DNA distribution (Marschner 1992). Interactive DNA image cytometry is particularly well suited for DNA measurements in urinary cytologic specimens, because it can be performed on existing smears or cytocentrifuge specimens that have already been evaluated by a cytologist. There is no need for special sample processing aside from Feulgen staining prior to the analysis. Relevant (dysplastic or tumor) cells can be measured within the normally diverse cell populations. After completion of the measurement, the results can be printed out in the form of a report sheet showing the DNA histogram and diagnostic analysis (Fig. 10.8).

10.5.2 Diagnostic Results

As stated in the section on the Biologic Principles of DNA Cytometry, this method is excellent for the identification of neoplastically transformed cells, for determining the benignancy or malignancy of cytologically or histologically diagnosed dysplasias (or borderline lesions), and for the grading of malignancies.

10.5.2.1 Identification of Neoplastic Urothelial Cells

Bass et al. (1989) found that the *sensitivity* of interactive DNA single-cell cytometry for the detection of malignant cells in urine samples was 88%, compared to only 58% with conventional Papanicolaou cytology (*n*=33). Even with grade 1 and 2 urothelial carcinoma, the sensitivity was still 86% (vs. 33% with cytology). These authors use the detection of cells >5c as markers for malignancy. The sensitivity of DNA cytometry for the identification of aneuploid cells in urothelial carcinoma, as in other tumors, is dependent on the grade of tumor malignancy. The lower the grade, the better the chance for the successful cytometric detection of aneuploidy.

The *specificity* of fluorescent image analysis was 96.7% in a study of 523 high-risk asymptomatic subjects (Parry and Hemstreet 1988). The rate of 3.3% false-positive diagnoses probably results from the approximately 8% prevalence of polyploid cells with >5c DNA content occurring normally in the urine (Biesterfeld et al. 1992). If the physiologic polyploidization of urothelial cells were taken into account in the interpretation of DNA measurements, the rate of false-positive diagnosis could probably be reduced. Koss et al. (1987) achieved a sensitivity of 60% with DNA single-cell cytometry in 30 tumor cell-positive urine samples versus only 23.2% with conventional Papanicolaou cytology. Because these authors

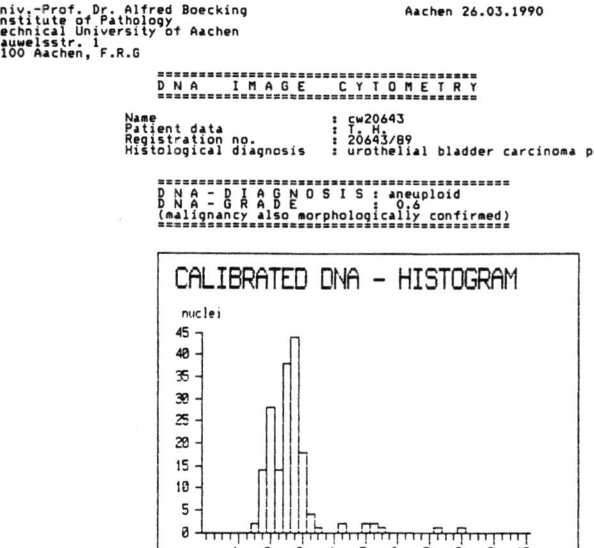

```
Univ.-Prof. Dr. Alfred Boecking                Aachen 26.03.1990
Institute of Pathology
Technical University of Aachen
Pauwelsstr. 1
5100 Aachen, F.R.G

==================================================
D N A     I M A G E     C Y T O M E T R Y
==================================================

Name                     : cw20643
Patient data             : T. H.
Registration no.         : 20643/89
Histological diagnosis   : urothelial bladder carcinoma pT2

==================================================
D N A - D I A G N O S I S : aneuploid
D N A - G R A D E         : 0.6
(malignancy also morphologically confirmed)
==================================================
```

CALIBRATED DNA - HISTOGRAM


```
The threshold for the assumption of malignancy is the detection of >= 3
aneuploid nuclei with a DNA content >= 5c, as you excluded :
            - polyploidization > 4c
            - cytologically detectable virus infection
            - irradiation or cytostatic therapy

In this case the DNA - diagnosis is  ANEUPLOID,
because at least three aneuploid cells with a DNA content >= 5c
were found (5cEE >= 3).
Within the relevant cell population nuclei were   n o t  selected.
The DNA-Grade of malignancy is based on the variance of the
tumor cell values around the normal 2c peak.
```

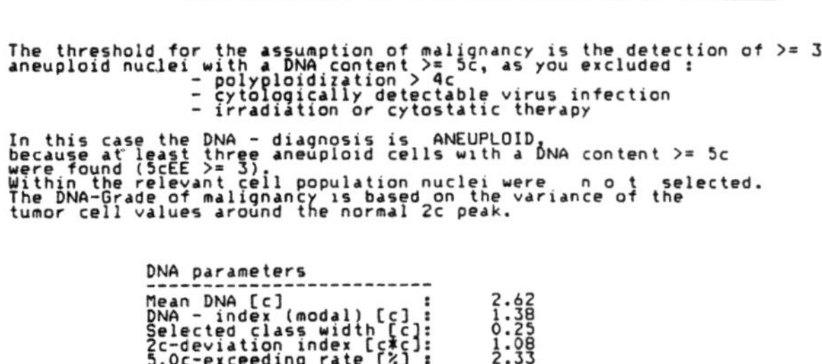

```
DNA parameters
--------------------------------
Mean DNA [c]              :    2.62
DNA - index (modal) [c]   :    1.38
Selected class width [c]  :    0.25
2c-deviation index [c*c]  :    1.08
5.0c-exceeding rate [%]   :    2.33
5.0c - exceeding events   :       4
9.0c - exceeding events   :       0
```

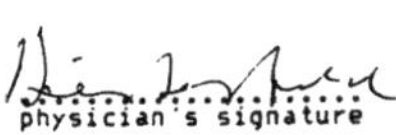

Fig. 10.8. **DNA cytometric printout for a grade 1 urothelial carcinoma, including the DNA histogram and diagnostic analysis:** aneuploid, DNA malignancy grade 0.6

used only the position of the stemline to detect aneuploidy ("diploid"/aneuploid) and did not also take into account individual aneuploid cells >5c, their sensitivity was significantly below that of Parry and Hemstreet (1988). Presumably the sensitivity and specificity of interactive DNA cytometry could be further improved by combining the single-cell and stemline detection of DNA aneuploidy and by taking into account the polyploidization of urothelial cells. We suggest that the identification of more than 10% cells >5c and/or a stemline outside the ranges of 2c, 4c, or 8c ± the standard deviation of the reference cell population should serve as the criterion for the DNA cytometric detection of aneuploidy, and thus of neoplasia, in urine samples. If one or both of the aneuploidy indices are positive, neoplasia may be assumed. Applied in this way, interactive DNA cytometry could increase the sensitivity of conventional urinary cytology.

10.5.2.2 Discriminating Benign and Malignant Dysplasias

Fossa (1977) found DNA aneuploidies suggestive of malignancy (cells >5c) in six of 12 moderate urothelial dysplasias in proximity to overt bladder carcinomas. In 50% of cases, then, malignancy could be diagnosed in the stage of dysplasia by the use of DNA cytometry. This suggests that dysplasias which exhibit an aneuploid DNA distribution (by single-cell or stemline interpretation) are prospectively malignant. We do not subscribe to the hypothesis that specific grades of dysplasia correlate with specific DNA distribution patterns, as there is no biological basis for this assumption. Rather, a dysplasia in which aneuploid cells are detected must be classified as prospectively neoplastic even though subjective morphology does not yet support that diagnosis. This is understandable when one considers that malignant transformation occurs at the DNA level and usually has chromosomal effects, whereas morphologic alterations appear later as epiphenomena. To date, however, there have been no precise studies on the prospective diagnosis and follow-up of malignancy in dysplastic urothelium.

10.5.2.3 Grading of Malignancy

Lederer et al. (1972) and Fossa (1975) were the first to document the suitability of the DNA distribution pattern as a grading parameter for bladder carcinoma. These authors used the position of the DNA stemline as their principal grading parameter but did not correlate their results with follow-up. Fossa et al. (1977) noted a statistically significant prolongation of survival in 63 patients with a "diploid" tumor stemline, compared with 60 patients with a "nondiploid" line. Hofstädter et al. (1984), in a study of 64 patients, found a close correlation of the stemline quotient SQ (equivalent to the modal DNA index) and diploid deviation quotient (DDQ; Fossa 1975) with the histologic grade according to Bergkvist et al. (1965) and the depth of tumor infiltration. The survival time of patients with "diploid" tumors (SQ =1.1) differed significantly from that of patients with aneuploid tumors

(SQ>1.1). It was found that the DDQ could be used to differentiate 30 groups of bladder cancer patients who had significantly different survival times. The frequency of tumor recurrence and the interval to recurrence also correlated significantly with the DDQ. Fossa and Kaalhus (1985) showed that the stemline quotient in 123 patients was prognostically relevant, regardless of the tumor stage. This means that the DNA determination had significant prognostic implications in every stage.

Böcking et al. (1990), in a study of 117 bladder cancer patients, investigated the prognostic influence of tumor stage (TNM), histologic grade (Mostofi et al. 1973), the subjective DNA histogram classification according to Fossa (1975), the mean DNA content, the differentiation of DNA stemlines into "diploid" versus aneuploid, and the DNA grade of malignancy (Böcking and Auffermann 1986). The TNM stage demonstrated the highest correlation with patient survival (Fig. 10.9). Multivariant regression analyses (Cox 1972) showed that histologic grade, another independent variable, demonstrated the second highest correlation with survival time (Fig. 10.10). When this variable was omitted from the models due to its insufficient interindividual reproducibility of only 62%, the DNA malignancy grade was introduced into the model as a further independent prognostic variable. Through DNA malignancy grading, it was possible to distinguish three patient groups with significantly different survival times (Fig. 10.11). The subjective histogram classification showed as little prognostic significance as the simple differentiation between "diploid" and aneuploid tumors. Nuclear size and its variability discriminated only two groups with significantly different lengths of survival. The interindividual reproducibility of DNA malignancy grading was investigated for 20 different tumors. With a correlation coefficient of $r=0.97$, it was significantly higher than that of subjective malignancy determinations, for which Ooms et al. (1983), for example, report a reproducibility of only 41%–58%. Thus, DNA malignancy grading provides a reproducible parameter that is equivalent to subjective morphologic grading in its prognostic relevance.

10.6 DNA Flow Cytometry

10.6.1 Measurement Systems

In DNA flow cytometry, isolated cells stained with a fluorescent dye that binds specifically to DNA (e.g., acridine orange, ethidium bromide, DAPI) flow in single file through a laser beam. The cells, enclosed by a fluid jet and accelerated under high pressure to a velocity of about 10 m/s, can be measured at a rate of at least 100 cells/s (e.g., FACStar plus, Becton-Dickinson, USA; Fig. 10.12). The minimum cell count in a measurable sample is 10 000, however, so hypocellular samples cannot be processed. The DNA content per cell corresponds to the level of emitted fluorescent light, which is measured by photomultipliers. The results are printed out as "scatter plots" and histograms and represent a summation result for the cell population as a whole, with no differentiation of specific subpopulations. *Advantages* of flow cytometry are that the results are available quickly (e.g., within 15 min) and are highly representative owing to the large number of measured cells. This results in a more precise determination of the DNA index of the tumor stemline. Another advantage is the ability to evaluate several parameters at once in the same sample, such as the nuclear DNA content, nuclear size, cellular protein content and, recently, various immunologic markers. This type of study utilizes several different labels in the same cell, which are measured at different wavelengths. Sophisticated systems (see Fig. 10.12) include a device for sorting out cells with specific detected properties, such as a DNA content >5c, for separate morphometric analysis in a TV image analysis system (Tanke et al. 1983). Flow cytometry also has several *disadvantages* compared with single-cell cytometry:

- The samples must contain *no fewer* than approximately 10 000 cells. Hypocellular samples must be analyzed by single-cell cytometry.
- Besides the routine cytologic specimens that are already available, an *additional sample* must be collected specifically for flow cytometry. To flow-process material from solid tumors, special cell recovery techniques must be used that employ mechanical and enzymatic measures (Vindelov et al. 1983; Hedley et al. 1983).
- Due to the lower resolution limit of approximately 1%, (tumor) cells that occur in the urine with a lower prevalence are not detected. Thus, *rare event detection is not possible*, and the sensi-

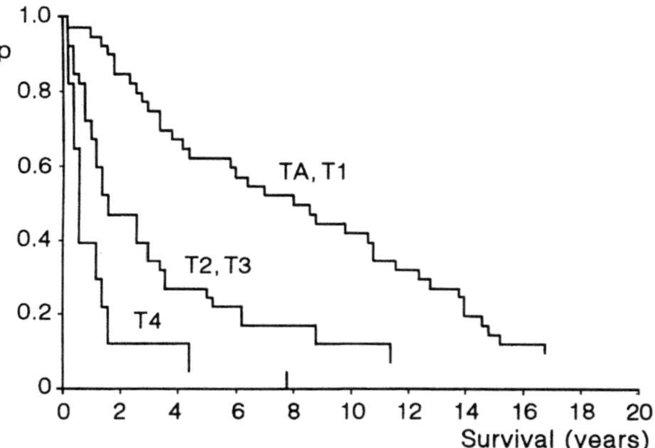

Fig. 10.9. Kaplan-Meier survival curves for 117 patients with bladder carcinoma, arranged according to pathologic stages pT1–4. (From Böcking et al. 1990)

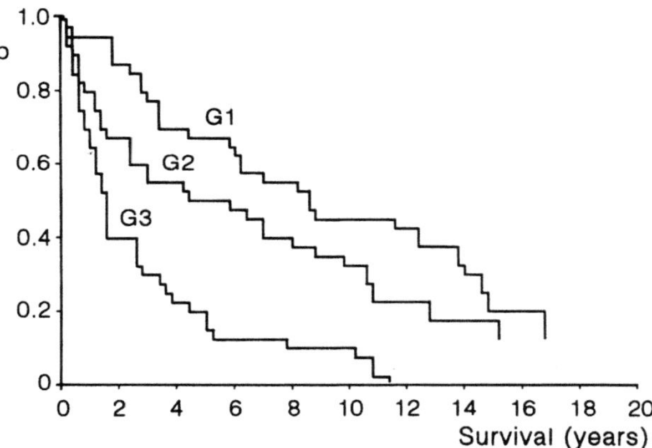

Fig. 10.10. Kaplan-Meier survival curves for 117 patients with bladder carcinoma as a function of the histologic malignancy grade according to Mostofi et al. (1973). (From Böcking et al. 1990)

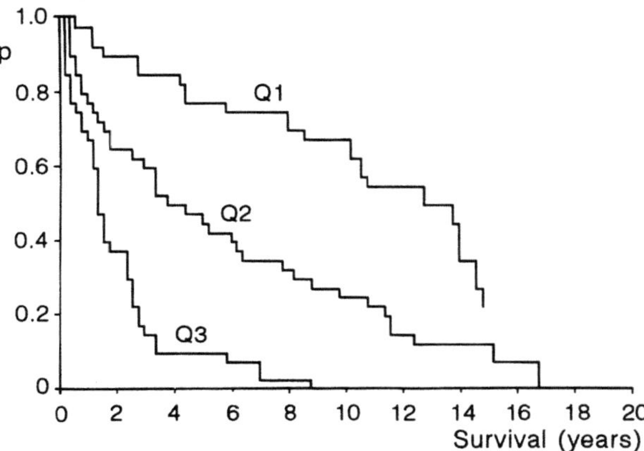

Fig. 10.11. Kaplan-Meier survival curves for 117 patients with bladder carcinoma as a function of DNA malignancy grade. (From Böcking et al. 1990)

tivity of the method is particularly limited by the low prevalence of atypical cells in low-grade and low-stage urothelial tumors.

- Because normal urothelial cells, squamous epithelial cells, inflammatory cells, and histiocytes are always measured in addition to questionable tumor cells and cannot be reliably differentiated from them, one cannot confidently assign the measurements to specific cell populations. Thus, DNA malignancy grading is not feasible in a mixed cell population. Because individual measured DNA values are not processed, there is no automatic accounting for smaller but prognostically relevant DNA values. Thus, DNA flow cytometry usually permits only a *rough* and biologically unwarranted *differentiation into "diploid" and "nondiploid" tumors.*

- Individual cells that have already been morphologically classified cannot be measured. Consequently, flow cytometry cannot be used for the DNA cytometric evaluation of urothelial dysplasias.

- *Artifacts* such as nuclear aggregates are indistinguishable from tumor cells.

- *Remeasurement* of the same sample for quality control is *not* possible due to the fading effect.

- Flow cytophotometers usually must be serviced and adjusted by *specially trained personnel.*

10.6.2 Diagnostic Results

10.6.2.1 Identification of Neoplastic Urothelial Cells

Klein et al. (1982) showed that DNA flow cytometry can detect aneuploidy 12–18 months before the appearance of a cystoscopically detectable tumor. These findings demonstrated the possibility of early cancer diagnosis by flow cytometry. Devonec et al. (1982), in tests on 110 bladder irrigation specimens from patients with conservatively treated low-stage bladder tumors, found that DNA flow cytometry was more sensitive than conventional subjective cytodiagnosis in detecting cytologic abnormalities. All 34 samples with positive cytology were diagnosed as atypical by flow cytometry, and there were an additional 39 cases in which only flow cytology was positive.

Generally, success in the detection of aneuploid DNA distributions correlates closely with the tumor stage and grade of malignancy: T0=0%, T1=27%, T2=71.4%, T3/4=75%; G0=0%, G1=30%, G3=77.0% (Chin et al. 1985).

Fig. 10.12. **FACStar Plus flow cytometer with cell sorter** (Becton-Dickinson, USA)

The average sensitivity of the DNA flow cytometric detection of aneuploidy in bladder irrigation specimens from bladder cancer patients as reported in the literature (Table 10.2) is 79%. Badalament et al. (1988) report an average sensitivity of 87% in a review of all 462 bladder carcinomas measured at the Memorial Sloan Kettering Cancer Center in New York. The average sensitivity of urinary cytology according to these studies is only 68.7%. By combining both methods, the sensitivity can be increased to 93%. The average specificity of DNA flow cytometry in bladder washings is 92% as reported in the literature, versus 97% for urinary cytology (see Table 10.2).

Thus, the sensitivity of urinary DNA flow cytometry exceeds that of conventional cytometry by more than 10%. Because the diagnostic capabilities of both methods are supplementary, they should be applied in an adjunctive rather than competitive fashion. Because the specificity of flow cytometry is compromised by a relatively high (8%) rate of false-positive findings, especially in patients with inflammations and stone disease, a positive DNA flow cytometric diagnosis should always be substantiated by another diagnostic technique. Tanke et al. (1983) suggested that DNA image cytometry would provide an accurate check when performed on abnormal nuclei that have been selected out by flow cytometry with a cell sorter, but this would be too costly for practical use.

10.6.2.2 Discriminating Benign and Malignant Dysplasias

Flow cytometry is poorly suited for this task because it does not permit the selective DNA measurement of cells previously diagnosed as dysplastic by morphologic evaluation. However, flow cytometry can often detect aneuploid cells in cystoscopically normal-appearing bladder mucosa that is located near an overt tumor (Hostmark et al. 1984; 54%). This provides an index for systemic tumor involvement of the urothelium.

10.6.2.3 Grading of Malignancy

Although DNA flow cytometry can scarcely do more than discriminate "diploid" from aneuploid tumors, this differentiation in itself has significant prognostic implications. Tribukait and Esposti (1978) found a correlation between histologic tumor grade, the occurrence of invasion, and the frequency of aneuploid cell populations. Later these authors (Tribukait et al. 1979) showed that the percentage of cells in the S phase is also prognostically significant. In 229 operatively treated patients with stage Ta/T1 bladder cancers, this same group (Gustafson et al. 1982a) showed that progression occurred exclusively in tumors with an "aneuploid" DNA distribution pattern, whereas none of the "diploid" tumors exhibited progressive behavior. Similar results were reported by deVere White et al. (1988a,b).

In patients with carcinoma in situ, more rapid progression was observed when more than one aneuploid stemline was detected (Gustafson et al. 1982b). There are no published reports of a correlation between DNA flow cytometric results and length of survival in patients with bladder carcinoma. This offers further evidence that DNA flow cytometry permits only a relatively gross prognostic evaluation of bladder carcinomas.

Bretton et al. (1989) used flow cytometry for monitoring therapeutic response in bladder cancer patients treated with BCG. While a negative examination was interpreted as evidence of therapeutic response, a positive result was considered a predictor of tumor progression.

Thus, rapid DNA flow cytometry appears to have value for bladder carcinoma screening in high-risk groups, for predicting invasive growth in Ta/T1 carcinomas, and for identifying tumor recurrence and confirming response to conservative treatment. However, all the prognostic information derived from DNA flow cytometry can also be furnished by DNA single-cell cytometry. The latter procedure, moreover, usually permits a finer prognostic evaluation.

Through the addition of further quantitative parameters to flow cytometry, it is hoped that the sensitivity and specificity of this procedure can be increased (Ley et al. 1989; Wright et al. 1989).

10.7 Automated Systems

It has been an expressed goal of cytoautomation to develop machines for the automatic prescreening of smears from the uterine cervix. With minor software modifications, these devices can also be used for the analysis of urinary sediment. Automated systems should assume the function of a cytologic assistant in the prescreening of cell smears, and do so at a consistently high performance level that is not degraded by fatigue. Additional requirements are a false-positive rate <10%, a false-negative rate <5%, the ability for a cytologist to check the machine diagnosis in the same specimen, a processing time of about 5 min or less per specimen, a processing rate of approximately 20 000 specimens per year, and a procurement cost below 500 000 DM (Sprenger 1985).

Koss' work with the TICAS system (Taxonomic Intra-Cellular Analytic System) achieved no better than a sensitivity of approximately 81% and specificity of approximately 78% using manual cell selection (Koss et al. 1975, 1978a,b, 1980). Similarly, use of the SAMBA system has not yet produced results that satisfy the requirements of an automated cytometric system (Brugal et al. 1986).

One development that could find use both for cytoautomation and for cytometrically assisted diagnosis is the CAESAR system (Gahm and Aikens 1990). Preliminary results in 63 urine samples indicate a sensitivity of 75% and specificity of 80%. This system employs a Kontron IBAS 2 image analysis system. The CAESAR system is not yet marketed commercially as a software-hardware unit.

The most advanced automated system at present is the LEYTAS II system (or MIAMED-ACA) manufactured by the Wild-Leitz Co. of Wetzlar, Germany (Ploem et al. 1979). Sensitivity/specificity levels of 100%/83.5%,

Table 10. 2. **Accuracy rates of the conventional and DNA flowcytometric diagnosis of urothelial carcinoma in spontaneously voided urine according to reports in the literature**

Authors	Cytologic sensitivity (%)	Flow cytometry		Cytology and flow cytometry sensitivity (%)
		Sensitivity (%)	Specificity (%)	
Klein et al. (1982)		93 ($n=208$)	98 ($n=100$)	
Murphy et al. (1986)	75 ($n=105$)	78 ($n=105$)	78.6 ($n=28$)	95 ($n=109$)
Badalament et al. (1986)	59.1 ($n=60$)	80.3 ($n=66$)		
Badalament et al. (1987)	61($n=70$)	83 ($n=70$)		
Jitsukava et al. (1987)	43 ($n=56$)	73 ($n=56$)		80 ($n=56$)

98.3%/88.4%, and 92.4%/96.7% have been determined for the automated screening of smears from the uterine cervix (Ploem 1989). Tanke et al. (1982) used the system for the automated screening of urine samples and achieved a sensitivity of 92.3% with a specificity of 84.1%. With the recent decision of the Wild-Leitz Co. to discontinue the development of image analysis systems for medicine, valuable developmental work invested in this outstanding system has been lost. There is little reason to expect that high-performance automated cytometric systems will be offered on the European market in the foreseeable future. The routine clinical use of automated systems is further constrained by the costly and involved preparatory procedures needed to produce cellular specimens compatible with automated analysis. As a result, routine smears cannot yet be processed on automated systems with satisfactory effectiveness. Until practical, cost-effective, automatic cyto-preparatory techniques become available, the routine use of automated cytometric systems for mass screening will remain an elusive goal.

Automated systems do have a limited role in analyzing the DNA distribution in tumor specimens for the purpose of determining prognosis (Stöckle et al. 1987). This task can be performed equally well by economical interactive systems, though these may require a somewhat greater investment of time. This drawback is offset, however, by the simpler technique of specimen preparation for interactive measurements. It is imperative that the pathologist be able to check at any time on the actual cell population that is being measured. Today these systems are incapable of identifying specific cell populations like normal urothelial cells or connective-tissue cells as such with any degree of accuracy. Automated systems are not yet able to process diagnostically questionable routine specimens with dysplasias or borderline lesions and classify them as malignant or benign by means of DNA cytometry.

10.8 Indications for Diagnostic DNA Cytometry

DNA measurements in urothelial carcinoma cells can provide more objective, more reproducible, and usually more prognostically relevant information than a conventional histologic or cytologic evaluation with malignancy grading. Thus, diagnostic DNA measurements are indicated whenever an accurate prognostic assessment can influence further diagnostic and therapeutic procedures. For example, the physician can use the DNA malignancy grade as a guide for determining further course of action following the transurethral resection of a superficial bladder carcinoma. If the DNA malignancy grade is low, a wait-and-see approach is justified, whereas a high DNA malignancy grade may warrant adjunctive chemotherapy. Today, DNA measurements for prognostic purposes are obtained most easily, economically, and precisely by means of interactive single-cell cytometry. This can be performed on any tumor-cell-positive urine sample or washing and even on biopsy material. Since modern TV image analysis systems can measure a cell nucleus in less than 1 s, larger numbers of cells, which previously were reserved for flow cytometers, can be measured within minutes. But TV image analysis is superior to flow cytometry in that the examiner knows which cell populations have been measured. The measurement of hypocellular samples (<10 000 cells) is also possible by single-cell cytometry.

In many cases of urothelial dysplasia, DNA cytometry also can establish a diagnosis of "prospective malignancy" by the detection of aneuploid cells. This provides a means for the further and possibly definitive investigation of cytodiagnostic problem cases. Only interactive DNA single-cell cytometry is appropriate for this task, as it enables the cytologist to individually identify and classify the questionable cells.

DNA flow cytometry offers a way to increase the generally unsatisfactory sensitivity of conventional urinary cytology. Centers that possess such a system and the personnel to operate and maintain it can use DNA flow cytometry as a rewarding adjunct to urinary cytology, especially for the screening of high-risk subjects. Due to the high rate of false-positive diagnoses in urine and bladder washings, however, the detection of aneuploidy by flow cytometry should always be corroborated by other investigations.

Automated systems for the automatic screening of urinary sediment are still in the developmental stage. They are inappropriate for the determination of prognostic DNA parameters in urothelial cells, which can be obtained more economically with interactive systems.

References

Aikens B, Liedtke C-E (1982) Untersuchungen zur Meßbarkeit und Objektivierbarkeit der Urinzytologie durch ein hochauflösendes mikroskopzytophotometrisches Meßsystem. Urologe [A] 21: 98–101

Atkin NB (1971) Modal DNA value and chromosome number in ovarian neoplasias. A clinical and histopathologic assessment. Cancer 27: 1964–1974

Auer GU, Caspersson TO, Wallgren AI (1980) DNA content and survival in mammary carcinoma. Anal Quant Cytol 2: 161–165

Auffermann W, Repges R, Böcking A (1985) Rapid diagnostic DNA-cytometry with an automatic microscope and a TV-image analysis system. Anal Quant Cytol 6: 179–188

Auffermann W, Urquardt M, Rübben H, Wohltmann D, Böcking A (1986) DNA grading of urothelial carcinoma of the bladder. Anticancer Res 6: 27–32

Bachmann P, Hinrichsen K (1979) Principles and methods of quantitative determination of Feulgen stained DNA with the Television Texture Analysis System (TAS). Histochemistry 60: 61–69

Badalament R-A, Gag H, Whitemore WF, Herr HW, Fair WR, Oeltgen HF, Melamed MR (1986) Monitoring intravesical bacillus Calmette-Guerin treatment of superficial bladder carcinoma by serial flow cytometry. Cancer 58: 2751

Badalament R-A, Hermansen DK, Kimmel M, Gay H, Herr HW, Fair W-R, Whitemore WF, Melamed MR (1987) The sensitivity of bladder wash flow cytometry, bladder wash cytology and voided cytology in the detection of bladder carcinoma. Cancer 60: 1423–1427

Badalament R-A, Fair WR, Whitemore WF, Melamed MR (1988) The relative value of cytometry and cytology in the management of bladder cancer: the Memorial Sloan Kettering Cancer Center-experience. Semin Urol 6: 22–30

Barlogie B (1981) Abnormal cellular DNA content as a marker of neoplasia in man. Eur J Cancer Clin Oncol 20: 1123–1125

Barlogie B, Drewinko B, Schumann J (1980) Cellular DNA content as a marker of neoplasia in man. Am J Med 69: 195–203

Bergkvist A, Ljungqvist A, Moberger G (1965) Classification of bladder tumors bases on the cellular pattern. Preliminary report of a clinical pathologic study of 300 cases with a minimum follow up of eight years. Acta Chir Scand 130: 371–378

Biesterfeld S, Gerves K, Fischer-Wein G, Böcking A (1992) Euploid polyploidization – a frequent phenomenon in nonneoplastic human tissues. J Clin Pathol (submitted)

Böcking A (1990) DNA-Zytometrie und Automation in der klinischen Diagnostik. In: Bonk U (ed) Aktuelle klinische Zytologie. Beitr. Onkol 38: 298–347

Böcking A, Auffermann W (1986) Algorithm for DNA cytophotometric diagnosis and grading of malignancy. Letter to the editor. Anal Quant Cytol Histol 8: 363

Böcking A, Adler CP, Common HH, Hilgarth M, Granzen B, Auffermann W (1984) Algorithm for a DNA-cytophotometric diagnosis and grading of malignancy. Anal Quant Cytol 6: 1–8

Böcking A, Auffermann W, Jocham D, Contractor H, Wohltmann D (1985) DNA grading of malignancy and tumor regression in prostatic carcinoma under hormone therapy. Appl Pathol 3: 206–214

Böcking A, Chatelain R, Auffermann W, Krüger GRF,

Asmus B, Wohltmann D, Schuster C (1986 a) DNA grading of malignant lymphomas I. Prognostic significance reproducibility and comparison with other classifications. Anticancer Res 6: 1205–1215

Böcking A, Chatelain R, Löhr GW, Reif M, Rosner R, Becker H (1986 b) DNA grading of malignant lymphomas II. Correlation with clinical parameters Anticancer Res 6: 1216–1223

Böcking A, Sanchez L, Stock B, Müller W (1987) Automated DNA cytophotometry. Lab Practice 36: 73–74

Böcking A, Chatelain R, Orthen U, Gier G, Kalkreuth G von, Jocham D, Wohltmann D (1988) DNA grading of prostatic carcinoma: Prognostic validity and reproducibility. Anticancer Res 8: 129–136

Böcking A, Chatelain R, Biesterfeld S, Noll E, Biesterfeld D, Wohltmann D, Goecke C (1989) DNA grading of breast cancer. Prognostic validity, reproducibility and comparison with other classifications. Anal Quant Cytol Histol 11: 73–80

Böcking A, Pollmann I, Biesterfeld S (1990) DNA grading of malignancy in urothelial carcinoma of the bladder: Prognostic validity, reproducibility and comparison with morphological parameters. 17th Meeting of the European Federation of Cytology Societies, 30 March–1 April 1989, Flims, Switzerland

Böcking A, Chatelain R, Biesterfeld S, Sanchez L, Kropff M, Stock B, Müller W (1992) MIAMED-DNA. Interactive system for rapid diagnostic DNA cytometry. Anal Quant Cytol Histol (submitted for publication)

Böhm N, Sprenger E, Schlüter G, Sandritter W (1968) Proportionalitätsfehler bei der Feulgen-Hydrolyse. Histochemie 15: 194–203

Bretton PR, Herr HW, Kimmel M, Fair WR, Whitemore WF, Melamed MR (1989) Flow cytometry as a predictor of response and progression in patients with superficial bladder cancer treated with bacillus Calmette Guerin. J Urol 141: 1332

Brugal G, Chassery J-M (1977) Un nouveau systeme d'analyse densitometrique et morphologique des preparations microscopiques. Histochemistry 52: 251–258

Brugal G, Quirion C, Vassilakos P (1986) Detection of bladder cancers using a SAMBA 200 cell image processor. Anal Quant Cytol Histol 8: 187–195

Chatelain R, Schunck T, Schindler EM, Schindler AE, Böcking A (1989 a) Diagnosis of prospective malignancy in koilocytic dysplasias of the cervix with DNA cytometry. J Reprod Med 34: 505–510

Chatelain R, Willms A, Biesterfeld S, Auffermann W, Böcking A (1989 b) Automated Feulgen staining with a temperature controlled staining machine. Anal Quant Cytol Histol 11: 211–217

Chin JL, Huben RP, Nava E, Rustum JM, Greco JM, Pantes JE, Frankfurt OS (1985) Flow cytometry analysis of DNA content in human bladder tumors and irrigation fluids. Cancer 56: 1677–1681

Coon GS, Schwartz D, Summers JL, Miller AW, Weinstein RS (1986) Flow cytometric analysis of deparaffined nuclei in urinary bladder carcinoma. Comparison with cytogenetic analysis. Cancer 57: 1594–1601

Cox DR (1972) Regression models and life tables. J Roy Statist Soc B 34: 187–200

Deeley EM (1955) An integrating microdensitometer for biological cells. J Sci Intrum 32: 263–267

Delgado R, Mikum G, Hofstädter F (1984) DNA Feulgencytophotometric analysis of single cells from paraffin-embedded tissue. Pathol Res Pract 179: 92–94

Devonec M, Darzynkiewicz Z, Kostyrka-Claps MC, Collste L, Whitemore WF, Melamed MR (1982) Flow cytometry of low stage bladder tumors: Correlation with cytologic and cystoscopic diagnosis. Cancer 49: 109–118

Farsund T, Hostmark J (1983) Mapping of cell cycle distribution in normal human urinary bladder epithelium. Scand J Urol Hepathol 17: 51–56

Feulgen R, Rossenbeck H (1924) Mikroskopisch-chemischer Nachweis einer Nukleinsäure vom Typus der Thymonucleinsäure und die darauf beruhende elektive Färbung von Zellkernen in mikroskopischen Präparaten. Z Physiol Chem 135: 203–248

Fossa SD (1975) Feulgen-DNA-values in transitional cell carcinoma of the human urinary bladder. Beitr Pathol 155: 44–55

Fossa SD (1977) DNA variations in neighbouring epithelium in patients with bladder carcinoma. Acta Pathol Microbiol Immunol Scand [A] 85: 603–610

Fossa SD, Kaalhus O (1976a) Nuclear size and chromatin concentration in transitional cell carcinoma of the human urinary bladder. Beitr Pathol 157: 109–125

Fossa SD, Kaalhus O (1976b) Computer assisted image analysis of Feulgen stained cell nuclei from transitional cell carcinoma of the human urinary bladder. Acta Pathol Microbiol Immunol Scand [A] 85: 590–602

Fossa SD, Kaalhus O (1985) The prognostic relevance of nuclear Feulgen DNA in transitional cell carcinoma of the urinary bladder. A long-term follow up study. Eur Urol 11: 418–421

Fossa SD, Kaalhus O, Scott-Knudsen O (1977) The clinical and histopathological significance of Feulgen DNA-values in transitional cell carcinoma of the human urinary bladder. Eur J Cancer Clin Oncol 13: 1155–1162

Gahm T, Aikens B (1990) CAESAR – a computer supported measurement system for the enhancement of diagnostics and quality in cytology. Micron Microsc Acta 1/2: 29–55

Graumann W (1953) Zur Standardisierung des Schiffschen Reagens. Z Wiss Mikrosc 61: 225–230

Gustafson H, Tribukait B, Esposti PL (1982a) DNA profile on tumour progression in patients with superficial bladder tumours. Urol Res 10: 13–18

Gustafson H, Tribukait B, Eposti PL (1982b) The prognostic value of DNA analysis in primary carcinoma in situ of the urinary bladder. Scand J Urol Nephrol 16: 141

Hedley DW, Rielander ML, Taylor IW, Rugg CA, Musgrove EA (1983) Method for analysis of cellular DNA-content of paraffin embedded pathological material using flow cytometry. J Histochem Cytochem 31: 1333–1335

Hedley DW, Philips J, Rugg CA, Taylor IW (1984) Measurements of cellular DNA content as an adjunct to diagnostic cytology in malignant effusions. Eur J Cancer Clin Oncol 20: 749–752

Heim S, Mittelman F (1987) Cancer cytogenetics. Liss, New York, pp 239–240

Hofstädter F, Jakse G, Lederer B, Mikuz G, Delgado A (1984) Biological behaviour and DNA cytophotometry of urothelial bladder carcinoma. Br J Urol 56: 289–295

Hostmark JG, Laerum OD, Farsund T (1984) DNA aberrations of bladder mucosa in patients with transitional cell carcinomas. Scand J Urol Nephrol 18: 113–120

Huber JC (1986) Numerische und strukturelle Chromosomenaberrationen bei gynäkologischen Malignomen. Thieme, Stuttgart

Jitsukava S, Tachibana M, Nakazomo M, Tazaki H, Addonizio JC (1987) Flow cytometry based on hetero-geneity index score compared with urine cytology to evaluate their diagnostic efficacy in bladder tumor. Urology 29: 218–222

Klein F, White FKH (1988) Flow cytometry deoxiribonucleic acid determinations and cytology of bladder washing: practical experience. J Urol 139: 275–278

Klein FA, Herr HW, Sogani PC, Whitemore WF, Melamed MR (1982) Detection and follow-up of carcinomas of the urinary bladder by flow cytometry. Cancer 50: 389–395

Koss LG, Bartels PH, Bibbo M, Freed SZ, Taylor J, Wied GL (1975) Computer discrimination between benign and malignant urothelial cells. Acta Cytol 19: 378–391

Koss LG, Bartels PH, Sychra JJ, Wied GF (1978a) Computer discriminant analysis of atypical urothelial cells. Acta Cytol 22: 382–386

Koss LG, Bartels PH, Sychra JJ, Wied GL (1978b) Diagnostic cytologic sample profiles in patients with bladder cancer using TICAS System. Acta Cytol 22: 392–397

Koss LG, Bartels PH, Wied GL (1980) Computer-based diagnostic analysis of cells in the urinary sediment. J Urol 123: 846–849

Koss L, Eppide EM, Melder KH, Wersto R (1987) DNA cytophotometry of voided urine sediment. Comparison of cytologic diagnosis and image analysis. Anal Quant Cytol Histol 9: 398–404

Lederer B, Mikuz G, Gütter W, Nedden G zur (1972) Cytophotometric investigations of the DNA content of benign and malignant transitional cell tumors of the bladder. Correlation of cytophotometric results with histological grading. Beitr Pathol 147: 379–389

Levi P, Cooper EH, Anderson CK, Patu MC, Williams RE (1969) Analysis of DNA content, nuclear size and cell proliferation of transitional cell carcinoma in man. Cancer 23: 1074–1085

Ley H, Valet G, Lehmer A, Hartung R (1989) Automatic identification of bladder tumor cells by multiple parameters in flow-cytometry. J Urol 141: 294 A (abstract 497)

Marschner S (1992) DNA-Malignitätsgrading des Mammakarzinoms. Prognostische Validität und Reproduzierbarkeit. Dissertation, RWTH Aachen

Montironi R, Scarpelli M, Pisani E, Ausuini G, Marinelli F, Marizzi G (1985) Noninvasive papillary transitional-cell tumors. Kayometric and DNA-content analysis. Anal Quant Cytol Histol 7: 337–342

Mostofi FK, Sobin LH, Torloni H (1973) Histological typing of urinary bladder tumors. International classification of tumors. WHO, Geneva

Murphy WH, Emerson LD, Chandlers RW, Moinuddin SM, Soloway MS (1986) Flow cytometry versus urinary cytology in the evaluation of patients with bladder cancer. J Urol 136: 815–821

Ooms EC, Anderson WA, Alons CL, Boon ME, Veldhiuzen RW (1983) Analysis of the performance of pathologists in the grading of bladder tumors. Hum Pathol 14: 144–150

Parry W, Hemstreet GP (1988) Cancer detection by quantitative fluorescence image analysis. J Urol 139: 270–274

Pauwels RP, Smeets WW, Geraedts JP, Debruyne FM (1987) Cytogenetic analysis in urothelial cell carcinoma. J Urol 137: 210–215

Pauwels RPE, Smeets AWGB, Schapers RWM, Geraedts JPM, Debruyne FMJ (1988) Grading in superficial bladder cancer. Cytogenetic classification. Br J Urol 61: 135–139

Pfitzer P, Vyska K, Stecher G (1976) Analysis of DNA histograms by computer. Beitr Pathol 159: 157–185

Ploem JS (31 8 1989) Das Leitz-MIAC-System; erweiterte

Fragestellungen. Arbeitstreffen: Bestandsaufnahme und Perspektiven der Zytoautomation. Institut für Physikalische Elektronik, University of Stuttgart

Ploem JS, Verwoerd N, Bonnet J, Koper G (1979) An automated microscope for quantitative cytology combining television image analysis and stage scanning microphotometry. J Histochem Cytochem 27: 136–143

Ploem-Zaaijer JJ, Beyer-Boon ME, Leyte-Veldstra L, Ploem JS (1979) Cytofluorometric and cytophotometric DNA measurements of cervical smears stained using a new bicolor method. In: Pressman NJ, Wied GL (eds) Automation of cancer cytology and cell image analysis. Tutorials of Cytology, Chicago, pp 225–235

Ratliff JE, Klein FA, White FKH (1985) Flow cytometry of ethanol-fixed versus fresh bladder barbotage specimens. J Urol 133: 958–960

Roels F (24 2 1990) Personal communication. CAAC-Meeting, Brussels

Rübben H, Rathert P, Roth S, Hofstädter F, Giani G, Terhorst B, Friedrichs R (1989) Exfoliative Urinzytologie, 4th edn. Harnwegstumorregister. Publication of the Fort- und Weiterbildungskommission der Deutschen Urologen. Arbeitskreis Onkologie. Sektion Urinzytologie, Essen

Sandberg AA (1986) Chromosomal changes in bladder cancer: clinical and other correlations. Cancer Genet Cytogenet 19: 163–175

Sandritter W (1981) Allgemeine Pathologie. Schattauer, Stuttgart

Sandritter W, Böhm N (1975) DNA in human tumors: a cytophotometric study. Curr Top Pathol 60: 151–219

Sandritter W, Carl M (1966) Cytophotometric measurement of the DNA-content (Feulgen reaction) of malignant human tumors. Acta Cytol 10: 20–30

Seidel A, Sandritter W (1963) Cytophotometrische Messungen des DNS-Gehaltes eines Lungenadenoms und einer malignen Lungenadenomatose. Z Krebsforschung 65: 555–559

Smeets AWGB (1987) Chromosome and flow cytometric studies of urinary bladder cancer. Dissertation, University of Maastricht

Sprenger E (1985) Automation in der Zytodiagnostik. In: Deutsche Forschungs- und Versuchsanstalt für Luft- und Raumfahrt eV (DFVLR) (ed) Automation der zytologischen Diagnostik. Verlag TÜV Rheinland, Cologne, pp 7–16

Sprenger W, Hilgarth M, Schaden M (1974) A follow up of doubtful findings in cervical cytology by Feulgen DNA cytophotometry. Beitr Pathol 152: 58–65

Stöckle M, Tanke HG, Mesker WE, Ploem JS, Jonas U, Hohenfellner R (1987) Automated DNA image cytometry: a prognostic tool in urinary bladder carcinoma? World J Urol 5: 127–132

Summers JL, Falor WH, Ward R (1989) A 10-year analysis of chromosomes in non invasive papillary carcinoma of the bladder. J Urol 125: 177–178

Tanke HJ, Ploem JS, Jonas U (1982) Kombinierte Durchflußzytometrie und Bildanalyse zur automatisierten Zytologie von Blasenepithel und Prostata. Aktuel Urol 13: 109

Tanke HJ, Driel-Kulker AMJ van, Cornelisse CJ, Ploem JS (1983) Combined flow cytometry and image cytometry of the same cytological sample. J Microsc 130: 11–22

Tavares AS, Costa J, Maia JC (1973) Correlation between ploidy and prognosis in prostatic carcinoma. J Urol 109: 676–679

Tavares AS, Costa J, Carvalho A de, Reis M (1986) Tumor ploidy and prognosis in carcinomas of the bladder and prostate. Br J Cancer 20: 438–441

Tribukait B, Esposti PL (1978) Quantitative flow-microfluorometric analysis of DNA in cells from neoplasms of the urinary bladder: correlation of aneuploidy with histological grading and cytological findings. Urol Res 6: 197–200

Tribukait B, Gustafson H, Esposti P (1979) Ploidy and proliferation in human bladder tumors as measured by flow-cytofluorometric DNA-analysis and its relations to histopathology and cytology. Cancer 43: 1742

Vere White RW de, Deitch AD, West B, Fitzpatrick JM (1988a) The predictive value of flow cytometric information in the clinical management of stage 0 (Ta) bladder cancer. J Urol 139: 279–282

Vere White RW de, Deitch A, Strand M (1988b) DNA flow cytometry using urine samples: diagnostic accuracy. J Urol 139: 321 A

Vindelov LL, Christensen IJ, Nissen NI (1983) A detergent-trypsin method for the preparation of nuclei for flow-cytometric DNA analysis. Cytometry 3: 323

Walker BE (1959) Polyploidy and differentiation in transitional epithelium of mouse urinary bladder. Chromosoma 9: 105–118

Wied GL, Bartels PH, Bahr G, Oldfield DG (1968) Taxonomic intra-cellular analytic system (TICAS) for cell identification. Acta Cytol 12: 180–204

Wied GL, Bartels PH, Dytch HE, Bibbo M (1983) Rapid DNA evaluation in clinical diagnosis. Acta Cytol 27: 33–37

Wright GL, Alexander JP, Konduba AM, Schlossberg SM, Schellhammer PF (1989) Multiparameter flow-cytometric analysis of low grade, low stage bladder washings and voided urin specimens. J Urol 141: 293 A

11 Immunocytology of Urothelial Tumors[*]

B. J. Schmitz-Dräger

CONTENTS

11.1 Introduction

In principle, urine is an ideal medium for the detection for urothelial tumors. This is based in part on the exclusive contact of the urine with the excretory portion of the urinary tract, which permits a given volume of urine to bathe the tumor for up to several hours. Other advantages are that specimens can be collected noninvasively and in virtually unlimited quantities. Urinary cytology and flow cytometry are techniques that exploit the advantages of this medium. Urinary cytology has become a particularly well established routine clinical modality in recent years and, together with cystourethroscopy, forms the second cornerstone in the diagnosis of transitional cell carcinoma (TCC) of the bladder. We know from experience, however, that the sensitivity of urinary cytology is not always adequate, especially in the detection of well-differentiated tumors (Dubernard et al. 1982; Koss et al. 1985; Zein et al. 1984). Moreover, the interpretation of specimens can vary greatly depending on the experience of the examiner.

The introduction of *hybridoma technology* has opened up entirely new approaches to the identification of tumor-associated parameters (Köhler and Milstein 1975). With this technique, it is possible to develop monoclonal, determinant-specific antibodies that can detect molecular changes in tumor cells at a stage when signs of malignant transformation are not yet visible by light microscopy. In past years a number of scientists have used the hybridoma technique as a means of identifying antigens that would be useful for the detection of urothelial tumor cells in urine (survey in Bander 1987).

11.2 Monoclonal Antibodies

A major concern in tumor immunology is the search for antigenic structures by which tumor cells can be differentiated from normal cells. Attempts to achieve this goal by the use of heterologous antisera have failed due to the inadequate specificity of the polyvalent antisera and the associated cross-reactions that occur. The knowledge that every plasma cell produces a unique antibody has prompted numerous attempts to immortalize B cells in cultures. There are two possible approaches to solving this problem:

– The permanent growth of B cells can be stimulated by transforming the B cells with mutagens or viruses.
– Immortalization can be accomplished by fusing B lymphocytes with a permanent cell line. This technique combines antibody production and perpetual growth in one hybrid cell.

Köhler and Milstein successfully implemented the second solution in 1975 and were the first to produce *monoclonal, determinant-specific antibodies* (MABs) to a desired antigen. At that time the cells were fused using inactivated Sendai virus, but today the cells are usually fused with polyethylene glycol (PEG), which is more efficient and easier to handle (Fig. 11.1).

[*] This study was funded by the Federal Ministry for Research and Technology (01 GA 8701/7) and the Ministry for Science and Research of the state of North Rhine-Westphalia.

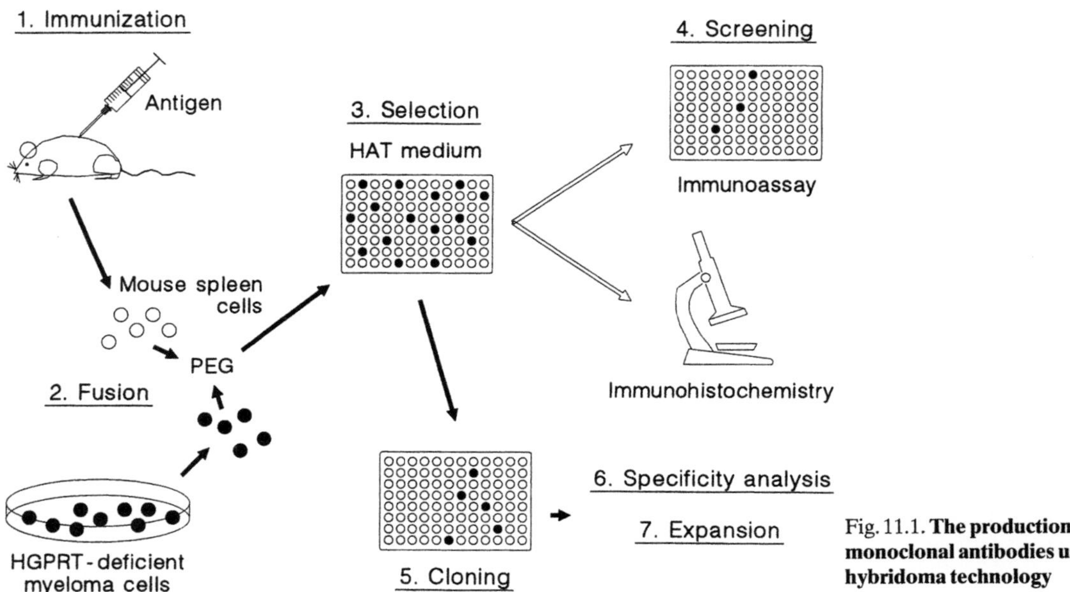

Fig. 11.1. **The production of monoclonal antibodies using hybridoma technology**

Since a fusion experiment cannot cover all the cells, and fusions between cells occur in a statistical manner, formation of the desired fusion products of lymphocytes and myeloma cells ("hybridoma" cells) is accompanied by the formation of additional products from the fusion of myeloma cells and lymphocytes. While unfused lymphocytes and lymphocyte hybrids cannot proliferate in vitro, unfused myeloma cells or fusion products from myeloma cells can grow indefinitely and overwhelm the hybridoma cells. Köhler and Milstein solved this problem by adding a step that selectively killed the myeloma cells. They used a myeloma cell mutant that did not express the enzyme hypoxanthine guanine phosphoribosyl transferase (HGPRT). The key enzyme of an accessory metabolic pathway in purine biosynthesis, HGPRT catalyzes the conversion of hypoxanthine to inosine monophosphate and of guanine to guanosine monophosphate. After the main metabolic pathway has been blocked by the cytostatic drug aminopterin, the nucleic acids adenosine triphosphate and guanosine triphosphate can form only in cells that possess the enzyme HGPRT. Because normal lymphocytes are HGPRT-positive, hybridomas, unlike myeloma cells, can grow in the presence of aminopterin, provided the substrate for HGPRT, hypoxanthine, is present in sufficient quantities. Since aminopterin also blocks the conversion of deoxyuridine monophosphate to deoxythymidine monophosphate, the presence of thymidine is necessary for the production of deoxythymidine triphosphate via an accessory metabolic pathway. Köhler and Milstein called their selective medium HAT because it was composed of *h*ypoxanthine, *a*minopterin and *t*hymidine. Culturing the fused cells in HAT medium ensured that only hybridoma cells would survive (see Fig. 11.1).

After the hybridomas have been grown, a preliminary *specificity analysis* (screening) is done to identify the hybridomas that produce a potentially useful antibody. This is an essential step, because several hundred hybridomas are generated in a single fusion experiment. As the culturing of so many hybridomas is prohibitive in terms of cost and labor, screening is done to eliminate hybridomas that definitely do not produce a useful antibody. Immunoassays or the immunohistochemical examination of selected tissue sections are the methods most commonly employed.

The hybridomas that are left *after screening* are cloned. *Cloning* (isolating) the cells ensures that the results of the screening tests are indeed based on the binding of one MAB and are not caused by a mixture of several antibodies. Cloning also serves another important function. Cellular fusion always yields a genetically unstable hybrid. The elimination of excess chromosomes or chromosomal fragments is a common event. If this process affects genes that play a key role in antibody production, the cell will suspend production of the antibodies. Such nonproducer cells require less energy to grow and thus can overwhelm hybridoma cells that continue to produce antibodies.

The final step in the production of MABs is an intensive immunohistochemical and immunocytologic *specificity analysis*. If the screening results are confirmed and the specificity tests indicate a useful antibody, the cells can now be grown in mass culture, and antibody-containing supernatant can be harvested from the cell cultures.

Hybridoma technology can be used to generate MABs to any desired antigens, thereby avoiding the problem of quality fluctuation in antiserum batches. The major impact of this technique on the biosciences in general was acknowledged by the awarding of the Nobel Prize for Medicine in 1984, just 9 years after the technology was first described.

11.3 Monoclonal Antibodies Directed Against Urothelial Tumors

Since the advent of hybridoma technology, a number of scientists have worked to develop monoclonal antibodies directed against urothelial tumors (Bander 1987). Information on some of these MABs is presented in Table 11.1. The first report on the production of MABs reactive with urothelial tumors was published by Grossman in 1983. The two antibodies, designated A8 and A80, were directed against antigens of the cell line RT4, which was used for immunization.

Shortly thereafter, Fradet et al. (1984) reported on the production of several monoclonal antibodies directed against urothelial tumors. Two of these antibodies reacted best with moderately or poorly differentiated TCCs.

A glycoprotein with a molecular weight of 85 kD, the antigen identified by MAB T43 was expressed not just on urothelial tumors but also on proximal tubular epithelium, epidermis, and lymphocytes. MAB T138 is also directed against a glycoprotein (gp 25). Cross-reactions were noted with endothelial cells of blood vessels and lymphatics.

Arndt et al. (1987) reported on MAB 486 P 3/12, which is directed against a tumor-associated antigen of bladder carcinoma. The percentage of antigen-positive cells in the specimen varies from tumor to tumor, showing no apparent correlation with the grade of tumor malignancy.

The antigen was not detected in only one of the 19 grade 2 tumors that were studied. The corresponding antigen is also present on some cells of the normal bladder mucosa and on other normal lymphatic tissues, granulocytes, gastrointestinal mucosa, and endometrium. Renal tumor cells and alimentary tract tumors also express the antigen identified by MAB 486 P. Biochemically, the antigen is a glycoprotein with a molecular weight of 200 kD that apparently belongs to the family of CEA proteins.

MAB Due ABC 3, produced by Schmitz-Dräger et al. (1988), likewise detects an antigen that is present on most cells in various urothelial tumors but also may occur on normal urothelial cells (Figs. 11.2, 11.3). The antigen-positive cells are present to varying degrees in the various tumors investigated. In contrast to the antigen recognized by MAB 486 P 3/12, the antigen detected by Due ABC 3 appears to have a greater degree of expression in poorly differentiated tumors.

The antigen is present on various normal tissues such as granulocytes, colon epithelium, proximal tubular epithelium, and various tumor cells of nonurothelial origin such as renal tumors and breast carcinoma. Biochemically, the antigen identified by MAB Due ABC 3 is a ganglioside (Decken et al. 1992).

Table 11.1. **Monoclonal antibodies directed against transitional cell tumors**

Antibody (subclass)	Immunogen	Antigen	Specificity (besides TCC)	Study
A2 (IgG1)	RT4	?	Exclusively RT4 cells[a]	Grossman (1983)
A80 (IgG1)	RT4	?	Various cell lines[a]	
T43 (IgG1)	T24	gp 85	Proximal tubular epithelium, epidermis, lymphocytes	Fradet et al. (1984)
T138 (IgM)	T24	gp 25	Vascular and lymphatic endothelium	
G4 (IgM)	?	gp 80	16 Normal tissues and 14 neoplastic tissues of various origins negative[b]	Chopin et al. (1985)
E7 (IgM)	?	?	16 Normal tissues and 14 neoplastic tissues of various origins negative[b]	
486 P 3/12 (IgM)	486 P	gp 200	Various normal and neoplastic tissues, occasionally normal urothelium	Arndt et al. (1987)
Due ABC 3 (IgM)	SW 1710	Ganglioside	Various normal and neoplastic tissues, occasionally normal urothelium	Schmitz-Dräger et al. (1988)
BLCA-8 (IgG3)	UCRU-BL-17CL	Glycolipid	Exclusively tumor cell lines of urothelial origin[a]	Walker et al. (1989)

[a] No immunohistochemical specificity analysis.
[b] Not further specified.

Monoclonal antibody BLCA-8 was recently described by Walker et al. (1989). The antibody was reactive with freshly cultured urothelial tumor cells and cell lines of urothelial origin but not with various cell lines of other origin. The article presents no further data on specificity, especially in tissue section. The authors report that the corresponding antigen is a glycolipid.

11.4 Immunocytology

Although a number of monoclonal antibodies directed against TCCs have been produced in recent years, only a few reports have been published on the use of such antibodies to diagnose TCC (Chopin et al. 1985; Huland et al. 1987, 1988; Sheinfeld et al. 1990; Walker et al. 1989). It is noteworthy that all of these reports concern the immunocytologic use of monoclonal antibodies. Apparently, the detection of tumor-associated antigens in the urine itself has met with formidable difficulties. Although such a test would be advantageous because it could be standardized, direct detection in fresh urine does not appear feasible for most of the antibodies cited above because of cross-reactions with other cells occurring in urine, such as granulocytes and macrophages.

Most of the antibody studies published to date have used *bladder washings* as the medium for immunocytologic analysis (Chopin et al. 1985; Huland et al. 1987, 1988; Sheinfeld et al. 1990). Compared with voided urine, bladder washings offer the advantages of better cellular preservation and a higher cell yield. Much as in flow cytometry, these parameters play an important role in immunocytology. On the other hand, the bladder must be catheterized to obtain the irrigation specimen, so this method, like cystourethroscopy, is an invasive procedure. It is especially burdensome for patients who, for clinical reasons, require serial cytologic monitoring. It remains to be determined, then, whether the advantages of examining bladder washings as opposed to voided urine will outweigh the disadvantages for the patient.

11.4.1 Bladder Wash Immunocytology

Chopin et al. (1985) first reported on the use of monoclonal antibodies to detect malignant urothelial cells in bladder wash specimens. Using the antibodies G 4 and E 7, they found immunocytologically positive tumor cells in 14 of 18 specimens from patients with known bladder tumors. They found no antigen-positive urothelial cells in washings from 13 of 15 controls.

Further observations on the use of monoclonal antibodies in urinary cytology were published by Huland et al. (1987). Although, as noted earlier, MAB 486 P 3/12 is not directed against a tumor-associated antigen in the true sense, an initial study in 40 patients with bladder tumors showed that use of this antibody can improve the sensitivity of the cytologic evaluation. Because the antigen gp 200 may occasionally be expressed on normal urothelial cells as well, it is considered normal for up to 30% of urothelial cells to exhibit a positive reaction.

Immunocytologic studies using MAB 486 P 3/12 yielded true-positive results in 36 of 40 patients with bladder carcinoma, whereas only 17 of these cases were positive by conventional urinary cytology. In another series of 104 patients, immunocytology correctly identified 60 of the 69 patients with bladder tumors (87%). A parallel determination of DNA content using flow cytometry and conventional cytologic analysis yielded positive results in 32 (46%) and 37 (54%) of the patients.

It is noteworthy that the *sensitivity* of immunocytology, unlike that of cytometry and conventional urine cytology, did not increase with the grade of tumor malignancy (see Table 11.4). Indeed, the accuracy of immunocytology, at 89%, was slightly better for the more highly differentiated grade 1 and 2 tumors than for grade 3 malignancies (78%). The *specificity* of the three methods was tested in 35 patients who did not have urothelial tumors. Immunocytologic and flow cytometric findings were suspicious for carcinoma in three patients, and cytologic findings were suspicious in four patients. This corresponds to a specificity of 91% for immunocytology and flow cytometry and 89% for conventional cytology.

The *Lewis X antigen* or *CD 15 antigen* is a cell-surface differentiation antigen that is carried on either protein or lipid moieties of various normal and malignant cells (Itzkowitz et al. 1986; Rettig et al. 1985).

Its immunogenic structure is the trisaccharide galactose B 1–4 (fucose 1–3) *N*-acetylglucosamine (Itzkowitz et al. 1986). Studies at the Memorial Sloan-Kettering Cancer Center in New York have shown that the Lewis X antigen occurs on transitional cell carcinomas but not on normal urothelial cells except for occasional umbrella cells (Cordon-Cardo et al. 1988). Figure 11.4 shows a MAB directed against the Lewis X antigen binding to tumor cells of a poorly differentiated TCC of the bladder.

Sheinfeld et al. (1990) investigated bladder wash specimens from 76 patients with histologically confirmed transitional cell carcinoma and 40 controls undergoing cystoscopy for reasons other than

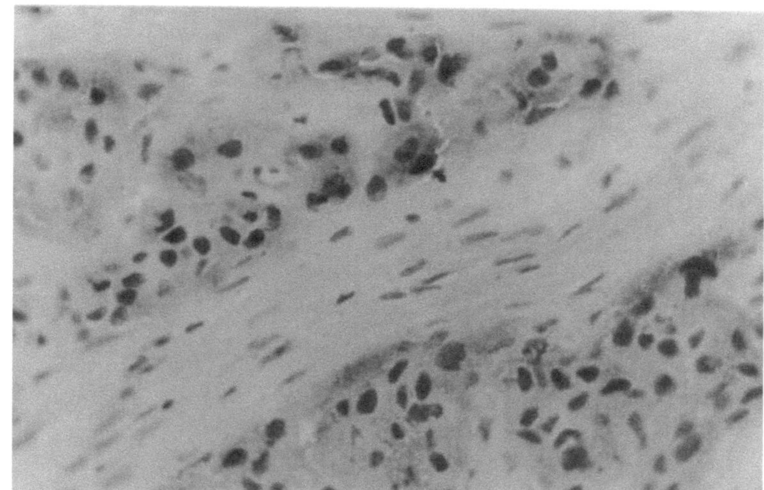

Fig. 11.2. **Immunoperoxidase reaction of monoclonal antibody Due ABC 3 on a frozen section from a moderately to poorly differentiated bladder carcinoma**

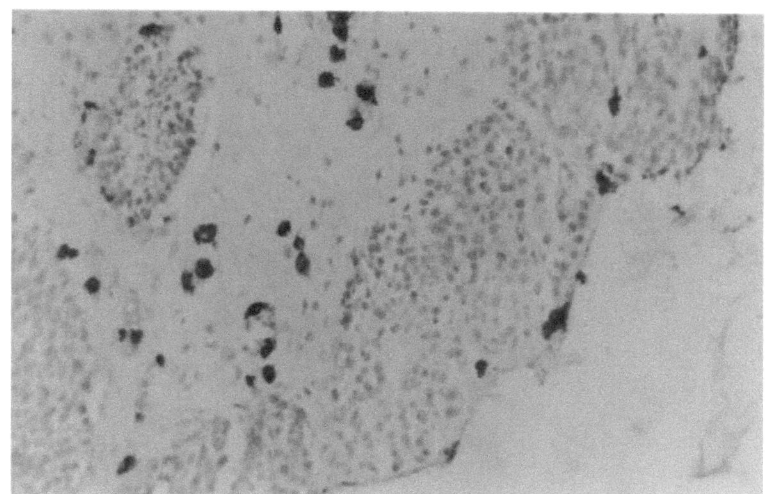

Fig. 11.3. **Immunoperoxidase reaction of monoclonal antibody Due ABC 3 on a frozen section from normal urothelial tissue.** The antibody reacts with granulocytes in the submucosa and with scattered superficial urothelial cells

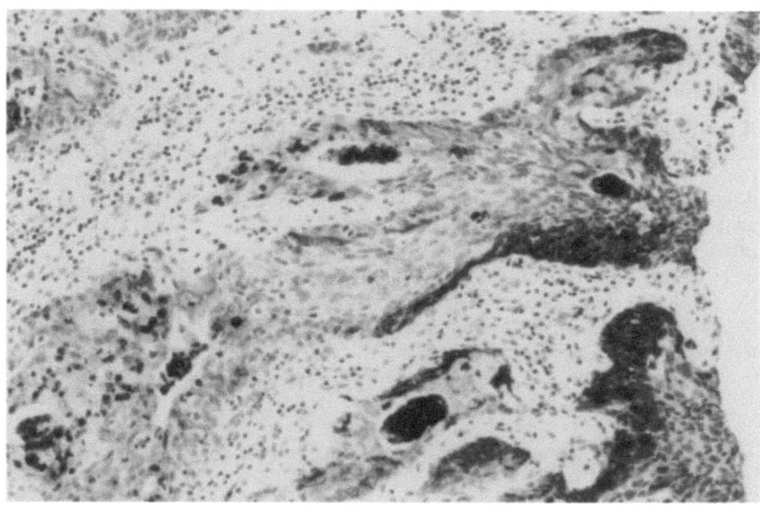

Fig. 11.4. **Immunoperoxidase reaction of monoclonal antibody P-12 on a frozen section from a poorly differentiated bladder carcinoma,** showing large areas of antigen-negative tumor cells. (Courtesy of Dr. V. Reuter, Memorial Sloan-Kettering Cancer Center, New York)

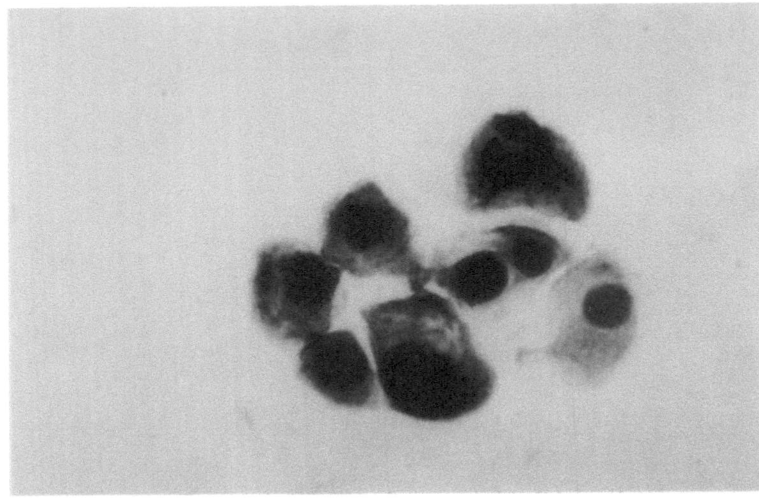

Fig. 11.5. **Immunocytology of a moderately differentiated bladder carcinoma with monoclonal antibody P-12.** (Courtesy of Dr. V. Reuter, Memorial Sloan-Kettering Cancer Center, New York)

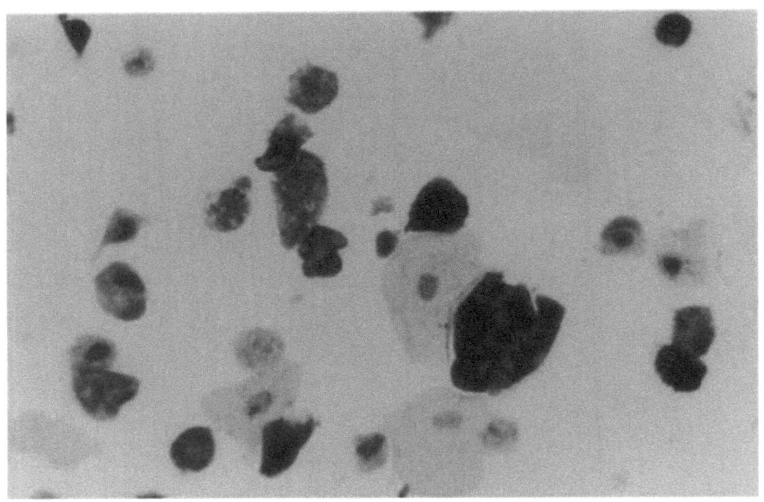

Fig. 11.6. **Immunocytology of a moderately differentiated bladder carcinoma with monoclonal antibody Due ABC 3**

bladder tumor. The anti-Lewis X *MAB P-12* was used for the immunocytologic analysis (Fig. 11.5).

A positive Lewis X result (specimen containing five or more antigen-positive cells) was noted in 67 of the 76 patients (88%) with TCC. Twelve of the 14 patients (86%) with well-differentiated TCC were immunocytologically positive, while urinary cytology was positive in only one of these cases (7%). Six of the 40 specimens from patients with no evidence of TCC were found to contain more than five antigen-positive urothelial cells (specificity=85%). It is noteworthy that five of the six patients who were immunocytologically positive had an adenocarcinoma of the prostate.

11.4.2 Voided Urine Immunocytology

Walker et al. (1989) published the first preliminary reports on the use of *voided urine* for immunocytologic analysis. Investigating eight specimens from patients with advanced TCC, they were able to detect antigen-positive urothelial cells in all the specimens using MAB BLCA-8. Since only a limited number of specimens from patients with advanced urothelial tumors were evaluated, further studies are needed to substantiate the value of this assay.

We recently conducted a *prospective study of 134 specimens* to assess the clinical value of MAB Due ABC 3 for the immunocytologic analysis of voided urine (Schmitz-Dräger et al. 1992). We investigated 74 specimens from patients with histologically proven TCC and 60 specimens from

Table 11.2. **Sensitivity and specificity of conventional cytology and immunocytology using monoclonal antibody Due ABC 3 in the diagnosis of urothelial tumors.** Analysis of 74 specimens from patients with urothelial tumors and 60 specimens from controls

Method	TCC (n=74) True-positive (%)	Control (n=60) True-negative (%)
Conventional cytology	35 (47)	55 (92)
Immunocytology	49 (66)	35 (58)
Conventional cytology and/or immunocytology	56 (76)	53 (88)

TCC, transitional cell carcinoma.

donors with no clinical evidence of urothelial carcinoma. A conventional cytologic evaluation accompanied the immunocytologic workup. Figure 11.6 shows the cytologic and immunocytologic specimen from a patient with a moderately differentiated bladder carcinoma.

11.4.2.1 Methods

Since urothelial cells become necrotic on prolonged contact with urine, morning urine should not be used, and the specimen should be *processed as rapidly as possible*. Immediately after collection, approximately 10 ml of urine is centrifuged for 5 min at 400 g. The supernatant is discarded, and the pellet is resuspended in RPMI 1640 culture medium with no serum added and cytocentrifuged (Hettich, Tuttlingen) for 3 min at 400 g onto a glass slide coated with poly-L-lysine (so the cells will not dislodge during further processing). The specimen is fixed with 1% formaldehyde and 4.5% acetone in phosphate-buffered saline (PBS) for 2 min. Preliminary tests showed that this fixation method does not cause destruction of the antigen. To inhibit endogenous peroxidases, the specimens are treated with 0.45% H_2O_2 in methanol for 30 min at 4 °C and then washed three times for 10 min in PBS. Nonspecific binding of the antibody is suppressed by treating the specimens with 5% rabbit serum in RPMI 1640 at room temperature for 10 min, followed by incubation with MAB Due ABC 3 (tissue culture supernatant) for 30 min at room temperature. The specimens are then washed three times for 10 min in PBS, and 0.1 ml of peroxidase-labeled rabbit antimouse IgG and IgM antiserum (Jackson Immune Research Laboratories, Westgrove, PA) is added to each. The antiserum is used in a dilution of 1:200 in PBS (with 10% fetal calf serum). The specimens are incubated at room temperature for 30 min. After washing again, 3,3'-diaminobenzidine (0.1% in Tris buffer containing 0.03% H_2O_2) is added as a substrate for the peroxidase. The enzyme reaction is run for 10 min in total darkness at room temperature. Then the specimens are counterstained with hemalum for 10 s. The slides are mounted with Depex, and the percentage of antigen-positive cells is determined by light microscopic examination. More than 100 cells, if present, are counted. To determine the actual percentage of antigen-positive cells, it is customary, for statistical reasons, to *count 100 urothelial cells*. Specimens with *fewer than 50 urothelial cells* are *not evaluated*. Hybridoma medium is used as a negative control, and serum from mice previously immunized with the bladder tumor cell line SW 1710 serves as a positive control. Preliminary tests are performed to establish a cutoff value for positive-negative discrimination. In the study described below, we used a cutoff value of 20%, meaning that specimens with more than 20% antigen-positive cells were classified as pathologic.

This method is very similar to that described by Huland et al. (1987) and differs from it only in the fixation and staining of the cells. Instead of peroxide-labeled antimouse Ig antiserum, Huland et al. (1987) used an alkaline phosphatase-coupled antiserum to demonstrate antibody binding. Sheinfeld et al. (1990) used a similar method, except that they smeared the sediment onto a slide; this is an option when a cytocentrifuge is not available.

Some antigens are denatured by the fixation process. If the selected MAB is directed against such an antigen, one can apply the technique of Chopin et al. (1985) in which the urine sediment is incubated with the MAB prior to cytocentrifugation and fixation.

The technique described by Walker et al. (1989) differs substantially from that of other investigators. In this method the centrifuged urine sediment is resuspended and embedded in agarose. A fluorescein-conjugated antiserum is used to demonstrate binding of the MAB. The advantage of this technique is that it provides a three-dimensional agar-embedded cellular array in which the examiner can focus on and evaluate individual cells. While this may be advantageous for scientific inquiries, however, screening a three-dimensional gel medium for individual urothelial cells would not be practical for routine clinical evaluations.

11.4.2.2 Results

In three of the 74 specimens (4%) from patients with TCC, we were unable to detect urothelial cells (Fig. 11.7). Sixteen of the 74 specimens (22%) were not evaluable by immunocytology due to an insufficient number of urothelial cells (<50), cellular degeneration, pyuria, or gross hematuria.

Five of the 60 cytologic specimens (8%) from the controls could not be evaluated (Fig. 11.8). Four specimens contained no urothelial cells, and in one case severe leukocyturia precluded adequate evaluation. Eighteen of the immunocytologic specimens (30%) could not be evaluated due to an insufficient number of urothelial cells (<50), cellular degeneration, severe leukocyturia, or gross hematuria.

We observed more than 20% antigen-positive urothelial cells in 49 of the 74 specimens (66%) from patients with TCC (Table 11.2). Figure 11.9 shows the immunocytologically determined percentage of antigen-positive urothelial cells in 58 evaluable specimens from patients with TCC and in 42 evaluable specimens from controls. Tumor cells were correctly detected by conventional urine cytology in 35 (47%) of these specimens.

Consistent with previous publications, we found that the sensitivity of conventional cytology correlated with the grade of tumor malignancy (Table 11.3) (Dubernard et al. 1982; Koss et al. 1985; Zein et al. 1984). The sensitivity of cytology was least satisfactory in the diagnosis of well-differentiated (grade 1) tumors. The sensitivity of immunocytology was comparable for grade 1 and grade 2 malignancies (57% and 53%, respectively) and was 79% for poorly differentiated grade 3 tumors (Fig. 11.10).

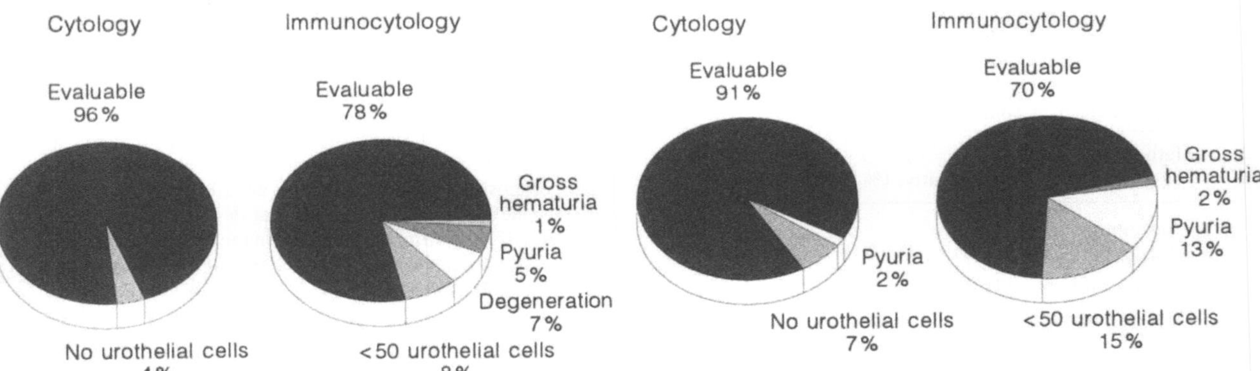

Fig. 11.7. **Prospective study of 74 cytologic and immuno-cytologic specimens from patients with urothelial tumors.** Nonevaluable specimens: percentage and reasons

Fig. 11.8. **Prospective study of 60 cytologic and immuno-cytologic specimens from patients with no clinical evidence of TCC.** Nonevaluable specimens: percentage and reasons

Cytologic examination did not reveal tumor cells in any of the 55 evaluable specimens from the controls, corresponding to a specificity of 92% (see Table 11.2). Immunocytology, on the other hand, showed an elevated number of antigen-positive cells in seven specimens. Since 18 additional specimens could not be evaluated, the specificity of immunocytology was 58%. The efficacy of both methods based on the calculated specificity, sensitivity, and incidence of disease in the study population was 67% for conventional cytology and 47% for immunocytology.

The sensitivity of cytology and immunocytology could be improved by combining the results of both techniques (see Tables 11.2, 11.3). By using this approach, we could correctly classify 34 of the 35 specimens (97%) from patients with poorly differentiated TCC and 13 of 21 specimens (62%) from patients with well-differentiated tumors. Overall, 56 of the 74 specimens (76%) from patients with TCC were positive by cytology, immunocytology, or by both methods. The combined specificity of cytology and immunocytology was 88% (53 of 60 specimens).

Table 11.4 summarizes the results in 100 specimens that could be evaluated by immunocytology. For comparison, the results for all 134 specimens and the results of Huland et al. (1988) and Sheinfeld et al. (1990) are also shown. Immunocytology showed a sensitivity of 84% in evaluating the 58 specimens from patients with TCC. When just these specimens are considered rather than all 74, sensitivity was only slightly improved by combining the results of cytology and immunocytology. This means that when all 74 specimens are considered, the improved sensitivity is based mostly on the large number of specimens not evaluable by immunocytology. Immunocytology showed a specificity of 83% in the 42 evaluable specimens from controls. The table shows that the results of the 100 immunocytologically evaluable specimens agree very closely with the results of Huland et al. and Sheinfeld et al.

11.5 Discussion

The increased risk of tumor recurrence in patients with a TCC calls for the use of sensitive diagnostic methods in the follow-up of these patients. Noninvasive techniques are desired, especially in cases where close surveillance is necessary on clinical grounds. A procedure that is accurate, noninvasive, and well-tolerated would represent a significant clinical advance in this area. Because cystoscopy is an effective but invasive procedure for verifying the diagnosis of bladder tumor, it is imperative that potential screening tests combine noninvasiveness with a high degree of sensitivity.

The studies by Huland et al. (1988), Sheinfeld et al. (1990), and our own study (Schmitz-Dräger et al. 1991) have consistently demonstrated that *immunocytology is a very sensitive test* (see Table 11.4). All three studies have shown that immunocytology is superior to conventional cytology in the diagnosis of well-differentiated malignancies. Combining the results can enhance the sensitivity of both methods without causing significant loss of specificity.

The three studies cited above have a number of features in common. The most striking is the similarity of the monoclonal antibodies that were used. None of the antibodies – 486 P 3/12, P-12, Due ABC 3 – are directed against tumor-specific antigens (Arndt et al. 1987; Cordon-Cardo et al. 1988; Decken et al. 1992). Specificity analysis and the characterization of the antigens indicate that they define related antigenic structures, apparently differentiation antigens that may be expressed in small amounts even on normal urothelial cells. This underscores the need for a quantitative method of immunocytologic evaluation.

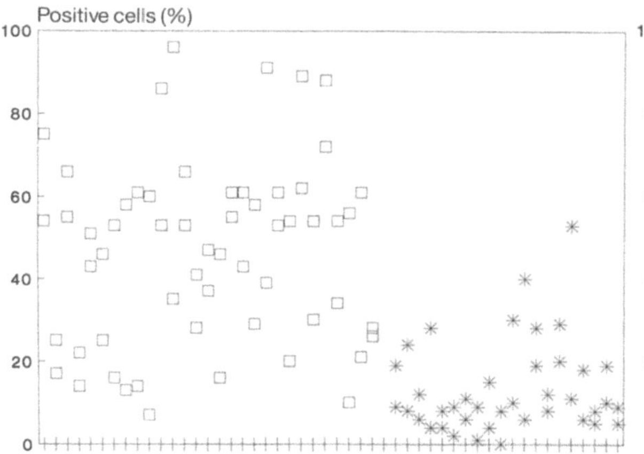

Fig. 11.9. **Immunocytologic results of 58 evaluable specimens from patients with urothelial tumors and 42 specimens from controls.** Percentage of antigen-positive urothelial cells. □, Specimens from patients with urothelial tumors; *, control group

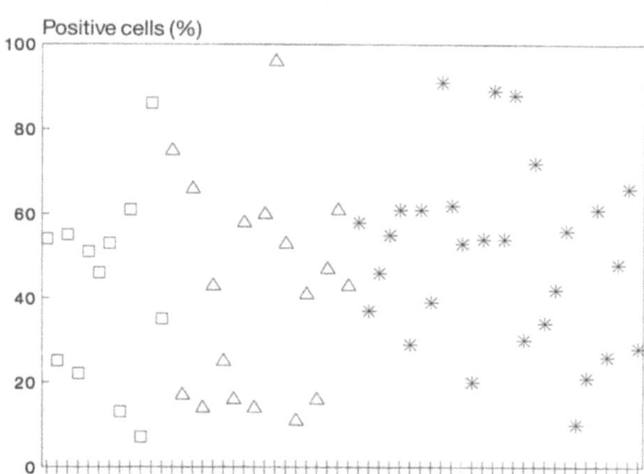

Fig. 11.10. **Immunocytologic results of 53 evaluable specimens from patients with various grades of urothelial tumor.** □, grade 1 TCC; △, grade 2 TCC; *, grade 3/4 TCC

Studies by Hakomori (1984) have shown that the transformation of urothelial cells is associated with changes in the glycosylation of membrane proteins or membrane lipids. The suppression of glycosyltransferases leads to a loss of normally expressed antigens (Liu et al. 1987). A familiar example is the loss of the ABO blood-group isoantigens in poorly differentiated TCC (Catalona 1981). Conversely, the activation of glycosyltransferases can lead to the expression of new or altered carbohydrate chains and, by altering membrane proteins or lipids, to changes in the membrane structure. The antigens defined by monoclonal antibodies 486 P 3/12, P-12, and Due ABC 3 appear to be carbohydrate structures whose expression is induced or amplified by the transformation of the urothelial cells (Arndt et al. 1987; Cordon-Cardo et al. 1988; Decken et al. 1992).

An interesting aspect of immunocytology is that the results are independent of the examiner. Evaluation of the same specimen by multiple independent examiners has shown a good *reproducibility of results* for monoclonal antibodies 486 P 3/12 and Due ABC 3. In the study by Huland et al. (1988) and in our own study (Schmitz-Dräger et al. 1991), the immunocytologic specimens were interpreted by personnel with no previous specialized training in pathology or cytopathology. Urothelial cells were simply differentiated from other cells such as leukocytes, erythrocytes, and squamous epithelium; they were not further differentiated into umbrella cells and cells of the superficial or deeper mucosal layers. This could partly account for the high cutoff value in these two studies compared with the study by Sheinfeld et al. (1990). Huland et al. (1988) regarded more than 30% antigen-positive cells as pathologic. while Schmitz-Dräger et al. (1991) used a threshold of 20%. In the study by Sheinfeld et al. (1990), the specimens were interpreted by a pathologist. Umbrella cells reactive with the MAB used were disregarded. Specimens containing five or more antigen-positive urothelial cells were classified as positive in this study.

Table 11.3. **Sensitivity of conventional cytology and immunocytology using monoclonal antibody Due ABC 3 in the diagnosis of various grades of urothelial tumors.** Analysis of 74 specimens from patients with histologically proven urothelial carcinoma

Method	Sensitivity (%)		
	Grade 1 (n=21)	Grade 2 (n=19)	Grade 3 (n=34)
Conventional cytology	3 (14)	8 (42)	24 (71)
Immunocytology	12 (57)	10 (53)	27 (79)
Conventional cytology and/or immunocytology	12 (57)	10 (53)	34 (97)
Total	21	19	34

The *disadvantages of immunocytology* are as follows:
1. Immunocytologic specimens are more costly to prepare than cytologic specimens.
2. A significant percentage of voided urine specimens are not evaluable.
3. A substantial cost factor would be associated with the broader use of monoclonal antibodies.

Table 11.4. **Sensitivity and specificity of immunocytology. Comparison of prospective studies**

Study	Sensitivity (%)				Specificity (%)
	G1	G2	G3	G1–3	
Huland et al. (1988)	16/18 (89)	33/37 (89)	11/14 (78)	60/69 (87)	32/35 (91)
Sheinfeld et al. (1990)	12/14 (86)	55/62 (89)		67/76 (88)	34/40 (85)
Schmitz-Dräger et al. (1990)					
134 Specimens[a]	12/21 (57)	10/19 (53)	27/35 (77)	49/74 (66)	35/60 (58)
100 Specimens[b]	12/14 (86)	10/16 (62)	27/28 (96)	49/58 (84)	35/42 (83)

[a] Of the 134 specimens that were analyzed in the prospective trial, 100 could be evaluated by immunocytology.
[b] Results of 100 immunocytologically evaluable specimens.

It should be considered, however, that immunocytology is still a very new method of diagnosing urothelial tumors that is still in the *clinical testing stage* and whose potential should not be underestimated. For example, the percentage of specimens nonevaluable due to insufficient cell numbers can perhaps be reduced or eliminated by processing larger urine volumes. Also, the development of *standardized test kits* with ready-to-use components should greatly simplify the relatively complex process of specimen preparation for immunocytologic analysis. Thus, solutions to the technical problems of immunocytology appear to be within reach.

Even with automation, the immunocytologic examination will probably always be significantly *more costly* than a conventional cytologic workup. To keep additional costs down, further prospective trials are needed to identify the specific patient groups that can benefit from an immunocytologic evaluation.

Evidence to date indicates that immunocytology will be a particularly useful adjunct to conventional cytology in patients with highly differentiated tumors. But even with moderately and poorly differentiated TCCs, the sensitivity of conventional cytology can be increased by adding an immunocytologic examination. The *combination* of both studies should significantly improve the noninvasive diagnosis of urothelial tumors.

11.6 Outlook

It is noteworthy that none of the MABs that have been widely used for immunocytology are "tumor-specific" in the true sense, yet the results achieved with these antibodies have been relatively good. It is reasonable to expect that even better results will be achieved by the development of new, more specific MABs. The production of such antibodies is not without problems, however. Attempts during the past decade to produce murine MABs directed against tumor-specific antigens have shown that the antibodies obtained are frequently directed against the strongly antigenic carbohydrate chains on the cell surface. It has not been possible to produce the desired tumor-specific MABs by the application of classic hybridoma technology.

Molecular biology appears to offer promising new approaches. Complicated techniques such as "subtractive hybridization" provide a means of identifying the mRNA of all proteins that are expressed in tumor cells but not in normal, nontransformed cells. Once this has been determined, monoclonal antibodies directed against likely target proteins can be produced by either the classic method or by genetic engineering.

Because immunocytology is examiner-independent, it should be possible to *automate* the test procedure. The MABs previously used in immunocytology all exhibit cross-reactivity with granulocytes. Automated immunocytology would require a means of discriminating between urothelial cells and other antigen-positive cells in the urine. Studies by Hijazi et al. (1989) suggest that this problem can be solved by employing flow cytometry and appropriate MABs. These authors analyzed the expression of cytokeratin 18 in urothelial tumors, using a MAB directed against cytokeratin 18 to investigate 76 tumor specimens and ten biopsies of normal bladder mucosa. The DNA content and malignancy grade of the tumor specimens were simultaneously determined. Hijazi et al. (1989) observed a significant correlation between the expression of cytokeratin 18, DNA content, and tumor grade. On the basis of these results, adaptation of this procedure to immunocytology appears feasible. Depending on the MABs selected, such a system would provide a valuable source of diagnostic and prognostic information.

References

Arndt R, Dürkopf H, Huland H, Donn F, Loening T, Kalthoff H (1987) Monoclonal antibodies for characterization of the heterogeneity of normal and malignant transitional cells. J Urol 137: 758–763

Bander NH (1987) Monoclonal antibodies: state of art. J Urol 137: 603–612

Baricordi OR, Sensi A, De Vinci C, Melchiorri L, Fabris G, Marchetti E, Corrado F et al. (1985) A monoclonal antibody to human transitional-cell carcinoma of the bladder cross-reacting with a differentiation antigen of neutrophilic lineage. Int J Cancer 35: 781–786

Catalona WJ (1981) Practical utility of specific red cell adherence test in bladder cancer. Urology 18: 113–121

Chópin DK, Kernion JB de, Rosenthal DL, Fahey JL (1985) Monoclonal antibodies against transitional cell carcinoma for detection of malignant urothelial cells in bladder washing. J Urol 134: 260–265

Cordon-Cardo C, Reuter VE, Lloyd KO, Sheinfeld J, Fair WR, Old LJ, Melamed MR (1988) Blood group-related antigens in human urothelium: enhanced expression of precursor, Le^x and Le^y determinants in urothelial carcinoma. Cancer Res 48: 4113–4118

Decken K, Schmitz-Dräger BJ, Rohde D, Nakamura S, Ebert T, Ackermann R (1992) Monoclonal antibody Due ABC 3 directed against transitional cell carcinoma. I. Production, specificity analysis and preliminary characterization of the antigen. J Urol 147: 235–241

Dubernard JM, Devonec M, Amiel J, Bouvier R, Fontaniere B, Faucon M (1982) Correlation between cytology and cystoscopy in the follow-up of patients with bladder tumours. Eur Urol 8: 5–8

Fradet Y, Cordon-Cardo C, Thomson T, Daly ME, Whitmore WF Jr, Lloyd KO, Melamed MR, Old LJ (1984) Cell surface antigens of human bladder cancer defined by mouse monoclonal antibodies. Proc Natl Acad Sci USA 81: 224–228

Grossman HB (1983) Hybridoma antibodies reactive with human bladder carcinoma cell surface antigens. J Urol 130: 610–614

Hakomori S-I (1984) Tumor-associated carbohydrate antigens. Ann Rev Immunol 2: 103–126

Hijazi A, Devonec M, Bouvier R, Revillard J-P (1989) Flow cytometry study of cytokeratin 18 expression according to tumor grade and deoxyribonucleic acid content in human bladder tumors. J Urol 141: 522–526

Huland E, Huland H, Arndt R, Baisch H, Klöppel G (1988) Urindiagnostik oberflächlicher Harnblasentumoren durch Zytologie, Immunzytologie und Flowzytometrie: Ergebnisse einer prospektiven vergleichenden Studie an 104 Patienten. Aktuel Urol 19: 13–17

Huland H, Arndt R, Huland E, Loening TE, Steffens M (1987) Monoclonal antibody 486 P 3/12: a valuable bladder carcinoma marker for immunocytology. J Urol 137: 654–659

Itzkowitz SH, Yuan M, Fukushi Y, Palekar A, Phelps PC, Shamsuddin AM, Trump BF et al (1986) Lewis^x- and sialyated Lewis^x-antigen expression in human malignant and nonmalignant colonic tissues. Cancer Res 46: 2627–2632

Köhler G, Milstein C (1975) Contionous culture of fused cells secreting antibody of predefined specificity. Nature 256: 495–497

Koss LG, Deitch D, Ramanathan R, Sherman AB (1985) Diagnostic value of cytology of voided urine. Acta Cytol 29: 810–816

Liu BC-S, Neuwirth H, Zhu LW, Stock LM, Kernion JB de, Fahey JL (1987) Detection of onco-fetal bladder antigen in urine of patients with transitional cell carcinoma. J Urol 137: 1258–1264

Rettig WJ, Cordon-Cardo C, Ng JS, Oettgen HF, Old LJ, Lloyd KO (1985) High-molecular-weight glycoproteins of human teratocarcinoma defined by monoclonal antibodies to carbohydrate determinants. Cancer Res 45: 815–822

Schmitz-Dräger BJ, Rohde D, Peschkes C, Ebert T, Ackermann R (1988) Monoklonale Antikörper gegen Harnblasenkarzinome – ein Beitrag zur Verbesserung der Diagnostik? Aktuel Urol 19: 117–123

Schmitz-Dräger BJ, Nakamura S, Decken K, Pfitzer P, Rottmann-Ickler C, Ebert T, Ackermann R (1991) Monoclonal antibody Due ABC 3 directed against transitional cell carcinoma. II. Prospective trial on the diagnostic value of immunocytology using monoclonal antibody Due ABC 3. J Urol 146: 1521–1524

Sheinfeld J, Reuter VE, Melamed MR, Fair WR, Morse M, Sogani PC, Herr HW et al. (1990) Enhanced bladder cancer detection with the Lewis X antigen as a marker of neoplastic transformation. J Urol 143: 285–288

Walker KZ, Russell PJ, Kingsley EA, Philips J, Raghavan D (1989) Detection of malignant cells in voided urine from patients with bladder cancer, a novel monoclonal assay. J Urol 142: 1578–1583

Zein T, Wajsman Z, Englander LS, Gamarra M, Lopez C, Huben RP, Pontes JE (1984) Evaluation of bladder washings and urine cytology in the diagnosis of bladder cancer and its correlation with selected biopsies of the bladder mucosa. J Urol 132: 670–671

12 Quantitative Immunocytology with Monoclonal Antibody (MAB) 486 p 3/12: Clinical Application

H. HULAND and E. HULAND

CONTENTS

12.1 Quantitative Analysis

Initially, the use of monoclonal antibodies (MABs) for clinical diagnosis met with serious difficulties. In the early 1980s we discovered that, contrary to several published reports, monoclonal antibodies were *not* tumor-specific, i.e., they did not react ideally with all the tumor cells in a given type of neoplasm (e.g., bladder carcinoma), and they were reactive to a degree with healthy cells (e.g., normal bladder) (Arndt et al. 1987). When we tested 15 of the best-documented MABs to bladder tumor antigen, we found that none of the antibodies tested were tumor-specific or organ-specific (Huland et al. 1991). To date, researchers have been unable to develop tumor-specific antibodies in man that react exclusively with the cancerous tissue in an organ and do not react with healthy cells. The few organ-specific markers, such as PSA, are not tumor-specific and additionally react with prostatic cancer cells and with cells of the BPH and normal prostate.

Another problem we noted at that time was the fact that MABs rarely reacted homogeneously with all the constituents of a bladder carcinoma (Fig. 12.1), but usually exhibited a heterogeneous labeling pattern (Fig. 12.2) (Arndt et al. 1987). Also, all the MABs were reactive to a degree with normal bladder cells (Fig. 12.3) (Arndt et al. 1987; Huland et al. 1991). This foiled initial attempts to use MABs for the urinary immunocytologic diagnosis of tumors, at least if the presence of antigen-positive cells is to be considered a positive finding and the absence of antigen-positive cells a negative finding.

Positive (MAB-binding) cells were commonly found in the urine of healthy subjects. The real breakthrough in the use of monoclonal antibodies in urinary cytology came when we detected a *quantitative* difference in the urine sediment specimens from tumor patients and controls (Huland et al. 1987) (Fig. 12.4). On comparing the specimens, we found that in virtually all cases less than 30% of the exfoliated transitional cells in specimens from healthy subjects reacted with MAB 486 p 3/12. At the same time, more than 30% of the counted transitional cells reacted with the MAB in 90% of all the tumor patients tested.

Years of experience with our MAB 486 p 3/12 and with MABs from other centers indicate that, for the present, our method of quantitative urinary immunocytology (QUIC) holds the only key to the successful use of MABs in the diagnosis of bladder tumors (Huland et al. 1987). Immunofluorescent techniques, though widely used to demonstrate antibody binding owing to their technical simplicity, are poorly suited for quantitative studies because they do not permit a morphologic evaluation of the labeled urothelial cells.

We have tested the majority of MABs directed against a bladder tumor-associated antigen that were described in publications up to 1986. Of the comprehensive comparative studies that we performed with these antibodies, here we shall present only the results of quantitative immunocytologic analyses (Fig. 12.5). Each of the 15 antibodies was tested on urine specimens obtained from 20 patients with bladder tumors, 14 patients with prostatic adenoma, and 13 patients with urolithiasis. We made the following discoveries (Huland et al. 1991):

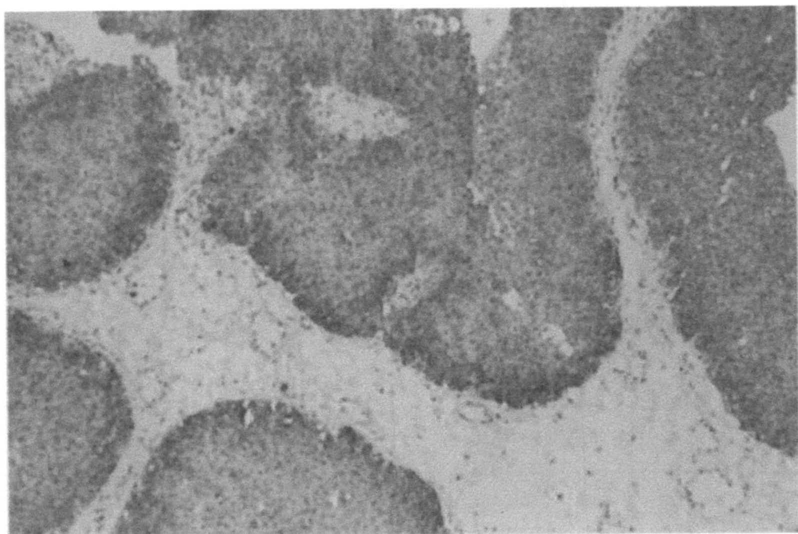

Fig. 12.1. **Immunohistologic staining of a grade 2 transitional cell carcinoma with MAB 486 p 3/12** (immunoperoxidase reaction): homogeneous staining of all tumor constituents

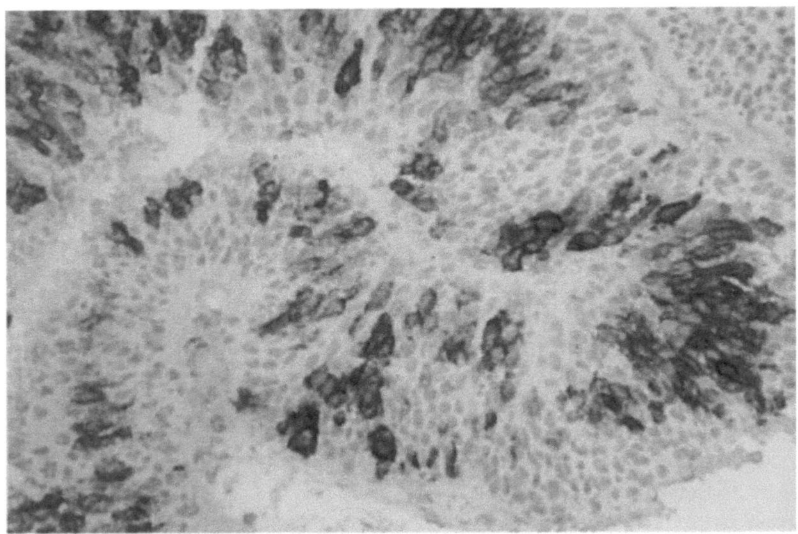

Fig. 12.2. **Immunohistologic staining of a grade 2 transitional cell carcinoma with MAB 486 p 3/12** (immunoperoxidase reaction): nonhomogeneous or heterogeneous labeling of the tumor constituents with antigen-positive and antigen-negative areas

1. None of the antibodies is tumor-specific.
2. Monoclonal antibodies are useful for the diagnosis of bladder tumors only if the principle of quantitative analysis is applied.
3. Several of the antibodies provide good quantitative discrimination between urine samples from bladder tumor patients (top row in Fig. 12.5) and patients without bladder tumors (middle row in Fig. 12.5).
4. Some antibodies concurrently stain urinary leukocytes, and others do not.
5. All the antibodies define different antigens.
6. All the antibodies are highly reactive with epithelial cells in urine samples from urolithiasis patients (bottom row in Fig. 12.5).

12.2 Practical Guidelines on the QUIC Technique

QUIC can be performed on a bladder wash specimen or spontaneously voided urine. The sample is centrifuged at 2000 rpm, washed once with PBS (phosphate-buffered saline), and cytocentrifuged. The cytocentrifuge specimen may be processed further immediately after air drying, or, if several specimens are to be stained at once, it can be frozen, freeze-dried (lyophilized), and vacuum-sealed in plastic for storage and later processing. This technique is particularly recommended in set-

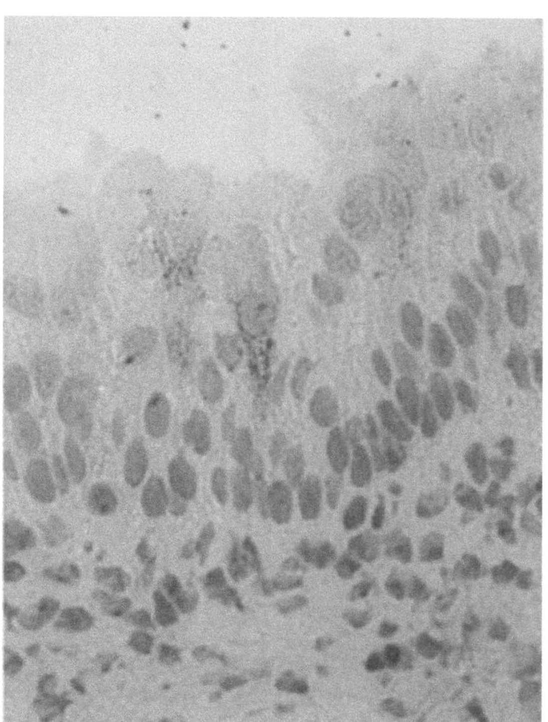

Fig. 12.3. **Immunohistologic staining of normal bladder mucosa with MAB 486 p 3/12:** immunoperoxidase reaction. Note the staining of isolated umbrella cells

Fig. 12.5. **Fifteen monoclonal antibodies:** percentage of antibody-binding urothelial cells in bladder washings from patients with a bladder tumor (n=20), prostatic adenoma (n=14), and urolithiasis (n=13)
▽

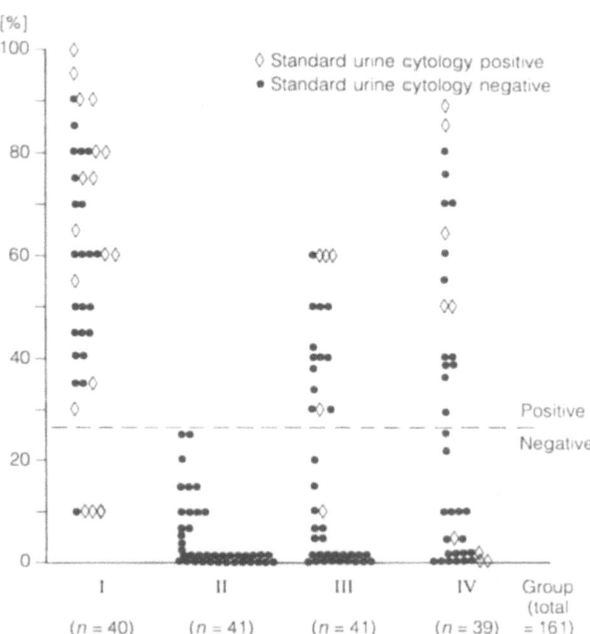

Fig. 12.4. **Quantitative immunocytology with MAB 486 p 3/12** (data in percent). Group I: 40 patients with transitional cell carcinoma of the bladder. Group II: 41 patients with urinary tract infection or benign prostatic hypertrophy. Group III: 41 patients who underwent complete removal of a bladder tumor, received mitomycin C instillation for prophylaxis, and had no evidence of local recurrence. Group IV: 39 patients who underwent removal of a superficial bladder tumor, received no prophylaxis, and had no endoscopic evidence of a recurrence

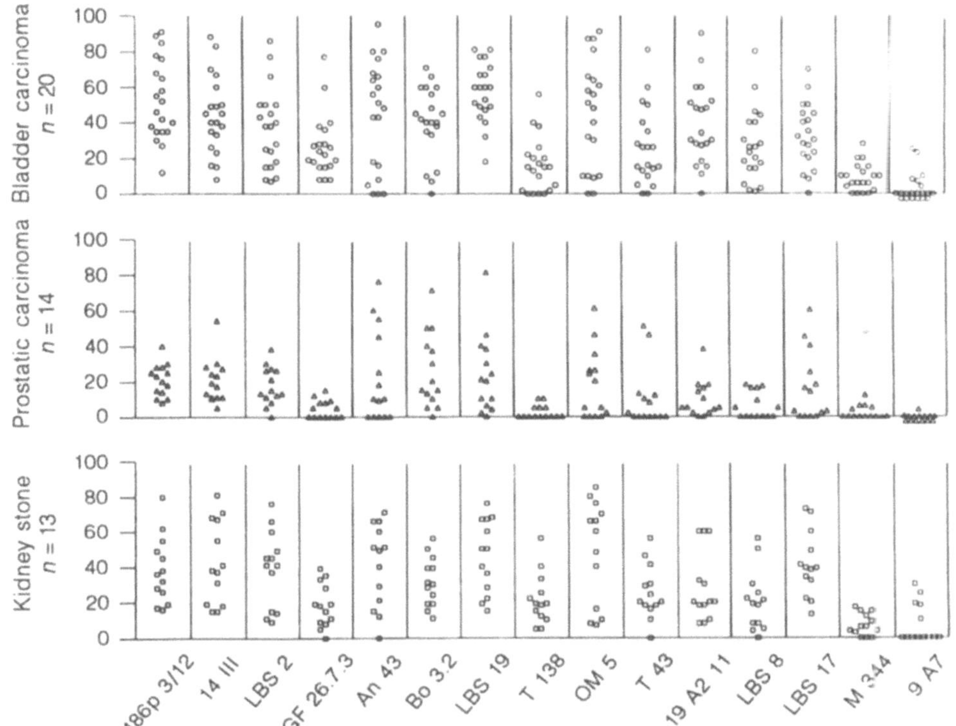

tings where large urine sample volumes are routinely processed (Table 12.1).

During preparation of the cytocentrifuge specimen, care should be taken that the cells are deposited in a monolayer, which is necessary for an accurate quantitative evaluation. This can be difficult in leukocyte-rich urine samples in which only a few urothelial cells are available for counting; quantification is uncertain when fewer than 100 urothelial cells are present. We use alkaline phosphatase-conjugated second and third antibodies as polyclonal indicator antibodies. This method is somewhat more complex than the peroxidase method, but we have found it to be more reliable and easier to interpret as it imparts a brilliant red stain to reactive cells (Figs. 12.6a,b, 12.7a–c). The interpretation itself is straightforward and does not require exceptional cytologic experience. The examiner must simply be able to distinguish among blood cells, transitional cells, and squamous epithelial cells. If the percentage of transitional cells staining with the MAB is 30% or more, a positive is declared. Rarely, all the transitional cells in the sediment are stained red (Fig. 12.6b), but usually a combination of antigen-positive and antigen-negative transitional cells is observed (Fig. 12.7). Only the transitional cells are counted, not the squamous cells.

When QUIC is performed with MAB 486 p 3/12, the concurrently stained granulocytes provide an internal "positive control." During each test a cytocentrifuged specimen of the same urine sample is also run as a negative control; instead of the first mouse monoclonal antibody (MAB 486 p 3/12), the specimen is incubated with antibody-free cell-culture supernatant. An ideal but more costly negative control would be to incubate the specimen with a mouse monoclonal antibody of the same subtype (e.g., IgM) but different specificity. The negative control serves to exclude nonspecific bindings that would produce a false-positive test. It is better to use $F(ab')_2$ fragments as polyclonal indicator antibodies (second and third antibodies) rather than complete antibodies. These fragments still exhibit specific binding but do not possess the Fc component or "sticky end" that very often leads to nonspecific, false-positive reactions.

Our quantitative findings fall into one of three patterns:

1. An obviously positive specimen with substantially more than 30% positive cells (Figs. 12.6, 12.7).

Table 12.1. **Quantitative immunocytology** (procedure)

1. Centrifuge urine sample at 2000 rpm (10 min).
2. Pipet off supernatant, swirl sediment with PBS.
3. Centrifuge sample at 2000 rpm.
4. Prepare cytocentrifuge specimens.
5. Air dry (approx. 2 h).
6. Fix in acetone for 5 min.
7. Rehydrate in rinsing buffer for 10 min.
8. Incubate for 30 min with 100 µl MAB 486 p 3/12 (humidified chamber) or with control antibody.
9. Wash slide in rinsing buffer for 10 min by repeated buffer changes in a cuvet.
10. Incubate for 30 min with 100 µl of 2nd antibody (humidified chamber).
11. Wash as in step 9.
12. Incubate for 30 min with 100 µl of 3rd antibody (humidified chamber).
13. Wash as in step 9.
14. Prepare substrate according to selected 2nd and 3rd antibodies.
15. Incubate substrate.
16. Wash as in step 9.
17. Counterstain with filtered hemalum for 15–30 s.
18. Blue in tap water for 10–15 min.
19. Mount with Kaiser gelatin.
20. Interpretation.
    Check for negative control staining (to exclude nonspecific staining).
    Count positive (antibody-binding) transitional cells (count at least 100 cells).
    Disregard squamous epithelial cells and blood cells.
    Positive test = 30 % or more positive transitional cells.
    Negative test = fewer than 30 % positive cells.
    Findings between 20 % and 30 % require further investigation.

2. An obviously negative specimen with fewer than 20% positive cells. This is seen in healthy patients, who tend to have hypocellular specimens.
3. Borderline findings around the 30% cutoff value. This pattern, especially common in the follow-up of bladder tumor patients, is difficult to classify as positive or negative and should prompt a repeat examination, preferably by bladder irrigation cytology using saline.

12.3 Sensitivity of QUIC with MAB 486 p 3/12

Our initial analysis of 40 patients with superficial bladder carcinomas showed that our method of quantitative immunocytology with MAB 486 p 3/12 has a sensitivity of 90% (Fig. 12.4): 36 of the 40 patients in group I had more than 30% positive cells with MAB 486 p 3/12.

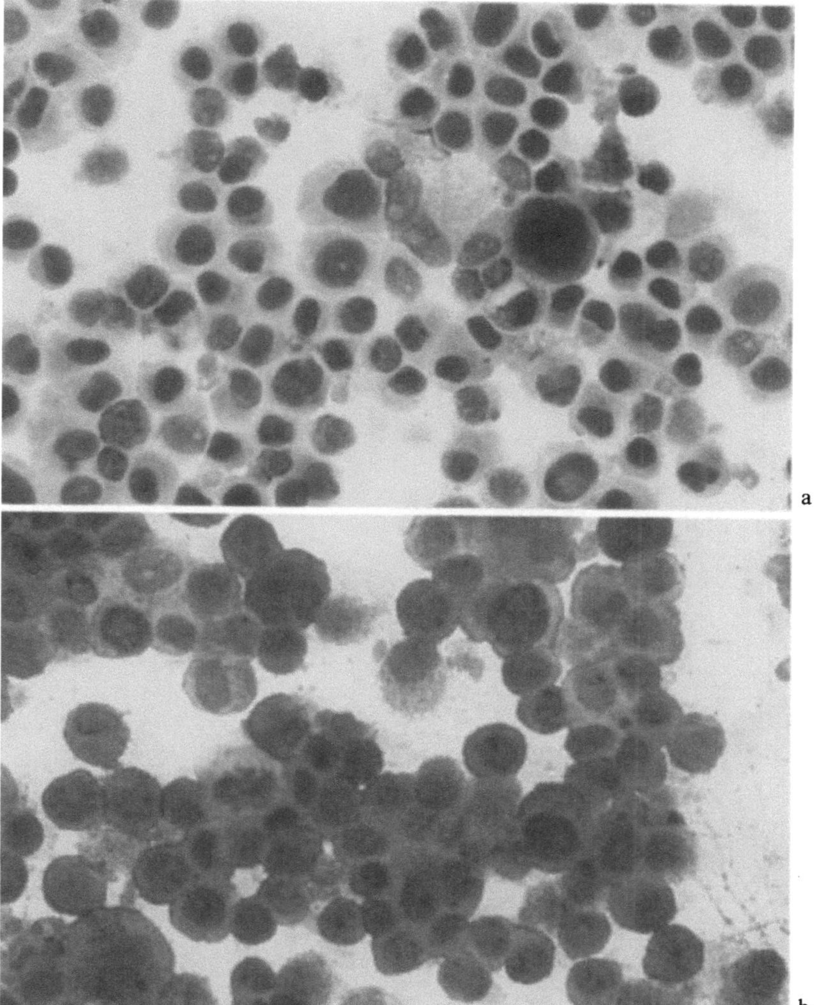

Fig. 12.6 a, b. **Immunocytology of a grade 3 transitional cell carcinoma (urine sediment) with MAB 486 p 3/12.**
a Example of direct stain.
b Example of homogeneous staining with MAB 486 p 3/12; alkaline phosphatase reaction

Later we conducted a more comprehensive comparative study in 69 bladder tumor patients whose urine samples were subjected to parallel evaluations by standard cytology, quantitative immunocytology, and flow cytometry. Those interpreting the tests were unaware of the patients' diagnoses (Huland et al. 1988).

The results are summarized in Table 12.2. For the first time we achieved an almost 90% accuracy rate with QUIC in the diagnosis of grade 1 tumors. This was comparable to the sensitivity achieved for grade 2 and grade 3 transitional cell carcinomas of the bladder. Conventional cytology and especially flow cytometry failed in the diagnosis of well-differentiated grade 1 tumors.

The superior results of urinary immunocytology using the QUIC method compared with standard cytology and flow cytometry are explained by the independence of QUIC on morphologic criteria. The transformation of a normal cell to a tumor cell is a gradual process. Changes in cellular morphology do not necessarily occur at the start of this cascade. Markers that detect cells which have already commenced malignant transformation but have not yet undergone appreciable morphologic change offer a crucial advantage in the early detection of malignant disease. MAB 486 p 3/12 defines an antigen that is already expressed on grade 1 tumors. These antigenic changes can be particularly useful in areas where conventional methods such as cytology have displayed inadequate diagnostic capabilities.

Only 27.7% of the grade 1 tumor specimens were positive by flow cytometry. This is consistent with the results of Oljanz and Tanke (1986), who had a 29% true-positive rate with grade 1 tumors,

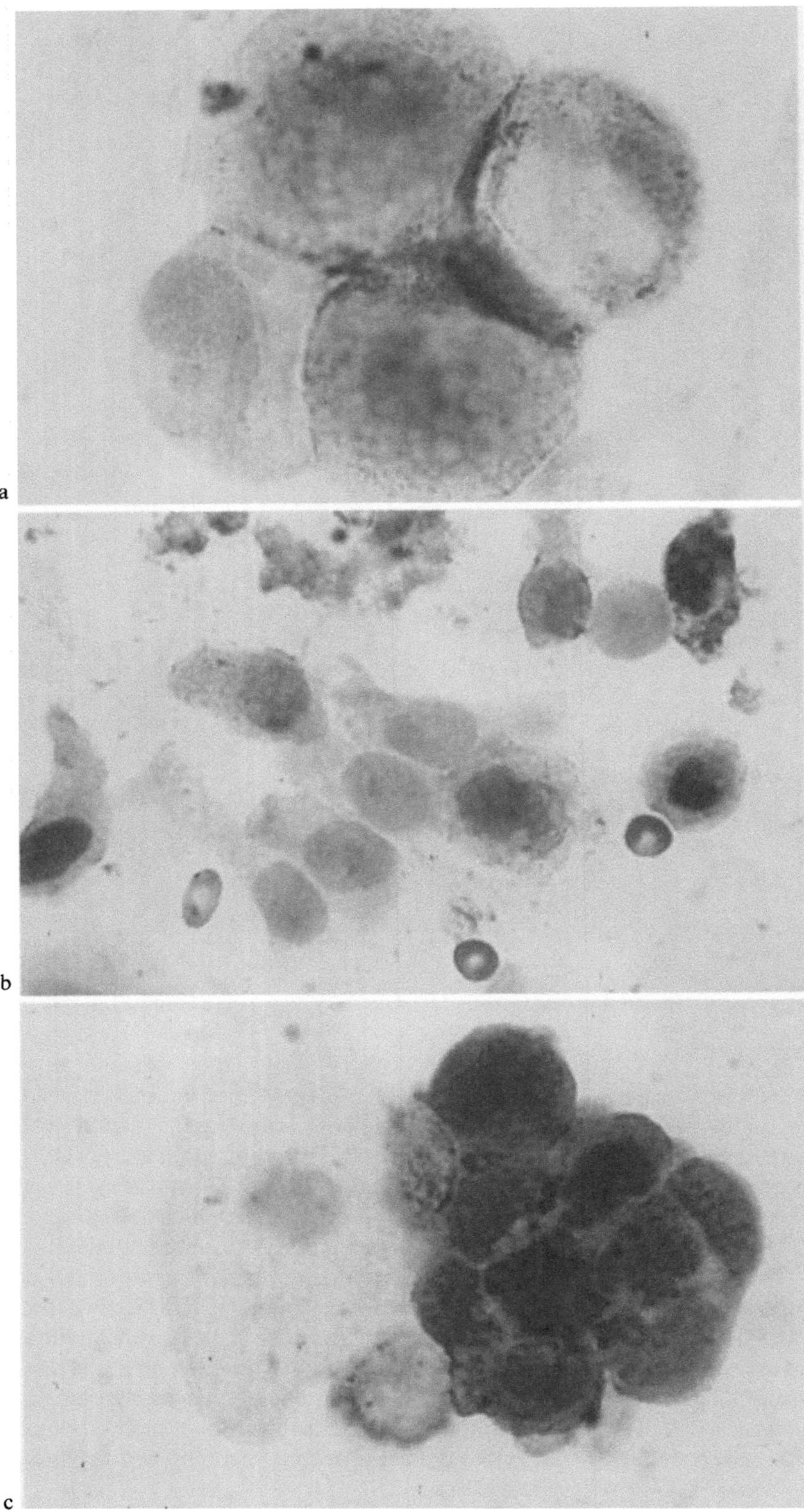

Fig. 12.7a–c. **Immunocytology of grade 3 transitional cell carcinoma (urine sediment) with MAB 486 p 3/12.** a,b Two examples of heterogenous staining. Alkaline phosphatase reaction. c The squamous epithelial cell is not stained

Table 12.2. Analysis of cytologic, immunocytologic, and flow cytometric findings in 69 patients with transitional cell carcinoma of the bladder, grouped according to tumor grades

Tumor grade	1		2		3		Total	
	(n)	$(\%)$	(n)	$(\%)$	(n)	$(\%)$	(n)	$(\%)$
Positive cytology (%)	8/18	44.4	19/37	51.3	10/14	71.4	37/69	53.6
Positive immuno-cytology (%)	16/18	88.9	33/37	89.2	11/14	78.5	60/69	86.9
Positive flow cytometry (%)	5/18	27.7	18/37	48.6	8/14	57.1	32/69	46.4

Table 12.3. Sensitivity of quantitative immunocytology and standard cytology in 241 bladder tumor patients

Tumor grade	1		2		3		Total	
	(n)	$(\%)$	(n)	$(\%)$	(n)	$(\%)$	(n)	$(\%)$
Totals	49		94		98		241	
Positive immuno-cytology	45	91.8	84	89.4	91	92.9	220	91.3
Positive cytology	29	59.2	60	63.8	83	84.7	172	71.4

and with the results of Melamed (1984), who could identify only 1/3 of papillomas by flow cytometry.

In a recent study we were able to reproduce the high sensitivity of QUIC with MAB 486 p 3/12 in 241 patients with bladder tumors (Klän et al. 1991). In this study we evaluated 49 patients with grade 1 tumors, 94 with grade 2 tumors, and 98 with grade 3 tumors. The results, which are compared with conventional cytology in Table 12.3, coincide with the results in Table 12.2.

QUIC with MAB 486 p 3/12 is currently the most sensitive marker for bladder carcinomas, achieving equal levels of sensitivity in the detection of grade 1, grade 2, and grade 3 neoplasms.

12.4 Specificity of QUIC with MAB 486 p 3/12

A marker must have high specificity if it is to be useful for screening. It should be positive in patients with a bladder tumor but negative in patients with other disorders and in healthy controls. None of the monoclonal antibodies that we have studied has a high specificity. The false-positive rate with QUIC is approximately 10% for nearly all the antibodies tested, and QUIC is almost always positive in patients with urinary stone disease (bottom row in Fig. 12.5). With our monoclonal antibody 486 p 3/12 and all the other MABs tested, there is always a high rate of positive quantitative immunocytologic findings in patients with kidney stones. We have proven that this is not caused by exfoliated renal epithelial cells. Among the 15 MABs tested, the antibodies that are not reactive with renal tissue gave a positive QUIC result in lithiasis patients just as frequently as antibodies that also react with, say, tubular epithelial cells. The cause of this phenomenon remains unclear. We have concluded

that none of the monoclonal antibodies currently known, including our MAB 486 p 3/12, is suitable for use as a screening tool.

Consequently, we do *not* recommend quantitative immunocytology for screening urine samples to exclude bladder cancer. The high false-positive rate would needlessly alarm the patient and physician. QUIC is not specific enough for this application.

12.5 QUIC with MAB 486 p 3/12 as a Marker for the Follow-up of Patients with Superficial Bladder Tumors

The high sensitivity of QUIC with MAB 486 p 3/12, especially in the detection of well-differented tumors, suggests that it would be excellent for the postoperative follow-up of patients with superficial bladder tumors.

Our years of research with quantitative immunocytology led us to believe that QUIC with MAB 486 p 3/12, used in patients who have undergone excision of a superficial bladder tumor, would become positive long before a recurrence of the tumor could be detected by endoscopy.

We tested the value of QUIC with MAB 486 p 3/12 in a prospective study of 55 patients with TAT1 superficial bladder carcinoma (Huland et al. 1990). None of the patients initially received antirecurrence prophylaxis. Urine samples were collected from all patients at 4-week intervals. In 40% of the patients ($n=22$), QUIC remained negative following transurethral resection (TUR) of the tumor. Only two of these patients developed recurrent tumors over an average follow-up period of almost 2 years (Table 12.4).

Apparently, QUIC provides a reliable means of identifying patients who will not develop recurrent disease.

As Table 12.5 indicates, the patients who remained immunocytologically negative were equally distributed between grade 1 and grade 2 tumors. All the patients with grade 3 tumors remained immunocytologically positive after TUR. The two patients who developed a recurrence despite negative immunocytology reinforce the original observation that MAB 486 p 3/12 labels only 90% of all bladder tumors. The sensitivity of QUIC in the follow-up of patients with superficial bladder tumors could be further improved by limiting the surveillance to patients in whom QUIC with MAB 486 p 3/12 was positive prior to operation. Based on our experience, this should be the case in 90% of all patients with superficial bladder tumors.

Thirty-three of the 55 patients (interestingly, the majority) were immunocytologically positive from the start of the follow-up period, i.e., immediately after the TUR. Fourteen of these patients subsequently developed a recurrence, although 15 of the remaining patients received prophylaxis following multiple positive immunocytologic tests; it is reasonable to assume that the actual recurrence rate in this group would have been even higher without prophylaxis.

In summary, QUIC with MAB 486 p 3/12 is of potential value for the follow-up of patients with superficial bladder tumors. This should be checked prior to tumor removal by determining whether QUIC is a good marker for the individual patient (QUIC should be positive before TUR). Following TUR, QUIC should be performed on voided urine samples collected at regular intervals (e.g., every 4 weeks).

If QUIC becomes or remains positive after TUR, the patient should be evaluated by cystoscopy, and a bladder wash specimen should be obtained; the higher cellular yield in the washing will make the specimen easier to evaluate by QUIC with MAB 486 p 3/12. If cystoscopy discloses a visible recurrence, the lesion must of course be removed by TUR. If a recurrence is not visible, the positive QUIC test indicates a strong propensity for recurrence and signals the need to initiate prophylaxis.

Currently our group is working to determine whether a positive QUIC after TUR can be used not only to assess the need for prophylaxis but also to establish its duration. This would give QUIC with 486 p 3/12 a role analogous to that of β-HCG or α-fetoprotein in patients with nonseminomatous testicular tumors. At present it is still unclear whether this will be possible.

Table 12.4. **Results of quantitative immunocytology with MAB 486 p 3/12** in the follow-up of 55 patients with superficial bladder tumors

Results	Patients (n) $(\%)$	Recurrent tumors (n)
Negative	22 40	2
Positive	33 60	14[a]

[a] Fifteen of the remaining 19 received prophylaxis.

Table 12.5. **Results of quantitative immunocytology** with MAB 486 p 3/12 according to tumor stage and grade

Results	TA	T1	Tis	G1	G2	G3	Total
Negative	9	13	0	11	11	0	22
Positive	8	22	3	8	17	8	33
Total	17	35	3	19	28	8	55

12.6 Outlook

Today, appropriately trained personnel must be available to interpret the result of quantitative immunocytology. This can be learned easily and quickly – far more easily than conventional cytology. As noted in Chap. 11, subjective impressions are not a factor in the QUIC method. The staining of cells in immunocytology is an objective criterion. In the future, efforts will be made to automate this method. This will require a means for eliminating or separately labeling granulocytes that stain concurrently with the MAB. Also, clumped cells must be individualized to produce monolayer suspensions suitable for machine evaluation. These problems are technically manageable. It is also conceivable that the test will utilize not just one MAB but two mutually supplementary MABs to further improve the sensitivity of the QUIC analysis. We doubt that the specificity of QUIC can be improved or that the test will ever be appropriate for screening. The ideal setting for QUIC is in the follow-up care of patients with superficial bladder carcinomas. QUIC with MAB 486 p 3/12 is the first tested bladder tumor marker that can help direct the postoperative management of these patients.

The idea is to evaluate urine samples by QUIC and limit endoscopy to those patients who remain or become QUIC-positive following TUR. It must be emphasized that positive QUIC does not necessarily mean that the patient has a pathologically demonstrable tumor recurrence that is already visible by endoscopy. QUIC with MAB 486 p 3/12 is such a sensitive marker that it signals the mere proneness or propensity to develop a recurrence.

Repeated positive QUIC tests are an indication for endoscopy to exclude a recurrent tumor. If a recurrence is not seen, prophylaxis can be instituted if QUIC continues to be positive.

Our policy is as follows: If a patient has two consecutive positive QUIC tests, we perform cystoscopy to check for a visible recurrence. If the positive immunocytology cannot be attributed to recurrent tumor, we initiate prophylaxis. We use QUIC, then, as a means of selecting patients who require postoperative prophylaxis. As noted in Sect. 12.5, it remains to be seen whether QUIC will be useful in establishing the duration of prophylaxis that is appropriate for the individual patient.

References

Arndt R, Dürkopf H, Huland H, Donn F, Loening T, Kalthoff H (1987) Monoclonal antibodies for characterization of the heterogeneity of normal and malignant transitional cells. J Urol 137: 758–763

Huland H, Arndt R, Huland E, Loening T, Steffens M (1987) Monoclonal antibody 486 P 3/12: a valuable bladder carcinoma marker for immunocytology. J Urol 137: 654–659

Huland E, Huland H, Arndt R, Baisch H, Klöppel G (1988) Urindiagnostik oberflächlicher Harnblasentumoren durch Zytologie, Immunzytologie und Flowzytometrie: Ergebnisse einer prospektiven vergleichenden Studie an 104 Patienten. Aktuel Urol 19: 13–17

Huland E, Huland H, Meier T, Baricordi O, Fradet Y, Grossmann HB, Hodges GM, Messing EM, Schmitz-Draeger BJ (1991) Comparison of 15 monoclonal antibodies against tumor-associated antigens. J Urol 146: 1631–1636

Huland E, Huland H, Schneider AW (1990) Quantitative immunocytology in the management of patients with superficial bladder carcinoma. I. A marker to identify patients, who do not require prophylaxis. J Urol 144: 637–640

Klän R, Huland E, Baisch H, Huland H (1991) Sensitivity of urinary quantitative immunocytology with mAk 486 p 3/12 in 241 unselected patients with bladder carcinoma. J Urol 145: 495–497

Melamed MR (1984) Flow cytometry of the urinary bladder. Urol Clin North Am 11: 599–608

Oljans PJ, Tanke J (1986) Flow cytometric analysis of DNA content in bladder cancer: prognostic value of the DNA-index with respect to early tumour recurrence in G 2 tumours. World J Urol 4: 205–210

13 Urinary Erythrocyte Morphology and Diagnosis of Hematuria

S. ROTH

CONTENTS

13.1 Introduction

One of the most common problems confronting the urologist is the investigation of microhematuria. This is due in part to the increased numbers of baseline, insurance and screening examinations that are being performed.

The search for an *effective and efficient primary diagnostic workup* of microhematuria was significantly advanced in 1979 when the Australian nephrologists Birch and Fairley first described characteristic morphologic changes of urinary erythrocytes in patients with glomerulonephritis.

Decades earlier, Larcom and Carter (1948) had noted the morphologic variability of red blood cells in urine sediment but were unable to establish a causal connection with specific disorders.

The ability to establish a glomerular source of bleeding by microscopic examination is of great importance for the urologist. After neoplasms and calculi, *renal parenchymal diseases* are recognized as the *third most common cause* of microhematuria, having a reported incidence of 2.5%–7.1% (Carson et al. 1979; Golin and Howard 1980; Davides et al. 1986; Olivo et al. 1989). Since experience indicates that renal parenchymal disease is treatable only in its early stage and that significant microhematuria is the only criterion for early diagnosis (Renner 1986), the *urologist* plays a *key role in directing patient management*. Urinary erythrocyte morphology can furnish the information needed to justify referral for appropriate nephrologic therapy.

13.2 Microhematuria

13.2.1 Definition of Microhematuria

Microhematuria is generally defined as the presence of red blood cells, detectable chemically or microscopically, in urine that appears grossly normal. Although the kidneys and lower urinary tract normally are not permeable to the passage of erythrocytes, red cells are present to a degree even in healthy urine, raising the problem of *where to draw the line between physiologic and pathologic microhematuria*. This problem is not only manifested in the difficult discernment of borderline pathologic values but also affects a variety of different methods of determination.

13.2.2 Diagnostic Evaluation of Microhematuria

13.2.2.1 Reagent Strips

Reagent strips (dipsticks) are widely used as a microhematuria screening test owing to their simplicity, and they hold considerable promise as an instrument for *home screening*. In 235 males over 50 years of age with no history of hematuria who tested their urine once a week at home for a period of 1 year, 31 patients had multiple positive tests prompting referral for urologic evaluation, which disclosed eight cases of urogenital carcinoma and seven cases of benign disorders necessitating treatment (Messing et al. 1989).

The principle of the reagent strip is based on the chemical reaction of organic hydrogen peroxide, which converts a highly sensitive chromogen to a blue-colored dye through the catalytic activity of hemoglobin. Since a positive test may be due to the presence of free hemoglobin and myoglobin as well as red blood cells, a *positive test should prompt a microscopic analysis of the urine sediment* to establish the presence of erythrocytes before additional diagnostic measures are performed.

The exclusive use of reagent strips is obsolete because of their inadequate specificity and sensitivity (hence the term "dipstick hematuria"). The false-negative rate can be as high as 16%–24% (Frölich and Sieck 1981; Arm et al. 1986), resulting in a *potentially low sensitivity*. The investigation of microhematuria with reagent strips alone would lead to an unacceptable diagnostic void. Figures on the false-positive rate range from 15% to 30% (Arm et al. 1986; Jardin and Madier 1987). Without microscopic verification, this *low specificity* would result in unnecessary, frequently invasive diagnostic measures.

13.2.2.2 Semiquantitative Techniques (Microscopic Field Counts)

In the semiquantitative techniques for the evaluation of microhematuria, the erythrocyturia is quantified independently of the factors of volume and time. An important argument favoring semiquantitative analysis is its practicality, since the *usefulness of a method is determined in part by the ease with which it can be performed* (Thiel et al. 1986).

In these techniques the examiner views the centrifuged urine sediment with a microscope and counts the number of red blood cells that can be seen per 400-power field (× 10 eyepiece, ×40 objec-

tive). Although there is *no definite consensus* with regard to normal values, many authors state that a count of 2–8 cells/field constitutes borderline or grade 1 microhematuria (Fig. 13.1), while ten or more cells/field signifies grade 2 microhematuria (Carson et al. 1979; Golin and Howard 1980; Thiel et al. 1986; Mohr et al. 1986; Conzelmann et al. 1988).

Grade 1: 2–8 cells/field
Grade 2: 8–30 cells/field
Grade 3: cells covering 3/4 of the field
Grade 4: cells covering the entire field

The count should be performed using phase contrast microscopy or on stained specimens during the oncologic cytologic investigation, since experience has shown that bright-field microscopy alone does not provide sufficient contrast or definition of the erythrocytes, leading to erroneously low counts.

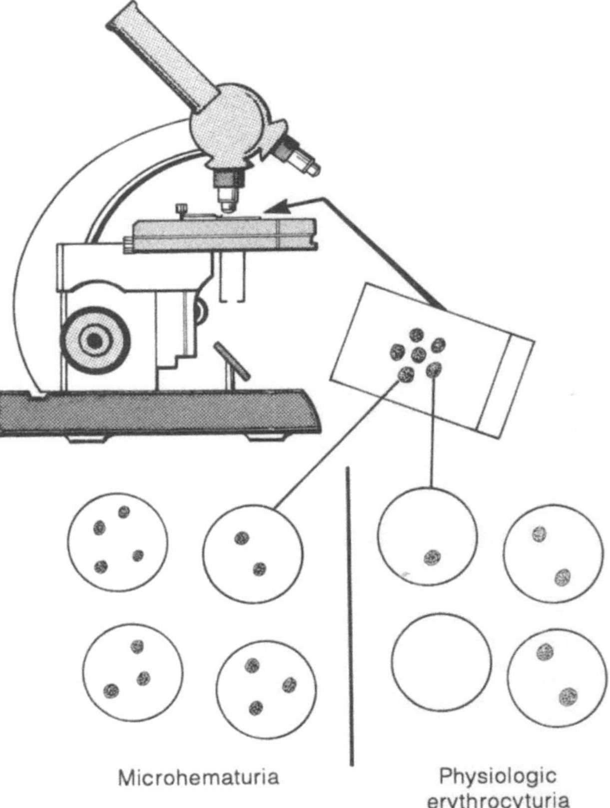

Fig. 13.1. **Microhematuria** is diagnosed if 2 or more erythrocytes are counted in multiple 400-power fields (× 10 eyepiece, × 40 objective). Otherwise **physiologic erythrocyturia** is present. (Modified from Thiel et al. 1986)

13.2.2.3 Quantitative Techniques

In quantitative techniques the uncentrifuged urine is examined in special *counting chambers* that permit the examiner to calculate the number of excreted urinary erythrocytes *per unit volume* of urine.

Addis in 1925 introduced the "Addis count" using a special counting chamber as a method of clinical examination. A later device, equally effective but easier to use, was the *Fuchs-Rosenthal chamber*, in which a drop of uncentrifuged fresh urine is placed into the chamber, which is divided into 16 squares. By counting the erythrocytes (and leukocytes), the investigator can compute the number of red cells per milliliter for the known chamber volume of 3.2 mm³. In contrast to the Addis count, a quantitative concentration is provided by the enlarged counting area of the device. Another variant is the *Neubauer chamber*.

Simpler yet more advanced is the *MD-KOVA system*, in which the centrifuged urine is deposited onto special disposable slides and counted. The defined volume of 6.6 µl per chamber permits an accurate quantification of urinary erythrocytes (Bauer et al. 1981).

The microsediment technique developed by *von Froreich* et al. (1974), which combines concentration of the sample with a direct coating of the counting surface, is simple and affords good results (Rathert and Lutzeyer 1974) but requires special apparatus in the form of an inverted microscope.

Adding the *time factor* to the quantitative determination (e.g., 12-h urine) in addition to volume is time-consuming and limits the usefulness of the method for routine evaluations. Another argument for an exclusively volume-oriented quantification is furnished by the studies of Teitel et al. (1964), who found that there were no significant differences in the urinary excretion of corpuscular elements during the course of a day.

The generally accepted *upper normal limit of microhematuria* in quantitative testing is 2000–3000 (maximum of 5000) red blood cells/ml (Olbing 1977; Althof and Ochs 1982; Bauer et al. 1981; Heintz and Althof 1989).

13.2.3 Incidence and Clinical Significance of Microhematuria

The *prevalence* of asymptomatic microhematuria ranges from 1% to 16% depending on the age of the population examined (Freni and Freni-Titulaer 1977; Sinniah et al. 1976; Vehaskari et al. 1979; Froom et al. 1984; Mohr et al. 1986).

The essential and still-controversial question, especially in patients under 40 years of age, concerns the *diagnostic implications* of microhematuria. How extensive an investigation is required, and how invasive must it be? Attitudes range from a complacent "diagnostic nihilism" to an aggressive "diagnostic overkill."

The basis for this controversy is the fact that the degree or severity of hematuria (microscopic or gross hematuria) does not correlate with the severity of the underlying disease (Messing et al. 1989). In a recent study of 1000 patients, Mariani et al. (1989) clearly demonstrated the *absence of a safe lower limit of hematuria*. Of the 86 patients who had a life-threatening cause of urinary tract bleeding, 16 (18.6%) had fewer than three erythrocytes/field in at least one urine sample. Three of these 16 patients had a renal cell carcinoma, and 13 had a urothelial cancer.

The absence of a "safe lower limit" justifies the claim that a *costly and precise quantification of hematuria* (e.g., using the counting chamber method) *does not have major practical implications*. It is far more critical to establish a reliable lower threshold.

Incidence of urologic carcinoma in patients with microhematuria: (regardless of age)

1. Greene et al. (1956)
   500 patients with microhematuria:
   11 carcinomas (2.2%)
2. Carcon et al. (1979)
   200 patients with microhematuria:
   25 carcinomas (12.5%)
3. Golin and Howard (1980)
   247 patients with microhematuria:
   24 carcinomas (10%)
4. Davides et al. (1986)
   150 patients with microhematuria:
   13 carcinomas (8.7%)
5. Mariani et al. (1989)
   1000 patients with microhematuria:
   86 carcinomas (8.6%)

Age Dependence of Carcinoma Incidence.
The opinion expressed in many textbooks that microhematuria should always be viewed as an ominous sign until proven otherwise poses a *danger of "diagnostic escalation,"* where every patient is placed through a gamut of diagnostic procedures until a tumor has been confirmed or excluded as the cause of the hematuria. The validity of this approach is qualified by the practical experience that, even in patients with microhematuria, a significant *age dependence* exists with regard to the *incidence of carcinoma.*

A screening study conducted in 10 050 young males disclosed microhematuria in 165 patients. Of the 24 patients who were thoroughly examined, one was found to have an exophytic bladder tumor (Richie et al. 1986). The authors' call for a full urologic investigation of all patients with microhematuria, regardless of age, prompted Jones et al. (1988) to conduct a prospective study of 100 patients under 40 years of age with microhematuria. While the authors were able to ascertain the cause of the microhematuria in 32 cases, there were only three cases in which cystoscopy was crucial in establishing a diagnosis. No case of microhematuria was found to be due to a malignant tumor. This is consistent with the findings of Froom et al. (1984), who found only one carcinoma in 1000 young members of the Israeli Air Force examined over a 15-year period.

These figures help us to understand the call of many urologists for a *rational diagnostic approach to microhematuria*, especially in younger patients. This basic workup would include:

History, for example,
- Exertion-dependent hematuria
- Familial microhematuria
- Medications (anticoagulants)

Laboratory data, for example,
- Coagulation disorder

Sonography, for example,
- Renal parenchymal tumor
- Renal cysts
- Congested kidneys

Urinalysis with
- Quantification of proteinuria
- Quantification of leukocyturia
- Oncologic urinary cytology
- Erythrocyte morphology

13.3 Nonglomerular and Glomerular Erythrocytes

13.3.1 Morphology

13.3.1.1 Nonglomerular Erythrocytes

The frequently cited link between the presence of nonglomerular erythrocytes in the urine and a postrenal bleeding source is misleading in that other renal processes such as cysts, tumors, and interstitial nephritis are also commonly associated with hematuria.

Despite the characteristic appearance of glomerular ("dysmorphic") erythrocytes, a description of the nonglomerular forms is necessary because these cells, too, are subject to morphologic variations due to differences in osmolarities, prolonged contact with the urine, or the mechanical stresses associated with extreme centrifugation or manual smearing (Figs. 13.2, 13.3). For this reason the *term "isomorphic" (morphologically uniform) is inappropriate for erythrocytes*, and the more neutral term "nonglomerular" is preferred (Thiel et al. 1986). A number of potential morphologic variants of nonglomerular erythrocytes may be observed:

Double-Rim Forms (Fig. 13.4a–d)
A double rim is commonly seen on erythrocytes that have been exposed to urine for several hours. The thinness of the double rim (Fig. 13.4a) distinguishes it from the annular shape of the glomerular erythrocyte (Fig. 13.4b,c). More importantly, the space bounded by the rim is still filled with cytoplasm and lacks the central "hole" characteristic of glomerular ring forms.

Spiked Forms (Fig. 13.5a–e)
Diffusion processes in hypertonic urine can cause shrinkage of the erythrocytes with the formation of spike-like projections on the cell surface (Fig. 13.5a). These pointed projections are not characteristic of glomerular erythrocytes, which frequently exhibit rounded, blister-like protrusions that may project toward the center of the cell (Fig. 13.5b) or outward (Fig. 13.5c,d), being attached to the cell body by a short, narrow stalk. In extreme cases the glomerular-type projections may appear as spheres adherent to the inside or outside of the cell.

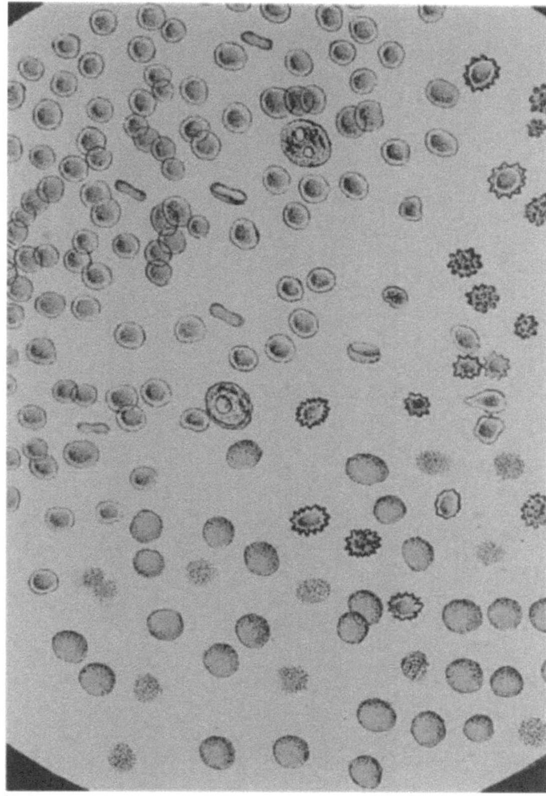

13.2

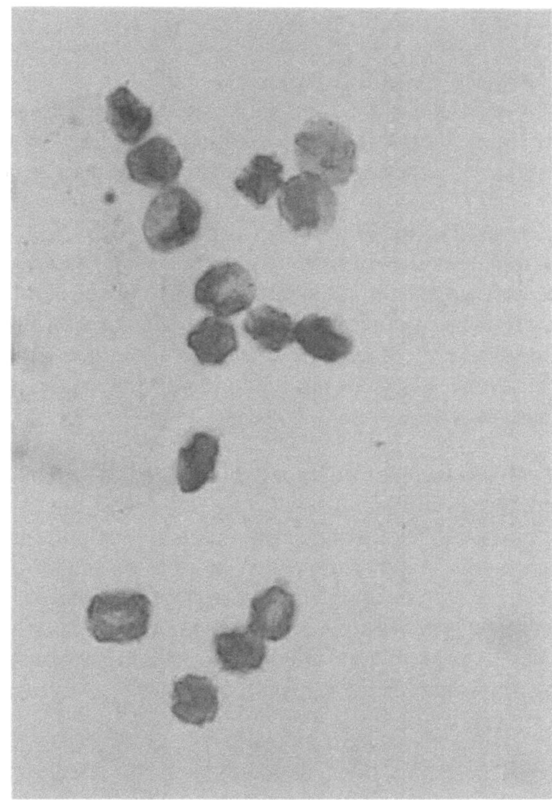

Figs. 13.2 and 13.3. As early as **1872**, Ultzmann and Hofmann showed in their *Atlas of Urinary Sediments* (Fig. 13.2) that nonglomerular (normal) erythrocytes are morphologically diverse (dysmorphic) as a result of different ambient conditions and positions on the optical axis (see text). The cells exhibit conspicuous spiked, biconcave, flattened, and dumbbell shapes. In Fig. 13.3 another specimen (Papanicolaou stain,× 850) clearly illustrates the dysmorphism of nonglomerular (normal) erythrocytes. Consequently, the term **dysmorphic erythrocytes** should **not** be equated with **glomerular erythrocytes** (see Sect. 3.1)

13.3

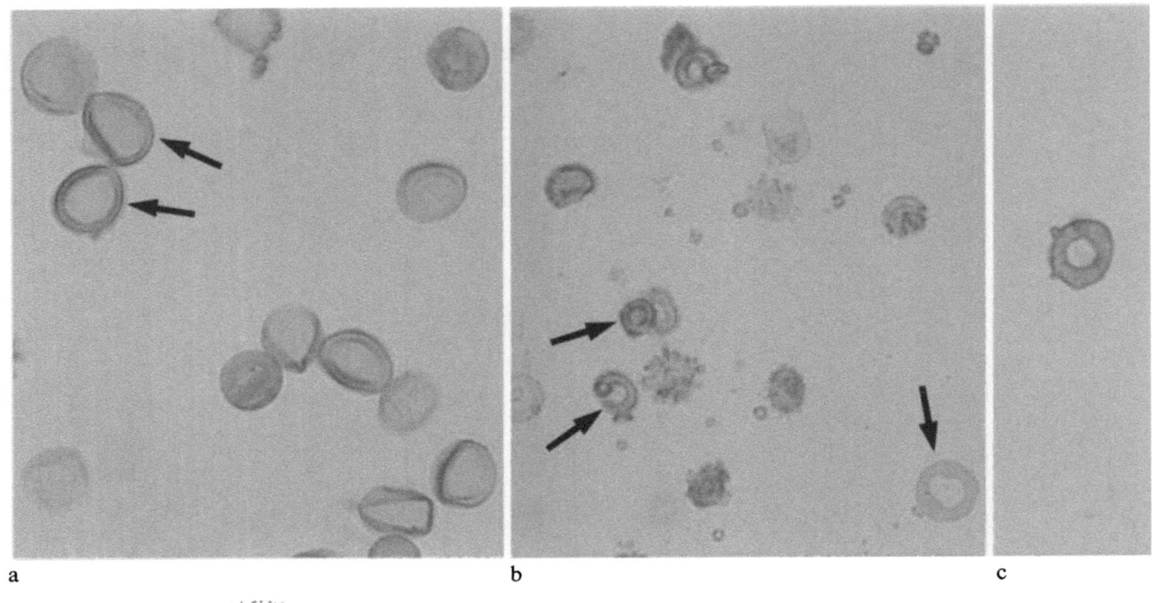

a   b   c

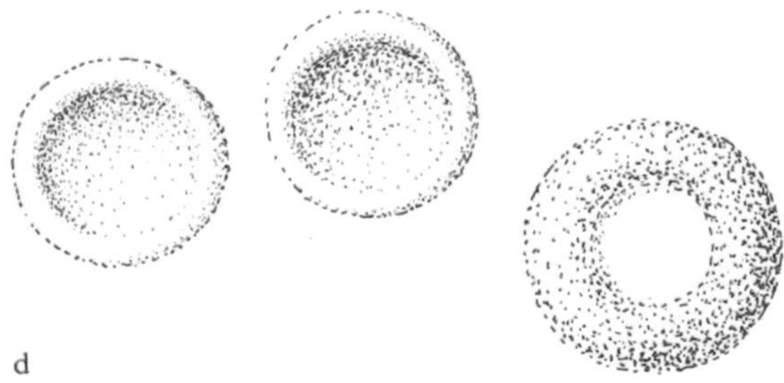

d

Fig. 13.4. a **Nonglomerular (normal)** erythrocytes (Papanicolaou stain, × 850) with osmotically induced double-rim structure (← ←). b **Glomerular erythrocytes** with typical annular structure (←) (Papanicolaou stain, ×850). c **Glomerular erythrocyte** (ring form) stained with methylene blue (× 850). d Schematic representation of **nonglomerular** (normal) erythrocytes with a fine double rim, contrasted with the "doughnut" appearance of a **glomerular** erythrocyte

Discoid Forms (Fig. 13.6)

In hypotonic urine, osmotically induced intracellular fluid uptake leads to enlargement of the erythrocytes, which appear as uniformly pale discs.

Creased Forms (Fig. 13.7)

Nonglomerular erythrocytes sometimes exhibit bizarre creases that radiate in a "Mercedes star" pattern from a central clear zone. Because of its translucency, the central zone may cause the cell to be mistaken for an annular glomerular erythrocyte.

Cells Viewed Obliquely or Edge-On (Fig. 13.8a,b)

Although most erythrocytes lie flat on the microscope slide, permitting a "planimetric" evaluation, some will occupy an oblique or edge-on position relative to the observer. An erythrocyte viewed edge-on will display a club or dumbbell shape (Fig. 13.8a) while an oblique cell will appear cap- or cone-shaped (Fig. 13.8a,b).

13.3.1.2 Glomerular Erythrocytes

Since glomerular erythrocytes were first described (Birch and Fairley 1979), a variety of morphologic classifications have been attempted. It is reasonable to object, however, that an attempt to designate individual forms with *specific names* (Hildebrandt et al. 1988) *needlessly complicates* what is basically a *simple morphologic phenomenon* (Thiel et al. 1986).

The following principal forms of glomerular erythrocytes are encountered:

Ring Forms (Fig. 13.9a–f)

In these forms, which are the most common morphologic variant of glomerular erythrocytes and are pathognomonic for glomerular hematuria, the entire cell volume is arranged in a peripheral ring, giving the cell a "doughnut" or "life preserver" appearance (Fig. 13.9a,c).

Although the central opening appears microscopically as a punched-out defect, it is actually covered by a stretched double thickness of "empty" membrane. This is made clear by the dumbbell

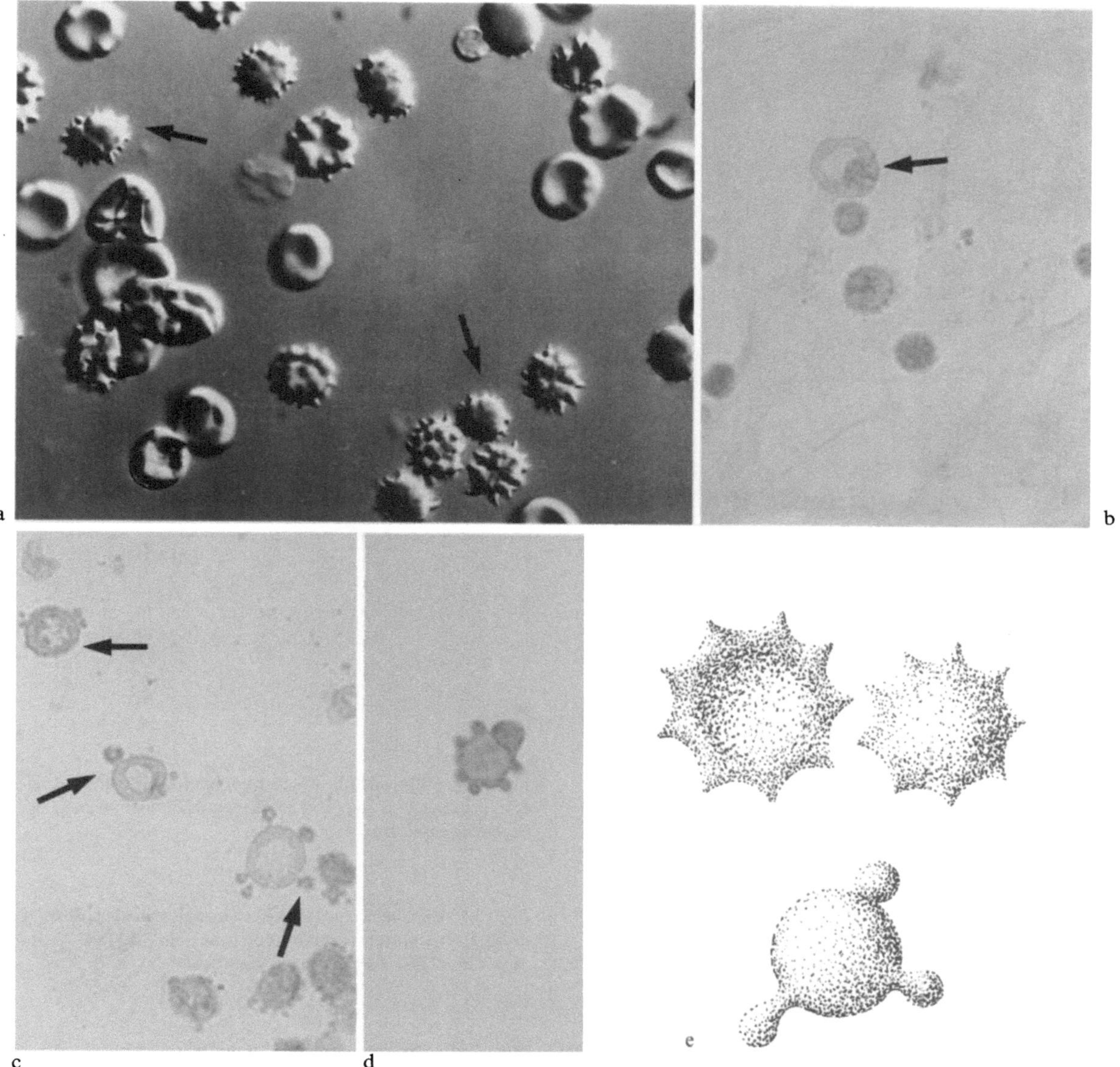

Fig. 13.5. a **Nonglomerular erythrocytes** with typical spiked surface projections (← ←) and biconcave appearance on interference contrast microscopy, × 765. (Photo courtesy of Priv.-Doz. Dr. H. Leyh of the Urologic Clinic and Outpatient Clinic of the Technical University of Munich, FRG). b **Glomerular erythrocyte** with an internal blister (← ←) alongside normal-appearing, nonglomerular erythrocytes (Papanicolaou stain, × 850). c **Glomerular erythrocytes** with vesicle-like surface projections (← ←) (Papanicolaou stain, × 850). d Methylene blue-stained **glomerular erythrocyte** with bulbous surface projections (←) (× 850). e Schematic contrast between **nonglomerular** spiked erythrocytes (*above*) and a **glomerular** erythrocyte with vesicle-like protrusions (*below*)

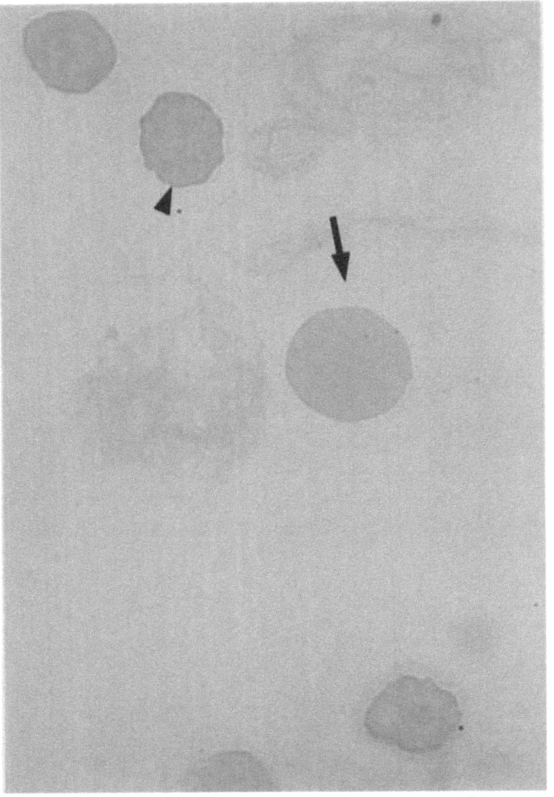

13.6

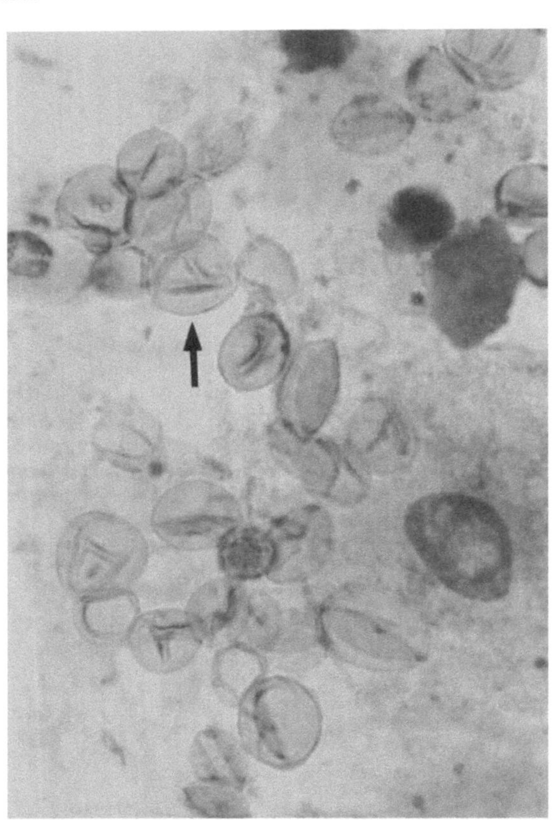

13.7

a

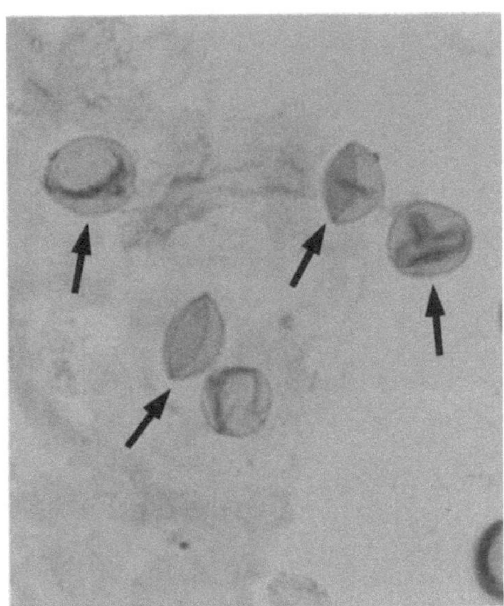

b
13.8

Fig. 13.6. Difference between hypoosmotically altered discoid erythrocytes (←) and erythrocytes of normal size (◄). Papanicolaou stain, ×850

Fig. 13.7. Bizarre creased forms of **nonglomerular** (normal) erythrocytes with an apparent central hole (◄) (see ← and ◄ in Fig. 12.8b). Papanicolaou stain, ×850

Fig. 13.8. a **Nonglomerular** (normal) erythrocytes viewed edge-on (dumbbell shape) and obliquely (cap shape). b Multiple cap-shaped, **nonglomerular** (normal) erythrocytes (← ←). Papanicolaou stain, ×850

Fig. 13.9a–f. a,b Ring-shaped **glomerular** erythrocytes ▷ (← ←) (a) compared with normal nonglomerular erythrocytes (b) on interference contrast microscopy, × 765. (Photos courtesy of Priv.-Doz. Dr. H. Leyh of the Urologic Clinic and Outpatient Clinic of the Technical University of Munich). c Typical ring-shaped ("doughnut") **glomerular** erythrocytes (← ←). Papanicolaou stain, × 850. d,e **Glomerular** ring forms viewed edge-on and in cross section (← ←); the cell contents are arranged in a peripheral ring, leaving a thin ("empty") central double membrane that looks like a hole when viewed from above. Papanicolaou stain, ×850. f Schematic illustration contrasting the **sharp inner contour of the glomerular ring form** (*right*) with a normal biconcave erythrocyte. The broken line marks the selected microscopic focal plane

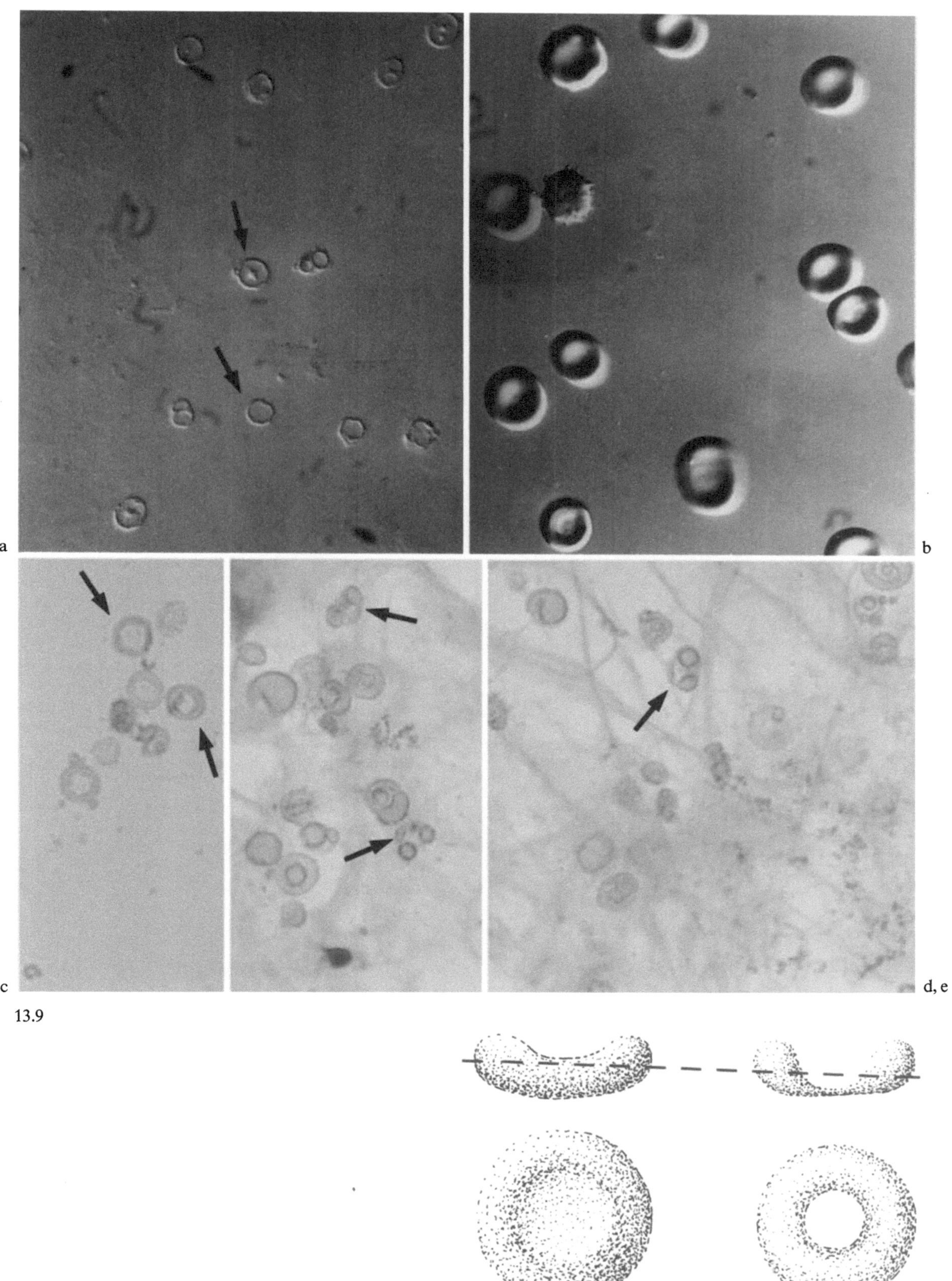

13.9

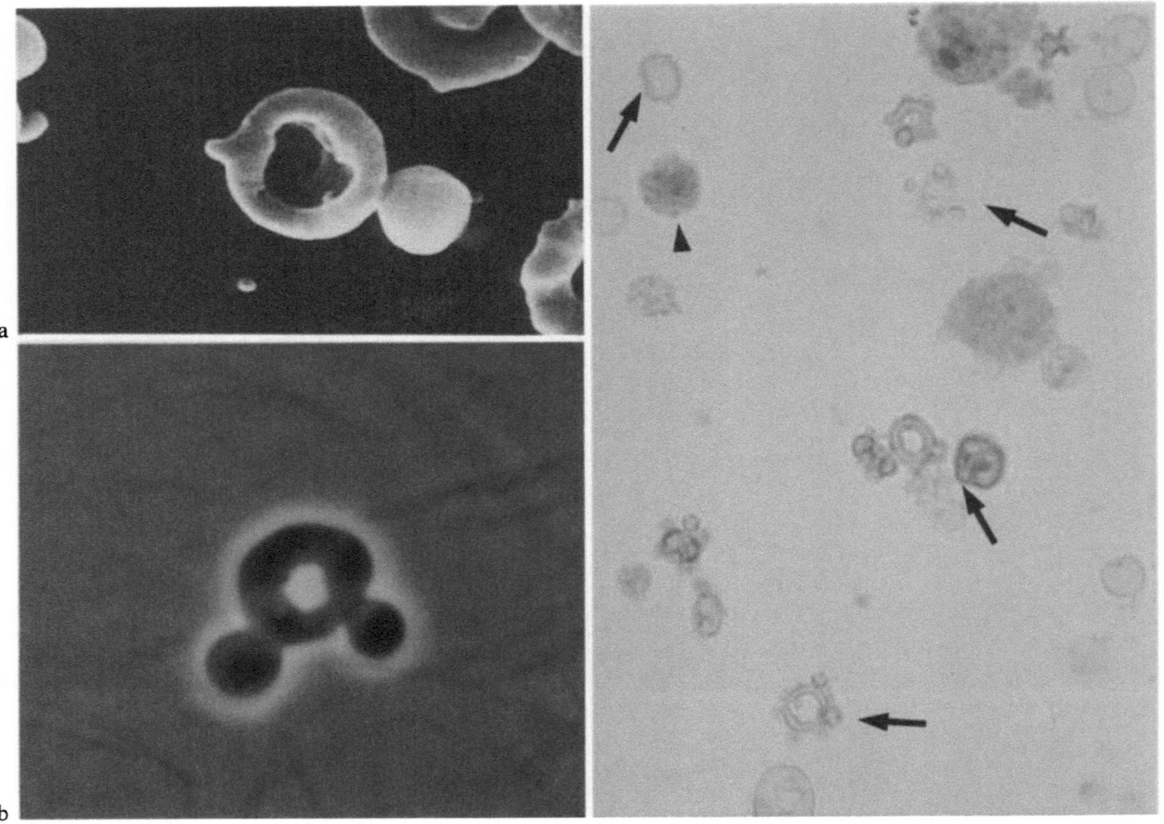

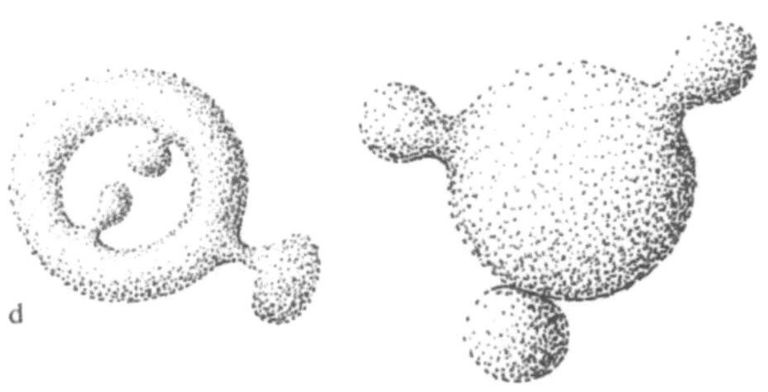

Fig. 13.10a–d. a,b **Glomerular** ring forms with vesicle-like protrusions (acanthocytes) as seen by scanning electron microscopy (a) and at × 1000 by phase contrast microscopy (b). (Photos courtesy of Prof. Dr. H. Köhler of the First Medical Clinic and Outpatient Clinic of the University of Mainz). c **Glomerular** erythrocytes with surface protrusions (acanthocytes, ← ←). Papanicolaou stain, × 850. d The **vesicles** may appear on ring forms as well as on erythrocytes without a central "hole." These blister-like protrusions are always interpreted as a sign of **glomerular** bleeding. Whether glomerular disease is present depends on the quantitative distribution pattern (see Sect. 13.3.3)

shape of the cell when it is viewed from the side (Fig. 13.9d–f).

These glomerular ring forms are alternatively known as annulocytes, Willisauer's rings (Thiel et al. 1986), or peppermint erythrocytes.

The *differentiating criterion of a sharp inner contour* readily distinguishes the glomerular ring forms from nonglomerular erythrocytes with a central clear zone (Fig. 13.9f). By adjusting the micrometer screw (focal plane adjustment), the examiner can define the inner contour of the ring with the same degree of sharpness as the opposite outer contour. This is not possible with a normal biconcave erythrocyte (Fig. 13.9b,f).

Vesicular Forms (Fig. 13.10a–d, see also Fig. 13.5b–d)

These cells are characterized by various degrees of membrane protrusions, which may occur in isolation or combined with glomerular ring forms. The vesicles are variable in their size and shape and may project from the outer surface of the cell or toward the center of an annular cell.

Ruined Forms (Fig. 13.11a,b)

These erythrocytes are severely distorted and may assume grotesque shapes. Often they appear crushed or shredded. The degree of destruction ranges from milder forms that can result from the prolonged storage of nonglomerular erythrocytes to severe forms that cannot be produced even by extremely long storage.

13.3.2 Sensitivity and Specificity

Today there is no longer any doubt concerning the *high diagnostic accuracy* of dysmorphic glomerular erythrocytes as an indicator of glomerular disease. Fasset et al. (1982), in a blind controlled study of 253 patients, correctly localized the bleeding source in 85% on the basis of erythrocyte morphology. One hundred fifteen patients with dysmorphic erythrocytes underwent renal biopsy, which histologically confirmed the diagnosis of glomerulonephritis in all cases.

Birch et al. (1983) examined 117 patients, 87 of whom had glomerulonephritis and 30 a nonglomerular disease. The glomerular disease was correctly diagnosed with a sensitivity of 99% (86/87) and the nonglomerular disease with a specificity of 93% (28/30). Rizzoni et al. (1983) obtained almost identical results in children, using red-cell morphology to diagnose glomerulonephritis with a sensitivity of 97% (63/65) and nonglomerular bleeding sources with a specificity of 95% (39/41). In 1989, Fünfstück and his group demonstrated a sensitivity of 94.8% (128/135) and a specificity of 95.5% (69/75).

13.3.3 Quantitative Limits

There continues to be controversy over the question of the *quantitative distribution pattern* of glomerular and nonglomerular erythrocytes that must be present in order to *make a diagnosis.*

In their original 1979 study, Birch and Fairley mentioned that erythrocytes with glomerular characteristics are found even in healthy patients. Thiel et al. (1986) note that this is not surprising since no other site in the urinary tract is as heavily perfused as the glomerular capillary filter. A rough estimate shows that of every billion erythrocytes that pass through a glomerulus, one physiologically crosses the glomerular barrier. This both documents the high quality of the glomerular filter and also accounts for the physiologic occurrence of glomerular erythrocytes in the urine.

With regard to the dividing line between glomerular and nonglomerular sources of bleeding, which different groups of workers have handled in different ways, it should be noted that ery-

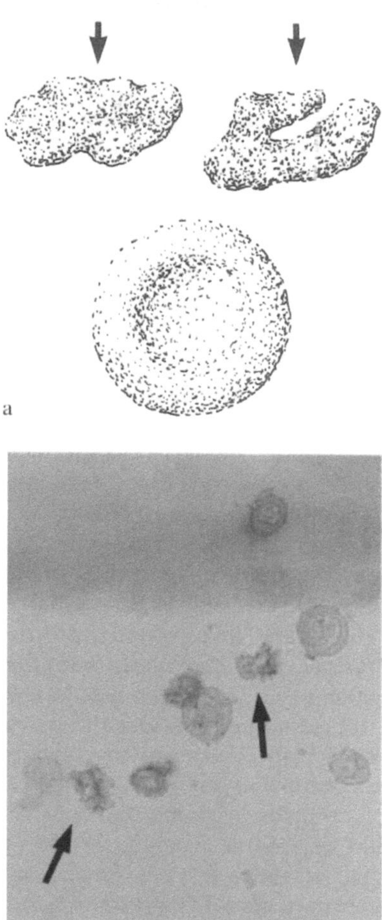

Fig. 13.11 a, b. Ruined **glomerular** erythrocytes ("dwarf" forms, ← ←), shown schematically and with Papanicolaou staining at ×850

throcyte morphology is *not* evaluated for the purpose of *definitive therapeutic planning* but "merely" as an *aid to diagnosis.* At the same time, since the certainty of the diagnosis determines the scope of the subsequent, frequently invasive diagnostic workup, one can appreciate the need for a quantitative orientation.

Investigators consistently report that the prevalence of dysmorphic erythrocytes may be as high as 20% even in the absence of glomerular disease. For example, Fünfstück et al. (1989) found 5.7% (± 5.4%) dysmorphic erythrocytes in 75 patients with urolithiasis, their statistical frequency distribution indicating that up to 17% dysmorphic erythrocytes may be found in patients with nonglomerular hematuria.

Reports vary concerning the critical *percentage limit* that must be present to establish a glomerular source of bleeding. While Thiel et al. (1986) believe that more than 30% glomerular erythrocytes signifies glomerulopathy (=more than two red cells per 400-power field), Rizzoni et al. (1983) defined a lower limit of 40% in pediatric patients.

Other groups have placed the limit even higher. De Santo et al. (1987) and Leyh et al. (1986) state that a percentage of red blood cells of 80% or more signifies glomerular disease, while Schramek and Schuster (1985) require 90%. Fünfstück et al. (1989) found a prevalence of 90.5% (± 7.4%) glomerular dysmorphic erythrocytes in patients with histologically confirmed glomerulonephritis, leading them to conclude that a 75% incidence of glomerular erythrocytes must be present to signify glomerular hematuria with reasonable confidence.

In a series of 300 patients, 120 of whom had glomerulonephritis and 54 interstitial nephritis, Müller et al. (1989) investigated the *factors that influence the changing quantitative limits* that characterize glomerular vs. nonglomerular hematuria. One of their findings was that the percentage of glomerular erythrocytes was inversely related to the degree of *hematuria*. Thus, while the percentage of these cells in glomerulonephritis patients with mild hematuria averaged 84.2% (± 15.2%) glomerular erythrocytes, it averaged only 50.9% (±5.2%) in patients with marked hematuria.

They also found that the percentage of glomerular erythrocytes declined with increasing *renal insufficiency*, averaging 67.6% (±25.3%) in patients with more than 5.5 mg% serum creatinine and 86.6% (±11.75%) in patients below that value.

A final parameter influencing the quantitative distribution was the *activity of glomerulonephritis* (ESR, proteinuria, electrophoresis, clinical presentation). As a result of this phenomenon, previously described by Iseghem et al. (1983), patients with a high glomerulonephritis activity had fewer glomerular red cells (74.6% ±23.3%) than patients with a low activity (86.0% ±12.3%).

Schuetz et al. (1985) found that water- or furosemide-induced *diuresis* in patients with glomerulonephritis also reduced the percentage of glomerular erythrocytes from an average of 64% to 19%.

Table 13.1. **Diagnostic options as a function of erythrocyte morphology**

Morphology	Option
Glomerular bleeding	Nephrologic workup
Mixed form	Nephrologic and urologic workup
Nonglomerular bleeding	Urologic workup

13.3.4 Causes of Erythrocyte Dysmorphism

The causes of glomerular erythrocyte distortion are not yet fully understood. Besides purely mechanical deformation of the cells on traversing the glomerulus, researchers have suggested toxic membrane damage by pathologic osmotic gradients or changes caused by lysosomal leukocytic enzymes that occur as mediators in renal parenchyma that has undergone inflammatory change (Fig. 13.12).

Lin et al. (1983) and Fasset et al. (1982) favor the theory of mechanical deformation, which postulates that the capillary spaces of the glomerular basement membrane are damaged in such as way that they permit an increased, shape-distorting passage of erythrocytes into the collecting system of Bowman's capsule. Lin et al. (1983) presented several electron micrographs illustrating the passage of erythrocytes through such capillary "holes," comparable to a traumatizing "passage through the eye of a needle" (Thiel et al. 1986).

Other authors (Birch et al. 1983; Schuetz et al. 1985) support the theory that tubular and glomerular transit leads to damaging effects on the erythrocyte membrane relating to *pathologic changes in pH and osmolarity*.

Madsen et al. (1982) showed that the cells of the postglomerular tubular system have a phagocytic potential. They believe that erythrocytes that have crossed the glomerular filter undergo membrane damage and morphologic change during tubular passage due to the action of secreted *lysozymes*.

Schramek et al. (1989) were able to generate dysmorphic erythrocytes in vitro that were identical to in vivo glomerular erythrocytes by exposing normal erythrocytes to various osmotic gradients. Glomerular changes were observed, however, only in cells that were additionally exposed to a hemolytic medium. Based on these results, the authors cited an *intrinsic factor* of the lytic erythrocytes as being responsible for the dysmorphism.

Table 13.2. **Quantitative limits of erythrocyte morphology** [a]

Less than 20 % glomerular erythrocytes	No glomerulonephritis
20 %–50 % glomerular erythrocytes	Questionable glomerulopathy
50 %–75 % glomerular erythrocytes	Strong suspicion of glomerulonephritis
More than 80 % glomerular erythrocytes	Definite glomerulonephritis

[a] The percentage of glomerular erythrocytes is decreased in the presence of:
- Strong diuresis
- Marked hematuria
- Increasing renal insufficiency
- High activity of glomerulonephritis.

Pathogenesis of dysmorphic
changes in glomerular erythrocytes

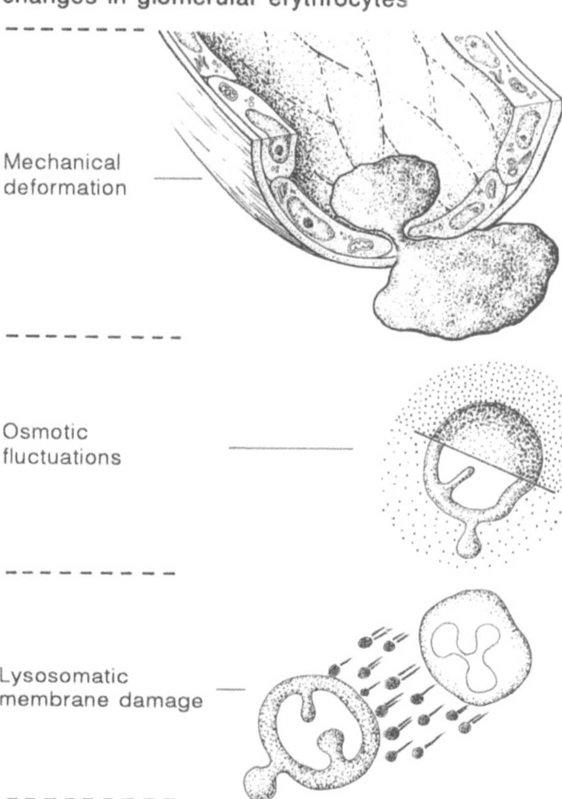

Fig. 13.12. Potential **causes** of the morphologic changes in **glomerular** erythrocytes (see Sect. 3.4)

13.4 Microscopic Examination of Glomerular Erythrocytes

13.4.1 Timing of Examination

In the past it was considered necessary to perform an *immediate* examination of the fresh urine, with no more than a 1 h delay, in order to detect glomerular erythrocytes. This policy was based on the fear that the autolytic potential of the urine could produce lytic changes in the glomerular erythrocytes and thus limit the validity of the examination.

In practice, however, this policy leads to *significant problems* because of the *urgency* that it imposes on the evaluation. As a result, it is not always possible to integrate the immediate microscopic examination into the working routine, and there is no opportunity to dispatch the sample for reference cytology.

Since the studies of Roth and Rathert (1989), it is known that urine samples can be *preserved* (Thiomersal and Carcyt "preservation-dispatch medium," Robapharm, see Sect. 8.4.1) for several days with no structural alteration in the glomerular erythrocytes. This makes it possible to defer the urinalysis or, if necessary, dispatch a sample for reference cytology.

13.4.2 Phase Contrast and Interference Microscopy

Owing to its relative simplicity, *phase contrast microscopy* has remained the primary method for the examination of glomerular erythrocytes (Fig. 13.13, see also Fig. 13.10b). The centrifuged urine is placed onto a glass slide, coverslipped, and analyzed at a magnification of 400 ($\times$ 40 objective, $\times$ 10 eyepiece) using a special objective and condenser (see Chap. 8).

The enhanced contrast produced by the separation of diffracted and undiffracted light causes the phase differences to become visible as differences in image brightness. Erythrocytes and other corpuscular elements in the urine appear surrounded by light "halos" that correspond to the contrast increase (see Figs. 13.10b, 13.13).

Interference contrast microscopy produces a relief-like image of the erythrocytes (see Figs. 13.5a, 13.9a,b). Although the information content of the image is comparable to that of the phase contrast image (Leyh et al. 1986), interference microscopy is reserved for suitably equipped institutions.

13.4.3 Rapid Staining Methods

The main disadvantage of phase contrast microscopy is the need for special microscopic equipment. Studies by Schuster et al. (1985) and Hauglustaine et al. (1982) have shown that glomerular erythrocytes can also be visualized by rapid staining methods, permitting the use of bright-field microscopy. This may be done using sediment stains (e.g., Sedicolor, MD-Kova dye solution, see Chap. 8), methylene blue stain (see Fig. 13.4c), or prestained slides (Testsimplets).

13.4.4 Alcoholic Stains (Papanicolaou)

The studies of Roth and Rathert (1989) have shown that, despite the familiar dehydrating effect of alcohol, glomerular erythrocytes can *retain their characteristic structural features on Papanicolaou staining*. Thus, Papanicolaou stain, still regarded as the standard stain for urinary cytology, is also suitable for identifying a glomerular source of urinary tract bleeding (Figs. 13.14–13.17). The *practical significance* of this is that the evaluation of erythrocyte morphology in the investigation of microhematuria can be performed during standard oncologic urinary cytology, without the need for a separate evaluation.

13.4.5 Summary

– In patients requiring investigation of microhematuria, the examination for glomerular erythrocytes can be integrated into the routine *oncologic urinary cytologic workup*. The erythrocyte evaluation may be done at once using a rapid stain, or the urine sample may be preserved and the evaluation deferred to facilitate integration into the standard laboratory routine.
– Because glomerular erythrocytes retain their structural features on Papanicolaou staining, permanent *filing* and *documentation* are possible. This not only has scientific implications in terms of possible later processing but also permits the documented monitoring of a patient's progress.
– The ability to preserve the urine sample and prepare permanent alcohol stains allows for *dispatch* and *reference cytology* when needed.

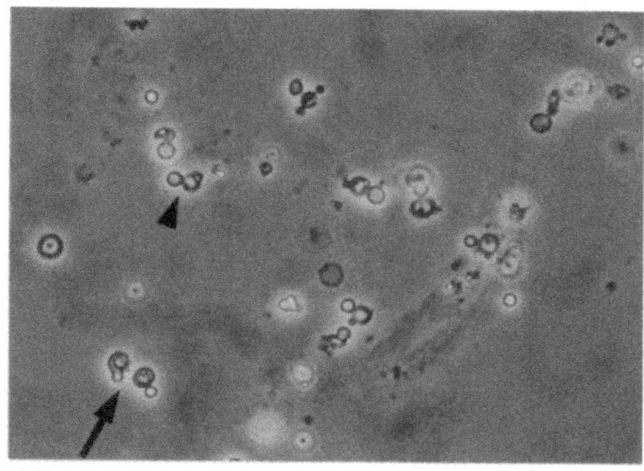

13.13

Fig. 13.13. **Phase contrast microscopy** (×340) of glomerular hematuria. The field contains ring forms with vesicle-like protrusions (←) and a glomerular ring form directly adjacent to a normal nonglomerular erythrocyte (◄). (Photo courtesy of Prof. Dr. H. Köhler of the First Medical Clinic and Outpatient Clinic of the University of Mainz)

Fig. 13.14 a, b. Dysmorphic **nonglomerular** (normal) erythrocytes. (a, ×340; b, ×850; Papanicolaou stain)

Fig. 13.15 a, b. Dysmorphic **glomerular** erythrocytes. (a, ×340; b, ×850; Papanicolaou stain)

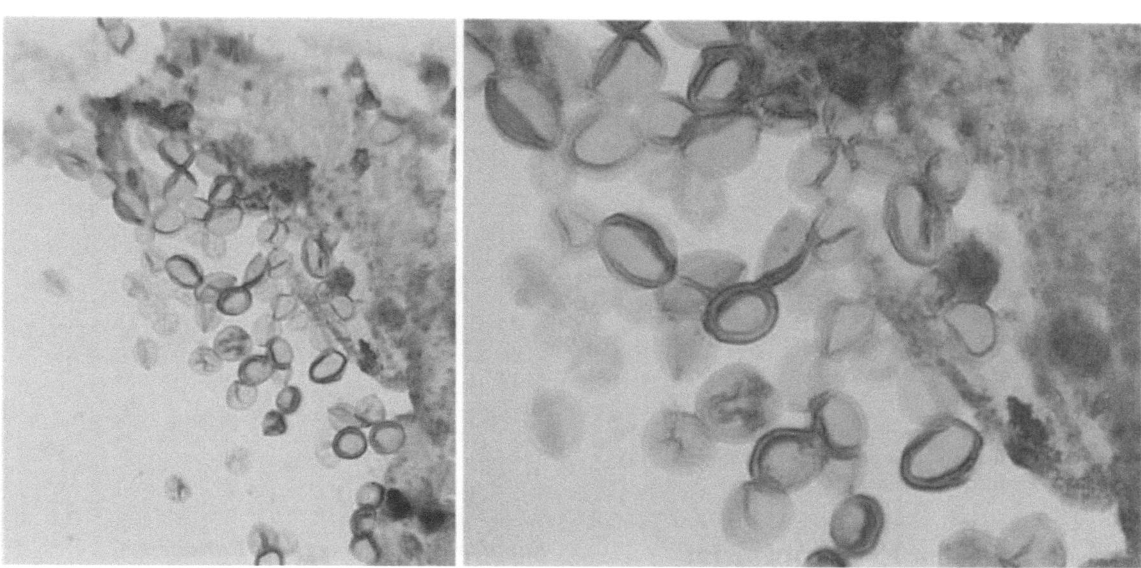

a   b

13.14

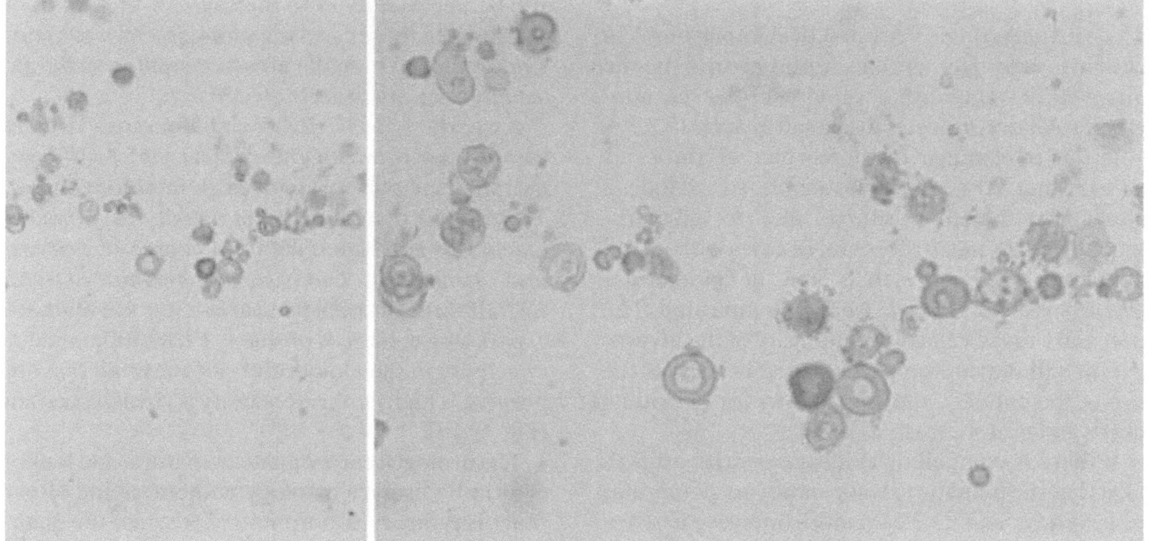

a   b

13.15

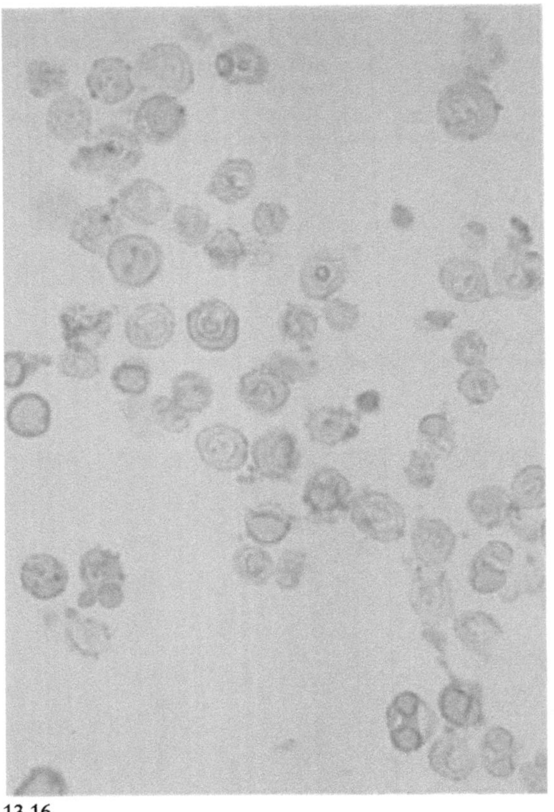

13.16

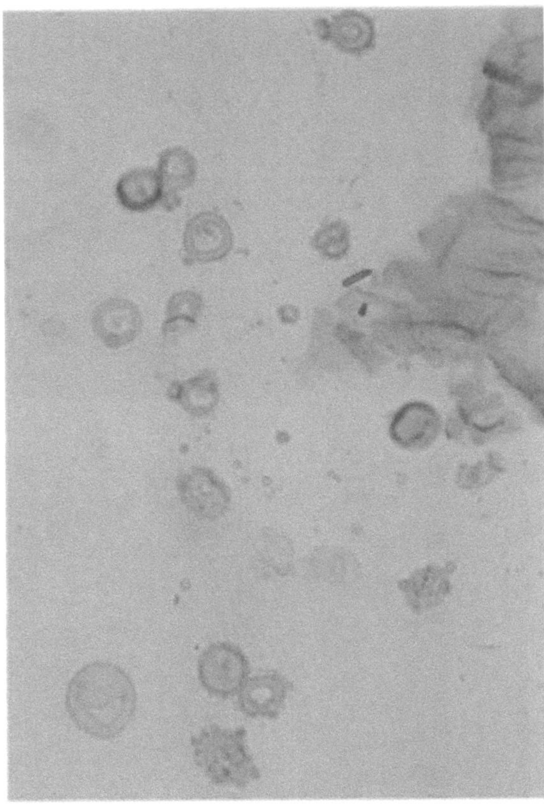

13.17

Figs. 13.16 and 13.17. Examples of **hematuria of glomerular origin** with a pathognomonic percentage of glomerular erythrocytes. Papanicolaou stain, ×850

13.5 Glomerular Erythrocyturia: Practical Implications

The *essential criterion* for practical implications in patients with glomerular erythrocyturia is the *quantitative relationship* of glomerular to nonglomerular erythrocytes discussed in Sect. 13.3.3.

If the percentage of glomerular erythrocytes is less than 20%, it is reasonable to exclude a glomerular bleeding source due to glomerulonephritis. In these cases and in cases with mixed hematuria (i.e., more than 50% nonglomerular erythrocytes), a *urologic workup* is indicated. The necessary invasiveness and intensity of the investigation will depend on various factors such as the age of the patient , allowing leeway for individual needs and preferences.

If there is compelling evidence or strong suspicion that the hematuria is secondary to glomerular disease (see Sect. 13.3.3), a more intensive *nephrologic workup* and therapy are indicated.

Supplementary Diagnostic Parameters
Despite the high sensitivity and specificity of erythrocyte morphology (see Sect. 13.3.2), there can still be uncertainty as to the cause of urinary tract bleeding. In these cases the urologist can work with a nephrologist to make an assessment of *nephritic activity* as an *aid to decision making.*

As early as 1957, Brod and Benesova drew a comparison between the clinical and histologic-morphologic activity factor in glomerulonephritis. They found that the clinical activity of *glomerulonephritis* correlated with the degree of *proteinuria, hematuria, leukocyturia, cylindruria,* and *ESR*. If three of these five parameters are markedly pathologic, there is probably a high inflammatory activity in the glomerulus; if four or all five are positive, a high nephritic activity is virtually certain (Fig. 13.18).

The urologist can evaluate hematuria and leukocyturia by urinary cytology without giving attention to erythrocyte morphology. Because the quantitative and qualitative assessment of proteinuria is of key importance in the further evaluation and treatment of glomerulonephritis, there is no real need to add an accurate quantification of hematuria and leukocyturia by the use of costly and time-consuming counting chamber methods (Lamberts 1982).

Important additional information can be provided by the detection of casts in the urine (sediment), since, except for isolated hyaline casts, cylindruria signals the presence of renal disease (Althof and Ochs 1989). Cylindruria requires the concomitant urinary excretion of proteins to serve as a "glue" for the clumped corpuscular elements. *Erythrocyte casts* are formed by the deposition of red cells onto a more or less dense protein-containing substance. They prove a renal origin of hematuria and are particularly common in glomerulonephritis and vascular disorders (Althof and Ochs 1982). Reports on the *incidence* of erythrocyte casts range from 22% to 66% (Fasset et al. 1982; Rizzoni et al. 1983; Birch et al. 1983).

13.6 Conclusion

A urologist faced with the problem of persistent microhematuria requires the least invasive, most revealing, and least expensive diagnostic plan that can be formulated. The general application of invasive and cost-intensive diagnostic procedures is neither possible nor prudent, especially in younger patients and children, and the possibilities of a basic, noninvasive workup should be fully exhausted. The urinary cytologic evaluation of *erythrocyte morphology* is of major important in this regard owing to its technical simplicity and its high sensitivity and specificity.

References

Addis T (1925) A clinical classification of Bright's disease. J Am Med Assoc 85: 163

Amrein R, Reber H, Widmer LK, Thölen H (1979) Normalwerte für die Ausscheidung von Erythrozyten und Leukozyten im 2-Stunden-Urin. Arztl Lab 25: 117–123

Althof S, Ochs H-G (1982) Mikroskopische Untersuchung des Harnsediments. In: Losse H, Renner E (eds) Klinische Nephrologie, vol 1. Thieme, Stuttgart New York, p 142–154

Arm JP, Peile EB, Rainford DJ, Strike PW, Tettmar RE (1986) Significance of dipstick haematuria. 1. Correlation with microscopy of the urine. Br J Urol 58: 211–217

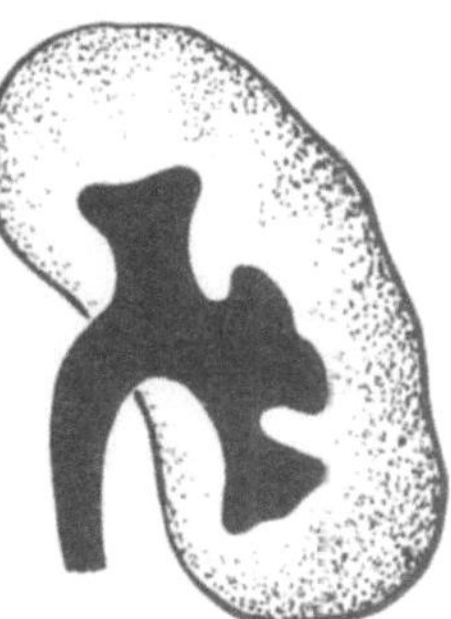

Fig. 13.18. Besides erythrocyte morphology, additional parameters can be determined to establish a possible glomerular origin for persistent microhematuria (see Sect. 13.5). The **"nephritic activity index"**, along with age, determines further nephrologic diagnosis and treatment

Bauer H-W, Haase W, Sieck R (1981) Neue diagnostische Möglichkeiten der Urinsedimentanalyse mit dem MD-KOVA-System. Urologe [B] 21: 295–299

Birch DF, Fairley KF (1979) Haematuria: glomerular or nonglomerular? Lancet 2: 845–846

Birch DF, Fairley KF, Whitworth JA, Forbes IK, Fairley JK, Cheshire GR, Ryan GB (1983) Urinary erythrocyte morphology in the diagnosis of glomerular hematuria. Clin Nephrol 20: 78–84

Brod J, Benesova DA (1957) Comparative study of funktional and morphological renal changes in glomerulonephritis. Acta Med Scand 157: 23

Carson CC, Segura JW, Greene LF (1979) Clinical importance of microhematuria. J Am Med Assoc 241: 149–150

Conzelmann M, Conen D, Besch W, Dubach UC, Thiel G (1988) Der Nachweis glomerulärer Erythrozyten im Urin mit Phasenkontrastmikroskopie: praktikabel in der ambulanten Praxis? Schweiz Med Wochenschr 118: 541–546

Davides KC, King L, Jacobs D (1986) Management of microscopic hematuria: Twenty-year experience with 150 cases in a community hospital. Urology 28: 453–455

Fasset RG, Horgan BA, Mathew TH (1982) Detection of glomerular bleeding by phase-contrast · microscopy. Lancet I: 1432–1434

Freni SC, Freni-Titulaer LWJ (1977) Microhematuria found by mass screening of apparently healthy males. Acta Cytol 21: 421–424

Fröhlich G, Sieck R (1981) Das MD-KOVA-System – Ein standardisiertes Verfahren zur Untersuchung des Urinsediments. Urologe [B] 21: 300–303

Froom P, Riback J, Benbassat J (1984) Significance of hematuria in young adults. Br Med J 288: 20–22

Froreich A von, Kaufmann J, Müller-Plathe O, Henze D (1974) Das Mikro-Sediment. Eine neue Methode der quantitativen Bestimmung cellulärer Harnbestandteile. Urologe [A] 13: 24–26

Fünfstück R, Schuster FX, Stein G, Beintker M, Schramek P, Jansa U (1989) Zur Bedeutung der Erythrozytenmorphologie bei glomerulären und nicht-glomerulären Hämaturien. Z Urol Nephrol 82: 85–91

Golin AL, Howard RS (1980) Asymptomatic microscopic hematuria. J Urol 124: 389–391

Greene LF, O'Shaughnessy EJ Jr, Hendricks ED (1956) Study of 500 patients with asymptomatic microhematuria. J Am Med Assoc 161: 610–613

Hauglustaine D, Bollens W, Michielsen P (1982) Detection of glomerular bleeding using a simple staining method for light microscopy. Lancet 2: 761

Heintz R, Althof S (1989) Das Harnsediment, 4th edn. Thieme, Stuttgart

Hildebrandt F, Fecht A, König B, Brandis M (1988) Exakte Beschreibung nicht-glomerulärer und glomerulärer Erythrozytenformen bei kindlicher Hämaturie. Monatsschr Kinderheilkd 136: 10–16

Iseghem P van, Hauglustaine D, Bollens W, Michielsen P (1983) Urinary erythrocyte morphology in acute glomerulonephritis. Br Med J 287: 1183–1186

Jardin A, Madier T (1987) Point de vue de l'urologue face à l'hématurie microscopique. Semin Uro-nephrol Salpetrière 6: 153–159

Jones DJ, Langstaff RJ, Holt SD, Morgans BT (1988) The value of cystourethroscopy in the investigation of microscopic haematuria in adult males under 40 years. Br J Urol 62: 541–545

Lamberts B (1982) Hämaturie, Hämoglobinurie, Chylurie. In: Losse H, Renner E (eds) Klinische Nephrologie, vol 1. Thieme, Stuttgart, pp 302–309

Larcom PC Jr, Carter GH (1948) Erythrocytes in urinary sediment: identification and normal limits. J Lab Clin Med 33: 875–888

Leyh H, Weitbrecht M, Schütz W (1986) Differentialdiagnostik glomerulärer und nicht-glomerulärer Hämaturien mit Hilfe der Interferenzkontrastmikroskopie. Urologe [B] 26: 260–264

Lin JT, Wada H, Maeda H, Hattori M, Tanaka H, Uenoyama F, Suehiro A et al. (1983) Mechanism of hematuria in glomerular disease. Nephron 35: 68–72

Madsen KM, Apolegate CW, Tisher CC (1982) Phagocytosis of erythrocytes by the proximal tubule of the rat kidney. Cell Tissue Res 226: 363–374

Mariani AJ, Mariani MJ, Macchioni C, Stams UK, Hariharan A, Moriera A (1989) 1000 hematuria evaluations including a risk-benefit and cost-effectiveness analysis. J Urol 141: 350–355

Messing EM, Young TB, Hunt VB, Wehbie JM, Rust P (1989) Urinary tract cancers found by homescreening with hematuria dipsticks in healthy men over 50 years of age. Cancer 64: 2361–2367

Mohr N, Offord KP, Owen R, Melton J (1986) Asymptomatic microhematuria and urologic disease. J Am Med Assoc 256: 224–229

Müller V, Siegel O, Siegel S, Flügel W, Göbel U, Natusch R, Pilz S (1989) Die Differenzierung der Hämaturie durch phasenkontrastmikroskopische Untersuchung von Harnerythrozyten. Z Urol Nephrol 82: 277–283

Olbing H (1977) Methoden zur Bestimmung von Hämaturien. Monatsschr Kinderheilkd 125: 759–760

Olivo JF, Guille F, Lobel B (1989) Hématurie microscopique. Valeur sémiologique en urologie. Conduite à tenir devant une hématurie microscopique. J Urol (Paris) 8: 453–458

Rathert P, Lutzeyer W (1974) Zytodiagnostik bei Harnwegserkrankungen. In: Kluthe R, Oechslen D (eds) Aktuelle Diagnostik von Nierenerkrankungen – 6th Freiburger Tagung über Nephrologie 1974. Thieme, Stuttgart, pp 22–30

Renner E (1986) Gutachterliche Stellungnahme zum Stellenwert der Vorsorgeuntersuchung auf das Vorliegen einer Mikrohämaturie – aus der Sicht der Nephrologie. Mitteilungen Dtsch Ges Urol 2: 77–81

Ritchie CD, Bevan EA, Collier StJ (1986) Importance of occult haematuria found at screening. Br Med J 292: 681–683

Rizzoni G, Braggion F, Zacchello G (1983) Evaluation of glomerular and nonglomerular hematuria by phasecontrast microscopy. J Pediatr 103: 370–374

Roth St, Rathert P (1989) Formstabilität glomerulärer Erythrozyten: Ergebnisse einer Studie und praktische Konsequenzen in der Hämaturiediagnostik. Urologe [A] 28 (Suppl): A 68, V 106

Santo NG de, Nuzzi F, Capodicasa G, Lama G, Caputo G, Rosati P, Ciordano C (1987) Phase contrast microscopy of the urine sediment for the diagnosis of glomerular and nonglomerular bleeding – data in children and adults with normal creatinine clearance. Nephron 45: 35–39

Schramek P, Schuster FX (1985) Persistierende Mikrohämaturie: Lokalisation der Blutungsquelle durch Beurteilung der Erythrozytenmorphologie. Urologe [A] 24: 216–220

Schramek P, Moritsch A, Haschkowitz H, Binder BR, Maier M (1989) In vitro generation of dysmorphic erythrocytes. Kidney Int 36: 72–77

Schuetz E, Schaefer RM, Heidbreder E, Heidland A (1985) Effect of diuresis on urinary erythrocyte morphology in glomerulonephritis. Klin Wochenschr 63: 575–577

Schuster FX, Schramek P, Schmidbauer CP (1985) Differenzierung glomerulärer und nichtglomerulärer Hämaturie im Hellfeldmikroskop. Aktuel Urol 16: 73–75

Sinniah R, Pwee HS, Lim CH (1976) Glomerular lesions in asymptomatic microscopic hematuria discovered on routine medical examination. Clin Nephrol 5: 216–228

Teitel M, Lambertson GH, Florman AL, Manhasset NY (1964) Filtration of urine for quantitation of cells and casts. Am J Dis Child 108: 19–27

Thiel G, Bielmann D, Wegmann W, Brunner FP (1986) Glomeruläre Erythrozyten im Urin: Erkennung und Bedeutung. Schweiz Med Wochenschr 116: 790–797

Ultzmann R, Hofmann KB (1872) Atlas der physiologischen und pathologischen Harnsedimente. Braunmüller, Wien

Vehaskari VM, Rapola J, Koskimies O, Savilahti E, Vilska J, Hallmann N (1979) Microscopic hematuria in schoolchildren: epidemiology and clinicopathologic evaluation. J Pediatr 95: 676–684

Völter D, Keller AJ (1983) Die Hämaturie. Schattauer, Stuttgart

14 An Expert System for Analysis and Standardization of Bladder Carcinoma Grading

R. Nafe, S. Roth, and P. Rathert

Correct evaluation of a urinary cytologic preparation is a matter of the cytologist's experience and precise visual assessment. Up to now, the necessary skills could only be learned through regular, intensive training at the microscope and, in a very limited form, from the literature, whereby the most serious problem is a lack of *standardized* criteria for grading urothelial carcinoma (see Chap. 7). This causes a high rate of diagnoses of grade 2 carcinoma of the bladder, which is clinically an inhomogeneous group with respect to prognosis and clinical outcome (Jordan et al. 1987). The necessity for an analysis and a better standardization of criteria for grading bladder carcinoma is thus evident.

Image analysis and cytophotometry can contribute to the quantification and standardization of the examination but cannot replace the knowledge of an experienced cytologist. Another standardization method is offered by so-called *expert systems,* special computer programs with four characteristic capabilities:

1. Large amounts of specialized data can be stored;
2. The expert's criteria can be analyzed according to the rules of artificial intelligence;
3. Conclusions can be drawn based on a specialized program and
4. The expert's diagnosis can be supported.

Contrary to a prejudice often voiced, the application of expert systems in medicine does not aim to replace the specialist and his or her knowledge in diagnosis, but rather to support the specialist, offering the possibility of analyzing large data pools containing quantitative or semiquantitative data. In contrast to "rule-based" systems, "inductive" expert systems are able to construct decision rules just be analyzing databases containing cases with known diagnoses and previously defined features (Curth et al. 1991). The rules are presented in the form of decision trees containing decision rules in the general form "If the value of feature X was a, then, if the value of feature Y was b, then . . . the diagnosis always was . . .". None of the trees contain redundant information, and the system is also capable of defining a rank order of the best discriminating features between two diagnoses (Table 14.1).

Therefore, the ideal field of application for those systems is the analysis of diagnostic decisions which cannot be defined by introspecitive qualitative analysis or by uni- and multivariate statistics. A good example, therefore, is the grading of urothelial carcinomas of the bladder in urinary cytologic tumor diagnosis.

Table 14.1. **Decision tree for the whole database** (202 preparations) containing seven different morphological features: nuclear-cytoplasmic ratio, nuclear shape, nuclear hyperchromasia, chromatin structure, membrane hyperchromasia, nucleolar shape, nucleolar size. Considerable variation of the cytomorphology of grade 2 carcinomas; less pronounced morphological variation of grade 1 and grade 3 tumors

```
Ncl-Cytopl-Rat??

small-norm: Ncl-Shape??
    round-oval:................................ no-Tumor
    part-irreg: Ncl-Hyperchr??
        not-present: Chromatin??
            fine:................................ no-Tumor
            mod-coarse:......................... G1
            dist-coarse:......................... /
        moderate:............................. G1
        distinct:............................. /
    pred-irreg:............................. /

var-small: Ncl-Shape??
    round-oval:............................. G1
    part-irreg:............................. G1
    pred-irreg:............................. G2

var-large: Nlo-Shape??
    round-oval:............................. G2
    part-irreg:............................. G2
    pred-irreg:............................. G3

large: Nlo-Shape??
    round-oval:............................. G2
    part-irreg: Nlo-Size??
        small:............................. G2
        varied:............................. G2
        large: Membrane-Hyperchr??
            not-present:..................... /
            moderate:........................ G2
            distinct:........................ G3
    pred-irreg:............................. G3
```

In order to provide an analytical and objective basis for grading bladder cancer and for its better standardization, we applied an inductive system (1st Class, BRAINWARE, Berlin, FRG) to 202 urinary cytologic preparations (69 free of tumor cells, 41 grade 1 bladder carcinomas, 66 grade 2 bladder carcinomas, 26 grade 3 bladder carcinomas) after defining 12 nuclear morphologic features semiquantitatively (Table 14.1). This was done with the goal in mind of compiling all criteria that might influence urothelial tumor diagnosis and grading and, thus, to minimize a subjective interpretation if at all possible. This analysis should contribute to the discussion of two questions: What are the most significant cytomorphologic criteria in differentiating urinary cytologic preparations free of tumor cells and those with tumor cells from grades 1, 2, and 3 bladder carcinoma, and 2. are the tumor grades clearly defined with respect to cytomorphology, or are grades 1, 2, and 3 tumors inhomogeneous groups?

For example, for 12 nuclear morphologic features entered into the system, the decision tree for the whole database contains only seven different features, which means that these seven criteria would have sufficed to achieve a correct and reliable grading for the 202 preparations. The best overall distinguishing feature between preparations free of tumor cells and the three grades of bladder carcinoma was the nuclear-cytoplasmic ratio. Depending on the results (small to normal; varying, generally small; varying, generally large; large), further criteria lead to a final diagnosis or tumor grading. Besides the tree for the whole database, further decision trees were developed by the inductive system based only on a pairwise comparison between two tumor grades or on the comparison between preparations free of tumor cells and grade 1 carcinomas.

The other deciding evaluation criteria, besides the nuclear-cytoplasmic ratio, are nuclear shape, chromatin structure, nucleolar-nuclear ratio, and nuclear hyperchromasia. The decision tree for differentiating grade 1 preparations and those classified as free of tumor cells contains only four different criteria, all of which refer only to the cell nuclei and not to the nucleoli. Only nuclear-cytoplasmic ratio and nuclear shape were decisive in differentiating between grades 1 and 2 preparations. Grades 2 and 3 preparations were clearly differentiated using four criteria, two characterizing the nucleoli and two the nuclei. Table 14.2 shows the five most important criteria for differential diagnosis in each case. The decision trees contain a large volume of information in compressed form, the complete verbal elaboration of which would go beyond the scope of this chapter. Particularly noticeable was the fact that favorable nuclear-cytoplasmic ratios and round to oval nuclear shape in no case led to suspicion of a tumor. Of the total of 69 preparations containing no tumor cells 67 were in this category of findings. The diagnosis "G1 tumor," on the other hand, was made in each case of partially deformed nuclear shape and moderate nuclear hyperchromasia. This was the most common constellation of findings among the grade 1 preparations. Other grade 1 preparations revealed moderately coarse chromatin structure and partially deformed nuclear shape. Even in cases of round to oval nuclear shape, the diagnosis "G1" was made whenever the nuclear-cytoplasmic ratio varied. Even preparations with generally small, but varying nuclear-cytoplasmic ratio were always given the "tumor" diagnosis. An interesting thing about the grade 2 tumors is the large variety of possible findings. The most common among them was a varying and, on the whole, large nuclear-cytoplasmic ratio combined with generally round to oval nucleolar

---

Table 14.2. The **most decisive criteria for the differentiation** between preparations free of tumor cells and grades 1, 2, and 3 bladder carcinomas in the order of their significance; criteria for the whole database and for pairwise comparisons

Whole database	"No suspicion of tumor" – grade 1
1. Nuclear-cytoplasmic ratio	1. Nuclear shape
2. Nuclear shape	2. Chromatin structure
3. Chromatin structure	3. Hyperchromasia of the cell nucleus
4. Nucleolar-nuclear ratio	4. Nuclear-cytoplasmic ratio
5. Hyperchromasia of the cell nucleus	5. Variations in size of nuclei
Grades 1–2	Grades 2–3
1. Nuclear-cytoplasmic ratio	1. Nucleolar shape
2. Nucleolar-nuclear ratio	2. Nucleolar-nuclear ratio
3. Hyperchromasia of the nuclear membrane	3. Nucleolar size
4. Variations in size of nuclei	4. Chromatin structure
5. Hyperchromasia of the cell nucleus	5. Nuclear-cytoplasmic ratio

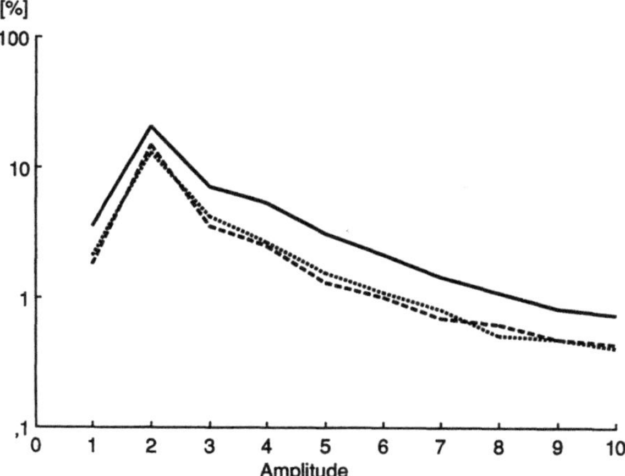

Fig. 14.1. **Fourier amplitudes (means) of urothelial cell nuclei** in preparations from washings from patients with reactive changes but no urothelial tumor *(dotted line)*, from grade 1 bladder cercinomas *(solid line)*, and from normal urothelial cell nuclei *(dashed line)*

shape (in 30 of 66 preparations). Particularly noticeable in differentiating between grades 2 and 3 tumors was the significance of nucleolus parameters, 22 of all 26 grade 3 preparations showing a predominance of irregularly shaped nucleoli and unfavorable nuclear-cytoplasmic ratios.

For the discussion of cytomorphologic criteria and degree of homogeneity of the three grades of bladder carcinoma, this inductive system is an adequate tool. Conventional multivariate statistics and so-called neuronal networks classify different entities or tumor grades on the basis of quantitative or semiquantitative data, but diagnostic decisions based on the rank order of the individual deciding features cannot be reconstructed using these methods.

The decision tree data include a large number of rules and information that would not be registered just by investigating the preparations. But information can easily be given a clear objective basis by comparing it with data on certain preparations already in the data base. For example, it was confirmed that grade 3 preparations differed in each case from grade 2 preparations in that they showed more pronounced irregularities in the nucleolus and less pronounced ones in the nucleus. In contrast, diagnosis of a grade 1 tumor was made almost exclusively on the basis of nuclear features. Immuncytochemistry (Chaps. 11, 12), flow cytometry (Chap. 10) and image analysis have been investigated as means of confirming the diagnosis of highly differentiated carcinomas. Our results suggest that it would be worthwhile to improve these

promising quantitative methods, since the deciding diagnostic features for grade 1 tumors (Table 14.2) are merely those of planimetry and densitometry. These can be detected by image analysis systems and by performing Fourier analysis of the outlines of the cell nuclei. It was possible to differentiate between preparations from washings with reactive changes in urothelial cells in the absence of a tumor and those from highly differentiated bladder cancer (Nafe et al. 1991). Higher Fourier amplitudes in grade 1 tumors indicate a more pronounced irregularity of tumor cell nuclei when compared with altered but not malignant cells and with normal urothelial cells from voided urine (Fig. 14.1). Nevertheless, it seems possible to standardize the criteria semiquantitatively too, such that a detailed description of the cytopathologic features and the consultation of the expert system might also improve the sensitivity of urinary cytologic examination for grade 1 tumors. Indeed, in our experience a detailed morphologic description of a preparation provides a better focus and overview. It does not seem impossible that by consulting an expert system even a specialist might improve the precision and the degree of standardization of diagnosis.

Another important result of the analysis of the database is the wide range of cytomorphologic variations in grade 2 carcinoma. Since they differ from grade 3 tumors, especially in the nucleolar features, it becomes obvious that morphology and number of nucleoli might become important in assessing cell proliferation in bladder cancer. This conclusion is supported by reports of a statistical correlation in the number of nucleolar organizer regions and proliferative activity of bladder cancer in humans and animals (Cairns et al. 1989; Takeuchi et al. 1990). More studies are needed to confirm whether these data can be used to improve the current system of grading and to avoid the prognostically inhomogeneous grade of 2. A further result of our analysis is that grade 1 and grade 3 tumors are more homogeneous with respect ot cytomorphologic variation between different preparations.

In summary, the analysis of the evaluation of urinary cytologic preparations by the expert system indicates that nuclear morphology is particularly important for detecting highly differentiated carcinomas, whereas nucleolar features may help in assessing the proliferative nature of the carcinoma. Although quantitative morphologic and immuncytochemical methods are available as adjuvant diagnostic and prognostic tools in urinary cytologic examinations, the goal of developing a grading system which rules out prognostically inhomogeneous tumor grades seems possible even when using an inductive expert system.

References

Cairns P, Suarez V, Newman J, Crocker J (1989) Nucleolar organizer regions in transitional cell tumors of the bladder. Arch Pathol Lab Med 113: 1250–1252

Curth M, Bölscher A, Raschke B (1991) Entwicklung von Expertensystemen. Hanser, Munich-Vienna, pp 9–25

Jordan AM, Weingarten J, Murphy WM (1987) Transitional cell neoplasms of the urinary bladder. Cancer 60: 2766–2774

Nafe R, Roth S, Rathert P (1991) Fourier analysis as a planimetric procedure – application to malignant and normal urothelial cells with reactive changes. Exp Pathol 43: 155–161

Takeuchi T, Tanaka T, Ohno T, Yamamoto N, Kobayashi S, Kuriyama M, Kawada Y, Mori H (1990) Nucleolar organizer regions in rat urinary bladder tumors induced by N-butyl-N-(4-hydroxybutyl) nitrosamine Virchows Arch [B] 58: 383–387

Subject Index

If you have any concerns about our products,
you can contact us on
ProductSafety@springernature.com

In case Publisher is established outside the EU,
the EU authorized representative is:
Springer Nature Customer Service Center GmbH
Europaplatz 3, 69115 Heidelberg, Germany

Printed by Libri Plureos GmbH
in Hamburg, Germany